Springer Oceanography

The Springer Oceanography series seeks to publish a broad portfolio of scientific books, aiming at researchers, students, and everyone interested in marine sciences. The series includes peer-reviewed monographs, edited volumes, textbooks, and conference proceedings. It covers the entire area of oceanography including, but not limited to, Coastal Sciences, Biological/Chemical/Geological/Physical Oceanography, Paleoceanography, and related subjects.

More information about this series at http://www.springer.com/series/10175

R. Venkatesan • Amit Tandon
Eric D'Asaro • M.A. Atmanand
Editors

Observing the Oceans in Real Time

Springer

World Meteorological Organization
Weather · Climate · Water

Editors
R. Venkatesan
Scientist G - Head Ocean Observation Systems Group
National Institute of Ocean Technology
Ministry of Earth Sciences
Chennai, TN, India

Eric D'Asaro
Physics Laboratory and School of Oceanography
University of Washington
Seattle, WA, USA

Amit Tandon
Department of Mechanical Engineering
University of Massachusetts Dartmouth
Dartmouth, MA, USA

M.A. Atmanand
Former Director
National Institute of Ocean Technology
Ministry of Earth Sciences
Chennai, TN, India

ISSN 2365-7677 ISSN 2365-7685 (electronic)
Springer Oceanography
ISBN 978-3-319-66492-7 ISBN 978-3-319-66493-4 (eBook)
DOI 10.1007/978-3-319-66493-4

Library of Congress Control Number: 2017955548

Printed on acid-free paper

This Springer imprint is published by Springer Nature
The registered company is Springer International Publishing AG
The registered company address is: Gewerbestrasse 11, 6330 Cham, Switzerland

Foreword

We no longer observe the ocean only from ship-based campaigns or with occasional glances from complex moorings in single locations. Modern observations are real-time, autonomous, distributed, and often networked. Dr. R. Venkatesan has collected chapters on the most recent methods of observations that combine these capabilities. These systems (gliders, floats, real-time reporting moorings, and networked radar and sea level systems) harness the latest in real-time distribution, onboard processing, satellite communications, and data networking. They are providing oceanographers with more information and more resolution of the time and spatial scales of critical processes in the ocean. The chapters give a basic overview of this new observational technology coupled with useful examples of their applications in the global ocean.

This book is particularly valuable for new oceanographers and resource managers. It will provide a basic understanding of the selection of sensors and systems that make observing systems possible and that provide enhancements to traditional ship-centric observing campaigns. Dr. R. Venkatesan has provided an international view of the establishment of several observing systems and uses these examples to demonstrate the affordability and utility of cutting-edge technology for nations with differing priorities. Observing in real-time will transform our understanding and management of our ocean.

Mark R. Abbott
Woods Hole Oceanographic Institution
Woods Hole, MA 02543, USA

Theresa Paluszkiewicz
Office of Naval Research
Physical Oceanography Program
Arlington, VA, USA

Preface

Advances in oceanography over the last few decades have been remarkable. Improved understanding of physical, biological, and chemical processes and improved numerical methods coupled to vastly faster computers and networks now place realistic maps of current ocean properties on the screens of anyone with a good web connection and allow plausible forecasts of future ocean conditions. But behind these numerical oceans are data from the real ocean, without which the models would uselessly drift away from reality, and behind the data are equally remarkable advances in ocean observing technologies. From almost entirely ship-based measurements a few decades ago, the ocean is now measured by a diverse combination of subsurface, surface, and space-based platforms using a diverse set of specialized oceanographic sensors. These are deployed operationally and globally to support the system of oceanographic models, regionally to address national and local resource issues, and in specialized arrays by researchers working to develop the next generation of scientific insights and the next generation of measurement platforms and sensors. We conceived this book to present the current state of these technologies, to guide those making ocean measurements, and to inspire those who will drive future technological advances.

The ocean transcends national boundaries and the study of the ocean has accordingly always been international. This is further emphasized by operational ocean modeling and prediction; accurate predictions even at the regional and local level can require knowledge of ocean properties well beyond national boundaries. Given the key role of technologies in ocean measurement, the most productive measurement programs are necessarily both international in scope and work to spread measurement technologies between international partners. This book was developed as part of collaborations between oceanographers based at institutions in India and in the United States to study monsoon air-sea interactions. Our collaborations highlighted the crucial role of modern measurement methods in developing new understandings of these processes and the importance of developing and maintaining the technological knowledge and skills necessary to make such measurements. We thus aimed this book at those who want to understand the fundamental measurement principles, the engineering approaches to implement these principles, and the techniques and strategies used to implement the measurements in the challenging ocean environment. Therefore, the final-year undergraduate students in ocean engineering, as well as other engineering students (mechanical, civil, electrical, and bioengineering) interested in specializing their skills toward modern ocean measurements, will find this book very helpful.

After introducing the state of the art in ocean measurements (Chap. 1), this book covers specialized in situ surface observations (Chaps. 2, 3, and 4) and subsurface observations (Chaps. 5, 6, 7, and 8), followed by remotely sensed observations (Chaps. 9, 10, and 11); data handling, optimization, and quality for meteorological and oceanographic applications (Chaps. 12, 13, and 14); and the societal applications (Chap. 15) by renowned experts in each of these topics.

This book is a result of OMM-ASIRI collaboration supported by the Indian Ministry of Earth Sciences (MoES) and the United States Office of Naval Research (ONR). Dr. Rajeevan, Dr. Shailesh Nayak, and Dr. Satheesh C. Shenoi are thanked for their constant motivation and support. We thank all the reviewers who contributed valuable comments to improve each chapter: Ken Jarrott, Dr. Ramesh Kumar, Dr. (Late) Vembu, Dr. Gopala Krishna, Dr. Biswajith Chakaraborthy, K. Ramesh, Dr. Latha, Dr. Jossia Joseph, Dr. Stephen Hall, N. Vedachalam, and Dr. Emily Shroyer. We are also grateful to organizations World Meteorological Organisation (WMO), Sea Bird Electronics for their inputs. K N Navaneeth is thanked for his contribution in compiling the chapters.

Chennai, Tamil Nadu, India R. Venkatesan
Dartmouth, MA, USA Amit Tandon
Seattle, WA, USA Eric D'Asaro
Chennai, Tamil Nadu, India M.A. Atmanand

Contents

About the Editors

R. Venkatesan is a Scientist G at the National Institute of Ocean Technology, Ministry of Earth Sciences, India, and served under the UNEP Regional Seas Programme. He received an M.Tech. from Karnataka Regional Engineering College and a Ph.D. from the Indian Institute of Science, India, and completed a course in marine environmental pollution and management. He is working on ocean observations and is also teaching ocean policy and ocean observation tools as an adjunct faculty. He has evolved and executed societal relevant projects and supports students. He is honored by receiving the MTS Lockheed Martin Award and National Geoscience Award and is serving as a member of the Steering Committee of GOOS/UNESCO/IOC, DOOS, and OceanSITES, the chair of the International Tsunameter Partnership and IEEE-OES India, and the founding chairman of MTS India.

Amit Tandon is a professor at the College of Engineering and the School for Marine Science and Technology at UMass Dartmouth. After obtaining his Ph.D. from Cornell University in 1992, Prof. Tandon was a UCAR postdoctoral fellow in ocean modeling and in climate system modeling.

Tandon, a Fulbright awardee and a Fulbright specialist, is a physical oceanographer whose research involves upper ocean dynamics, from mixing at microscales and Langmuir cells to frontal sub-mesoscale and mesoscale processes in the ocean, and their impact on air-sea interaction and the large-scale ocean circulation. His current research involves investigating the role of these sub-grid processes in global oceans.

Eric D'Asaro is a senior principal oceanographer at the Applied Physics Laboratory, University of Washington, and a professor of oceanography at the University of Washington. He received an M.S. in applied physics from Harvard University and a Ph.D. from the joint program in oceanography of the Woods Hole Oceanographic Institution and the Massachusetts Institute of Technology. He received the Sverdrup Gold Medal of the American Meteorological Society for pioneering instrumental, observational, and analytical progress in understanding upper ocean responses to atmospheric forcing and is a fellow of the American Geophysical Union and a member of the National Academy of Sciences.

M.A. Atmanand Former Director of the National Institute of Ocean Technology, Ministry of Earth Sciences, India, received his M.Tech. and Ph.D. from the Indian Institute of Technology Madras. He has done pioneering work in the area of deep sea technologies in India and spearheaded many indigenization programs like autonomous underwater profilers, drifters, and technologies for the extraction of polymetallic nodules from deep sea. He has published 90 papers in international journals, international conferences, national conferences, and others. He received the IEEE-OES Presidential Award, National Geoscience Award, and ISOPE Ocean Mining Symposium Award. He is the founding chairman of the IEEE Oceanic Engineering Society of India, the chair of IEEE Madras Chapter, and an associate editor of an IEEE journal.

Part I
Introduction

Recent Trends in Ocean Observations

R. Venkatesan, Amit Tandon, Debasis Sengupta, and K.N. Navaneeth

Abstract This chapter focuses on recent progress and emerging directions in ocean observations. The importance of sustained observations and status of the global ocean observing network are covered. Emerging trends include many exciting developments, such as Bio-ARGO floats, autonomous or remotely-operated instruments and platforms, and new measurement capabilities focused on fundamental ocean processes. Numerical models that integrate and assimilate multi-scale observations of the atmosphere, land, ice and ocean will lead to new science as well as improved forecasts of great societal value.

1 Introduction

Throughout history, the oceans have been a vital source of sustenance, transport, commerce, growth, and inspiration. However, our knowledge of the ocean is still quite limited. Thus, Ocean Observations have a central role to play in delivering ocean-related services to the society. Ocean Observational Systems (OOS) typically consist of (a) in-situ measurements using sensors mounted on ships, buoys, moorings and coastal stations to capture changes in properties with time and depth at specific points or tracks and (b) remote sensing systems such as satellites, aircraft and radar to capture the spatial and temporal variations of the ocean properties synoptically. However, the copious amounts of raw data produced by such systems need to result in quality-controlled processed data, which need to be further translated into user-friendly ocean information services.

Ocean observations serve many useful purposes and provide economical and societal benefits to the public. These benefits include safe and efficient marine oper-

R. Venkatesan (✉) • K.N. Navaneeth
National Institute of Ocean Technology, Chennai, India
e-mail: dr.r.venkatesan@gmail.com

A. Tandon
University of Massachusetts, Dartmouth, MA 02747, USA

D. Sengupta
Indian Institute of Science, Bangalore, India

R. Venkatesan et al. (eds.), *Observing the Oceans in Real Time*, Springer Oceanography, DOI 10.1007/978-3-319-66493-4_1

ation, improved marine commerce, coastal hazard mitigation, sustained marine resources and ecosystem management, reducing public health risk, and improve national security. A well-designed observation system should be meet users' needs, providing end-to-end products, and adaptive to changes in user requirements and technology advances. An observation system and any one of its components need to go through several development phases including the initial research and development (R&D), pilot demonstrations, and pre-operational tests and validation prior to becoming operational.

The environmental variables measured by OOS can be classified into physical (such as water level, surface waves, currents, bathymetry, water temperature, salinity), meteorological (such as wind, temperature, pressure, visibility, humidity), biological (such as ocean color, fish species and abundance, zooplankton species and abundance, phytoplankton species), chemical (such as contaminants, dissolved oxygen, nutrients), terrestrial (such as river discharge), and human health (such as seafood contamination, concentration of human pathogens).

Accurate and reliable environmental information can be derived by combining field observations with predictive operational models. Many technologies come together to accomplish these tasks. Data can be measured either directly using in-situ sensors or remotely via remote-sensing techniques such as satellite, airborne, or land-based or ship-based instruments. Predictive models expand the observation coverage and provide forecast capability, provided they have been validated and meet accepted standards prior to becoming an operational tool.

The expanding needs of users and new instrumentation technology have both led to several notable trends in the evolution of observational systems. These include real-time or near real-time data reporting, faster data telemetry via satellite during significant events (such as storm surge or a tsunami), water-level measurements with absolute reference, coastal wave, and water quality monitoring (such as nutrients from watershed, dissolved oxygen), integration of satellite sensing data and increasing spatial resolution of remotely sensed data for coastal applications, developing HF radar for surface current mapping, wireless data telemetry (such as Iridium and Argos satellites data transmission, cellular phone, underwater acoustic modem, spread spectrum modem), and system integration.

2 Recent Trends in Ocean Observations

Recent trends in ocean observations rely on measuring essential ocean variables, using global observing networks, transmitting the data, and applying quality control on the resulting data. We discuss each of these facets next.

(a) Essential Ocean Variables (EOVs)

Ocean Observations are expensive due to the vast, harsh, and remote oceanic environment. The international Global Ocean Observing System (GOOS) has set up the framework and requirements for an effective and efficient ocean observing system. The expert panel of GOOS has identified essential ocean

Table 1 Major physical and biogeochemical EOVs and their measuring platforms (http://www.goosocean.org)

SI No.	Essential oceanic variables (EOV) (physical and biogeochemical variables)	Observing platforms
1	Sea state	Satellite altimeters, moored buoys, drifting buoys, and coastal radars
2	Surface stress	Buoys, ships, and satellites
3	Sea surface height (SSH)	Satellite measurements, in-situ global sea level observing system (GLOSS) water level gauges
4	Sea ice	Passive microwave satellite sensors, e.g. SMOS
5	Sea surface temperature (SST)	Satellite (infrared and microwave radiometers), in-situ observations from ships, moorings, drifting buoys, floats
6	Subsurface temperature	Moorings, gliders, ARGO floats, ship-based Conductivity Temperature and Depth (CTDs)
7	Surface currents	Moorings, HF radars, satellite altimetry, Lagrangian drifting buoys
8	Subsurface currents	ADCP, ARGO floats, gliders
9	Sea surface salinity (SSS)	Satellite, floats and drifting buoys
10	Subsurface salinity	Moorings, gliders, ARGO floats, CTDs
11	Oxygen	Moorings, profiling floats, gliders, ships
12	Nutrients	Ship-based observations
13	Inorganic carbon	Surface moorings, drifting buoys, ships
14	Transient tracers (CFC, sulfur hexaflouride, tritium)	Ship-based hydrography
15	Suspended particulates	Moorings, satellite, floats, ship-based samplings
16	Nitrous oxide	Ship-based hydrography
17	Stable carbon isotopes	Ships, ship-based time series, repeated hydrography
18	Dissolved organic carbon	Ship-based time series, repeated hydrography
19	Ocean color	Satellite

variables in physics (10), biogeochemistry (9), and biology and ecosystem (7), based on their relevance, feasibility of measurement, and cost effectiveness. The major physical and biogeochemical EOVs and their measuring platforms are summarized in Table 1. (1-10 are physical parameters and 11-19 are biogeochemical parameters).

(b) Global Observing Networks: The persistent, core elements of the current global ocean observing system are the Argo float network, the moored buoy network, the drifting buoy network, voluntary observations from ships (VOS) and the global sea-level observing system (GLOSS). These are supplemented by data available from national observation assets, research activities and special projects.

- Ship Observations Team-VOS: About 4000 operational ships with approximately 240 automatic weather stations (AWS) from 25 active countries are part of the Voluntary Observing Ship (VOS) scheme. These observations are concentrated along major shipping routes, primarily in the North Atlantic

Fig. 1 Oceanographic data collection using various platforms such as ships, moored buoys, gliders, Argo floats, remote sensing, and drifting buoys

and the North Pacific. Four hundred new AWS are being installed as part of VOS by a European joint project.

- Argo: The Argo network consists of robotic profilers in the ocean, deployed around the world. This network consists of more than 3800 floats at a spatial resolution of 3 × 3 grids, with about 800 deployed every year, and each float lasting about 3–4 years. Presently, 3829 operational floats are active, contributed by 30 countries. The current network covers about ~70% of the global ocean, allowing observations of oceanic heat gain with unprecedented accuracy. These floats have provided more than 100,000 temperature and salinity profiles of the upper 2000 m of the ocean. The Argo data have been extensively used by the researchers to study the regional and global changes in the ocean temperature and ocean heat content, large-scale circulation [5, 9, 13], salinity and freshwater content [4, 6], upper ocean temperature and salinity structure during cyclones [2, 3, 10], initialization of ENSO forecasts, and sea-level studies [1, 7, 18].
- Moored buoy network: There are currently 318 moored buoys (Fig. 2) which provide 83% of the coastal/national moored wave measurements data to the countries subscribing to the GTS. Moored buoy metadata is collected in a common format and is available to the community to improve forecasts. In the near future, 35% of moored buoys (tropical and coastal/national moored buoys) will report data on the GTS in binary universal form for the representation of meteorological data (BUFR) format. This will allow integration of the moored buoy metadata system into the Joint Technical Commission for

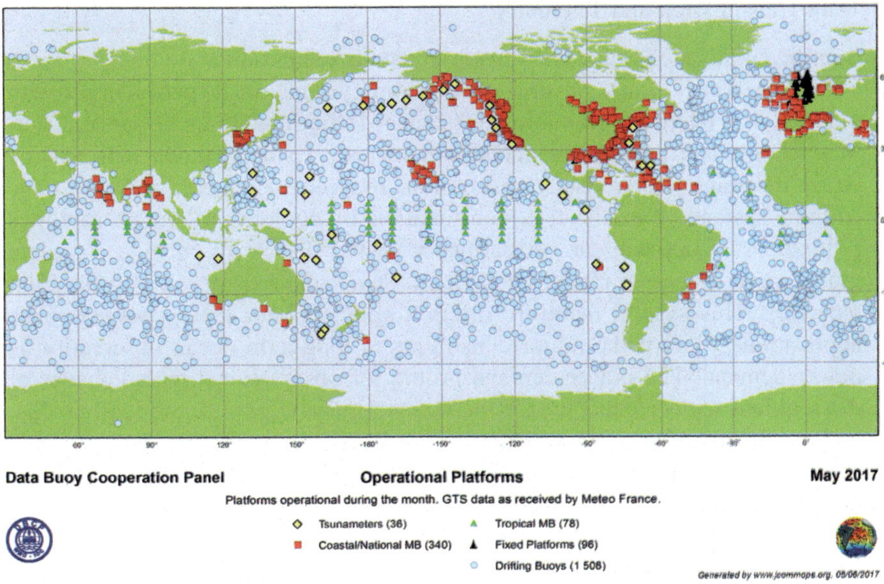

Fig. 2 Status of buoy network in the global ocean (source: http://www.jcommops.org)

Oceanography and Marine Meteorology in-situ Observing Programmes Support Centre (JCOMMOPS). The Global Tropical Moored Buoy Array (GTMBA) includes Tropical Atmosphere Ocean/Triangle Trans-Ocean Buoy Network (TAO/TRITON) in the Pacific, the Prediction and Research Moored Array in the Tropical Atlantic (PIRATA), and the Research Moored Array for African-Asian-Australian Monsoon Analysis and Prediction (RAMA) in the Indian Ocean [11]. Besides the GTMBA, several national networks are maintained by agencies such as National Institute of Ocean Technology (NIOT, India) [14] and National Data Buoy Center (United States). In the Indian Ocean, the moored buoys maintained by RAMA and NIOT provide invaluable data sets for understanding of the Madden-Julian Oscillation (MJO), Indian Ocean Dipole (IOD), intraseasonal variability of currents at the equator [8], and cyclones [10, 15].

- Drifting buoys: There are 1508 drifting buoys which provide data to the GTS from 11 countries. The data from 44% of the drifting buoys is reported to the GTS in less than 1 h. All of the drifting buoys report data to the GTS in BUFR format, with 57% of the drifting buoys reporting barometric pressure data to the GTS. The future plan for the drifter network calls for increasing the number of drifting buoys with air pressure measurements, improving the drifting buoys density distribution coverage, and improving the data reporting timeliness to GTS.
- Global Sea-Level Observing System (GLOSS): This network consists of about 290 sea-level stations around the world. There are roughly 161 stations reporting data, 52 stations with some data, and 77 stations not reporting . The new technologies will use microwave sensors at tide gauges. The University of Hawaii Sea Level Center produces quality-controlled data sets.

(c) Data Transmission and Telemetry

Significant advances have been made in data transmission and telemetry technology in recent years. Several observation systems are experimenting with wireless data telemetry technologies (such as Iridium satellites, IP modem, cellular phone, and underwater acoustic modems). These methods will provide wider coverage, faster data transmission, reduced field work and cost. The Iridium systems consists of a constellation of 66 low-earth orbit satellites (780km vs 35,000 km for geostationary orbits). Its merits include complete global coverage, less transmission delay, small and low power consumption receivers. This system therefore offers a good tool for emergency response applications. One of the major constraints in long-term ocean observation is power demand. The power demand profile for observational tools consists of the required battery capacity for the defined mission period, the load profile, battery chemistry, ambient temperature, self-discharge, terminal voltage, and the efficiency of the battery under the envisaged load profile. The peak power demand that occurs during the data transmission to the INMARSAT satellite transmission is 32 W, which is considerably higher than the 5 W required for an Iridium satellite communication.

(d) Quality Control and Corrections

Real-time observations of oceanographic and meteorological parameters are vital to study various short-term and long-term climatic events like cyclones, monsoons, as well as for weather forecasts. Satellites, ship-based observations, moored buoys, drifting buoys, and profiling floats are the various platforms used in modern times to get the real-time ocean and meteorological observations. These platforms sometimes transmit erroneous/bad data. Hence, to ensure the credibility of data, quality control procedures have to be adopted. Various agencies such as NDBC [7] and Quality Assurance of Real Time Ocean Data (QARTOD) have published manuals for the quality control of data sets. Simple automatic QC procedures at the data-processing centers include gross-range check, spike check, stuck value check, and climatological limits. Gross-range check looks for the values outside the range of a sensor and can measure and flag the data. The spike test looks for the difference between the present value and the average of the previous day, and if the difference is greater than the threshold value, it is marked as spike. Some parameters have a large range of variability within a day, and it is really very hard to find the real spikes in data. The wind speed during cyclonic storms increases very rapidly and crosses the threshold set for a spike. In such cases, other tests such as a correlation test with related parameters such as sea-level pressure (SLP) are used to flag the data. Stuck values or unchanging values arise due to failure of sensors or the communication system. A climatological check looks for the data points which are falling outside the seasonal expectations. Higher level delayed QC involves comparison of data with nearby or colocated observations from different sources. For example, the temperature and salinity data from buoys or floats can be validated against a CTD cast to understand drift in the sensors. Also, multiple sensors utilizing different technologies for measuring a parameter can be colocated, and the data obtained can be compared. This close neighbor approach would provide the ultimate QC check, but cost prohibits such a deployment in most cases. Density inversion tests are performed in the case of subsurface temperature and salinity measurements. Densities are compared at consecutive levels from top to bottom profile. If the density calculated at the greater pressure is less than that calculated at the lesser pressure, both the temperature and salinity values should be flagged as bad data. In delayed mode QC, the data obtained from in-situ platforms can also be compared with model simulations.

3 Emerging Trends

Oceans have been continuously explored by humans for many years. Ships, satellites, and Argo floats have been the backbone of ocean observation systems for the past many years. These are now networked and being used simultaneously for greater understanding, along with the development of new technologies. Figure 3 shows an example of a recent Indo-US collaboration OMM-ASIRI process cruise which used multiple air–sea flux moorings and a series of autonomous observational tools such as wire walkers, gliders, a robotic oceanographic surface sampler, and a large number of drifting buoys. Examples of rapidly innovative technologies include cabled observatories, animal-borne instruments, wire walkers, semi-autonomous robotic surface vehicles, and gliders, which will help to improve our measurements. An array of various sensors connected to the Internet through underwater cable will allow us to detect various energetic phenomena such as big earthquakes and volcanic eruptions remotely, thereby bringing the ocean to the desktop. Cabled observatories are yet another novel technology which has solved the power demand for long-term continuous measurements of the oceans.

While the Argo floats have increased the data density in the top 1500 m, there is a need for increasing observations to deeper depths. Long-term time series of deep observations are necessary to examine the heat budget of the oceans and monitor the long-term changes in the heat capacity of the ocean. To be able to forecast the basin-scale heat budget changes, deep ocean measurements are required. Thus, new designs such as the spherical glass deep Argo floats are emerging to enable these measurements.

Ecosystem monitoring and forecasts require in-situ observations of biogeochemical variables. Therefore, Argo floats incorporating chlorophyll, oxygen, nitrogen, backscatter, color-dissolved organic matter (CDOM), and irradiance measurements have been developed, which are called Bio-Argo (biogeochemical Argo) floats.

Numerical models have improved as faster computers have allowed better resolution, and they are important and widely used tools for oceanographers and meteorologists. Numerical modeling uses mathematical tools (differential equations), which are solved in time and space to simulate the oceanic and atmospheric conditions.

Fig. 3 Ocean data collection during the joint Air Sea Interaction Research Initiative (ASIRI) and Ocean Mixing and Monsoons (OMM) cruise in the Bay of Bengal, showing the US research vessel Roger Revelle and the Indian research vessel Sagar Nidhi

Accurate information about the marine environment is essential for shipping and navigation, defense purposes, fishing, and renewable energy. However, while solving these complex partial differential equations, a number of approximations are made. This leads to generation of errors, which will eventually grow with the model's integration time. To eliminate these errors, the models need to be checked against data, and, during the evolution, corrections need to be applied to the models. Data assimilation, the combining of model output and observations, is necessary to reduce the errors or uncertainties in forecasts. With the availability of the observations from global networks and local data sets discussed in Sect. 2, the process of data assimilation is getting increasingly more detailed. One of the recent challenges is to assimilate high-resolution data sets into numerical models. The current generation of nowcast and forecast models is capable of resolving mesoscale eddies in the oceans. Several forecast systems developed by different countries provide global ocean coverage and provide global analyses and medium and extended range forecasts of 7–18 days depending on the system, and long-range forecasts of 7 months [17]. These are increasingly done using cloud server technologies and parallel computation. Another direction of advancement is numerical investigations of biogeochemical processes to understand the oceanic ecosystem. Models like MEDUSA

(Model of Ecosystem dynamics, nutrient Utilization, Sequestration and Assimilation) and Nucleus for European Modeling of the Ocean (NEMO) are used to study the physical processes that control the distribution and activity of marine ecosystems and also to forecast the oceanic acidification and its effect on marine organisms. New mathematical models to resolve tides and waves, as well as various coupled models (ocean–wave–atmosphere–ice), are being developed [17].

Fine resolution models conducting large eddy simulations and process studies are emerging to focus on the processes which occur at scales too fine to be resolved in global, basin scale, and regional models. These process studies are then used to develop parameterizations for large-scale models. As the supercomputing capability increases in future due to faster electronics and new chip designs, future developments should also consider user efficiency, higher temporal and spatial resolution, as well as improved parameterizations of unresolved processes.

A combination of ocean observations and model simulations allows characterization of individual processes that contribute to recent changes in the ocean. The topic of in-situ ocean measurements is an evolving field, and there is continued interest in adapting new technologies to this topic. Oceanic observations are the backbone for any kind of operational services (operational forecasts, cyclones, storm surges, monsoon variability, tsunami), research and development including validation of satellite sensors, and parameterizing key processes for models and verifying model simulations. First, ocean events are dynamic and require rapid response to measure the phenomena such as eddies and other ocean currents that need to be observed in hours or days. Second, the ocean environment requires study of subtle signatures that challenge in situ instrumentation to accurately distinguish between event types. Third, since the ocean presents a hostile environment, designing and deploying autonomous underwater vehicles, gliders, and drifting buoys presents a range of engineering and technological challenges. Finally, observing systems are required to be operated in extended periods autonomously considering the remoteness of such ocean observations.

The accumulation of climate-relevant time-series parameters, for example, surface ocean currents and deep ocean temperature, will support progress of climate sciences in the years to come. Contributions from citizen-based science (CS) are a novel opportunity to fill observational gaps and increase environmental stewardship among the public (Fig. 4). CS is being used from data collection through to hypothesis generation [11].

4 Concluding Remarks

These are exciting times for future of ocean observations and their integration into predictive and ocean process models. While the ocean observatories are challenging to set up, the research community recognizes that technical ambition and scientific

Fig. 4 Futuristic forecasts will include integrating multiple networked platforms along with the "Citizen in Science" participation

excellence must be balanced with a promise of societal value. New technologies enable multiple networked tools to monitor our oceans economically and often provide greater coverage than traditional shipboard surveys. Modeling to integrate observations as well as diagnose and understand key processes is also a very important component of modern oceanographic research.

Currently using data (i.e., direct and remote measurements, model results) allows ocean scientists to accurately describe the processes occurring in the marine environment and make predictions. The future heralds several new directions, including the role of Citizen Scientist, a systemic approach to address industry needs, and research into biodiversity and climate change. Other developing areas include coastal tourism, biodiversity forecasts, and applications to marine infrastructure developmental projects. Another emerging trend would be the ability to run task-specific ocean models on mobiles and tablets.

Acknowledgments The authors thank the financial support extended by Ministry of Earth Sciences, Government of India, and the Office of Naval Research, USA, for their support during the Indo US Collaborative program (OMM-ASIRI) to conduct field experiments in the Bay of Bengal which led to the idea of writing this chapter.

References

1. Cazenave A, Dominh K, Guinehut S, Berthier E, Llovel W, Ramillien G, Ablain M, Larnicol G (2009) Sea level budget over 2003–2008: a reevaluation from GRACE space gravimetry, satellite altimetry and Argo. Glob Planet Chang 65:83–88
2. Balaguru K et al (2012) Ocean barrier layers' effect on tropical cyclone intensification. Proc Natl Acad Sci 109(36):14343–14347
3. Balaguru K et al (2014) Increase in the intensity of postmonsoon Bay of Bengal tropical cyclones. Geophys Res Lett 41(10):3594–3601
4. Durack PJ, Wijffels SE, Matear RJ (2012) Ocean salinities reveal strong global water cycle intensification during 1950 to 2000. Science 336(6080):455–458
5. Gille ST (2008) Decadal-scale temperature trends in the Southern Hemisphere ocean. J Clim 21:4749–4765
6. Joseph S, Freeland HJ (2005) Salinity variability in the Arabian Sea. Geophys Res Lett 32:4
7. Leuliette EW, Miller L (2009) Closing the sea level rise budget with altimetry, Argo, and GRACE. Geophys Res Lett 36:L04608
8. Masumoto Y, Horii T, Ueki I, Hase H, Ando K, Mizuno K (2008) Short-term upper-ocean variability in the central equatorial Indian Ocean during 2006 Indian Ocean Dipole event. Geophys Res Lett 35:L14S09
9. McCarthy GD, Smeed DA, Johns WE, Frajka-Williams E, Moat BI, Rayner D, Baringer MO, Meinen CS, Collins J, Bryden HL (2015) Measuring the Atlantic Meridional overturning circulation at 26°N. Prog Oceanogr 130:91–111
10. McPhaden MJ, Foltz GR, Lee T, Murty VSN, Ravichandran M, Vecchi GA, Vialard J, Wiggert JD, Yu L (2009a) Ocean-atmosphere interactions during cyclone Nargis. Trans. Am. Geophys. Union (EOS), 2009, 90, 53–54
11. McPhaden MJ et al (2009b) The Global Tropical Moored Buoy Array. Community White Paper, Oceanobs'09
12. NDBC (2009) NDBC Technical Document 09–02, Handbook of Automated Data Quality Control and Checks and Procedures
13. Roemmich D (2007) Physical oceanography – super spin in the southern seas. Nature 449:34–35
14. Venkatesan R, Shamji VR, Latha G, Mathew S, Rao RR, Muthiah A, Atmanand MA (2013) In situ ocean subsurface time-series measurements from OMNI buoy network in the Bay of Bengal. Curr Sci 104(9):1166–1177
15. Venkatesan R, Mathew S, Vimala J, Latha G, Arul Muthiah M, Ramasundaram S, Sundar R, Lavanya R, Atmanand MA (2014) Signatures of very severe cyclonic storm Phailin in met–ocean parameters observed by moored buoy network in the Bay of Bengal. Curr Sci 107(4):589–595
16. Fritz S, Fonte CC, Linda (2017) The Role of Citizen Science in Earth Observation, remote sensing, licence from MDPI, Basel, Switzerland 2017, pp 1–13 http://creativecommons.org/licenses/by/4.0/
17. Tonani M et al (2015) Status and future of globals and regional ocean prediction systems. J Oper Oceanogr. https://doi.org/10.1080/ 1755876X.2015.1049892
18. Willis JK, Fu LL (2008) Combining altimeter and subsurface float data to estimate the time-averaged circulation in the upper ocean. J Geophys Res Oceans 113:C12017

Part II
Surface Observations

Observing Surface Meteorology and Air-Sea Fluxes

Robert A. Weller

Abstract The focus of this chapter is on unattended observations of surface meteorology made from surface buoys. Also discussed is the computation of the air-sea fluxes of heat, freshwater, and momentum from the surface meteorological data obtained from surface buoys. The computation of these fluxes requires measurements of wind speed and direction, air temperature and humidity, barometric pressure, incoming shortwave radiation, incoming longwave radiation, sea surface temperature and salinity, and surface currents. Progress has been made in recent years in developing low power, robust, and accurate instrumentation. The present state of the sensors and related hardware is summarized. Typical buoy installations are described, and a summary of where these observations have been made is presented. The accuracies of the observations of surface meteorology are discussed. These accuracies are used together with the bulk formulae and intercomparisons to estimate the uncertainties in the flux observations. In many cases, it is now possible to make measurements of the mean net air–sea heat flux to an accuracy of 8 W m^{-2}. Challenges remain, and these challenges are also discussed.

1 Introduction

Observing the surface meteorology and the air-sea fluxes of heat, freshwater, and momentum is an essential component of the global ocean observing system and of research studies aimed at improving our understanding of the coupling between the ocean and atmosphere. It is these vertical exchanges across the air-sea interface that couple the ocean to the atmosphere. The sea surface is where the exchanges occur and that provides the bottom boundary to the atmosphere over approximately 70% of the earth's surface. At the same time, the sea surface is the top of the ocean that provides large storage capacity for heat and other properties.

For Observing the Oceans in Real Time – Instruments, Measurements and Experience
March 31, 2016

R.A. Weller (✉)
Woods Hole Oceanographic Institution, Woods Hole, MA, USA, 02543
e-mail: rweller@whoi.edu

R. Venkatesan et al. (eds.), *Observing the Oceans in Real Time*, Springer Oceanography, DOI 10.1007/978-3-319-66493-4_2

The benefits of observing the surface meteorology and air-sea fluxes are apparent over a wide range of phenomena and timescales. Latent heat flux that provides energy from the ocean to hurricanes is seen as a cause for rapid intensification of hurricanes [10]. Improved observations, understanding, and thus representation in models of the air-sea fluxes are seen as necessary for improved prediction of the Asia-Pacific monsoon [24]. Dramatic impacts on regional weather, rainfall, and drought are associated with El Niño, and El Niño monitoring and prediction efforts depend upon the observations of the surface winds and air-sea coupling across the tropical Pacific Ocean [15]. More generally, ocean models require accurate surface forcing; and considerable effort (e.g., [12]) has been directed at developing accurate global surface forcing data sets.

Past climate research programs had identified targets for the accuracies sought in surface meteorological observations and in the air-sea fluxes of heat, freshwater, and momentum. In reviewing these, Weller and Taylor [21] found that the World Ocean Circulation Experiment (WOCE) had set accuracy requirements for surface meteorology as follows: surface pressure – 0.3 hPa, precipitation – 0.4 mm day^{-1} or 15 cm yr.$^{-1}$, sea surface temperature – 0.25 °C, air temperature – 0.25 °C, sea surface salinity – 0.02 PSU, specific humidity – 0.25 g kg^{-1}, surface wind speed – larger of 2% or 0.2 m s^{-1}, surface wind direction – 2.8°, incoming shortwave radiation – 10 W m^{-2}, incoming longwave radiation – 10 W m^{-2}. The WOCE requirements on air–sea fluxes were set for 30-days; spatial averages over a 2° latitude × 10° longitude area; total heat flux – 10 to 15 W m^{-2}, and wind stress – ±10%. WOCE targets were set from the perspective of improved understanding of observed change in ocean surface mixed layer temperature and atmospherically forced ocean dynamics. The Tropical Ocean Global Atmosphere (TOGA) research program established similar accuracy targets and also pursued efforts to close upper ocean heat and freshwater budgets, testing the accuracy of observed surface heat and freshwater fluxes against those surface fluxes inferred from closing the budgets in a volume of the upper ocean. The TOGA Coupled Ocean Atmosphere Response Experiment (COARE) [22] in particular had a focus on improving surface meteorological and air-sea flux observations through deployment of multiple platforms and dedicated periods of in-situ intercalibration. TOGA COARE sought to observe net air-sea heat flux to ±10 W m^{-2} [20]. More recent work on the earth's energy balance (e.g., [18]) has heightened attention to the uncertainties in various estimates of heat fluxes and points to the value of accurate in-situ air-sea flux estimates from buoys in assessing the realism of models used to examine global energy budgets.

The goals of this chapter are to outline what surface meteorological variables need to be observed and what challenges are faced in making those observations unattended at sea, to describe the present state of the art in sensors and related electronics and packaging, to show an example of at-sea data logging and telemetry hardware, to review how the observations are used to calculate the air-sea fluxes of heat, freshwater, and momentum and also to describe the uncertainties associated with these estimates, and finally discuss future work on observing surface meteorology and air-sea fluxes. An additional resource for those seeking to make meteorological and air-sea flux measurements at sea can be found in the technical report by Bradley and Fairall [3].

2 The Required Observations and the Challenges

The observations of surface meteorology to be made include sampling sea surface temperature, surface wind velocity, sea-level barometric pressure, air temperature and humidity, precipitation, incoming shortwave radiation, and incoming longwave radiation. With these observations, the bulk formulae can be used to compute the air-sea fluxes of momentum, heat, and freshwater. In the bulk formulae, it is the wind velocity relative to the surface current that is required; so, a surface or, more practically, a near-surface current observation that can be used as an estimate for surface current is needed. Similarly, a sea surface skin temperature is needed, for which a measurement at the near-surface depth is made and used to estimate skin temperature. Because of the dependence of the surface saturation specific humidity on ocean salinity, the surface salinity or a proxy is also sought as an observation.

The bulk formula for vertical flux of horizontal momentum or wind stress is

$$\tau = \rho C_D \left|U_S - U_z\right|\left(U_s - U_Z\right) \tag{1}$$

where U_s is the surface current vector, U_z is the wind velocity vector at height z, ρ is air density, and C_D is the drag coefficient which varies with stability and height. For the latent heat flux, Q_H, and the sensible heat flux, Q_B, the corresponding formulae are as follows:

$$Q_H = \rho C_H \left(q_s - q_z\right)\left|U_s - U_z\right| \tag{2}$$

$$Q_B = \rho c_p C_B \left(T_s - T_z\right)\left|U_s - U_z\right| \tag{3}$$

where q_s is the saturation-specific humidity at the sea surface, q_z is the specific humidity observed at height z, C_H is the transfer coefficient for moisture known as the Dalton number, c_p is specific heat of air, T_s is the skin temperature of the ocean, T_z is the air temperature at height z, and C_B is the transfer coefficient for temperature known as the Stanton number.

These formulae make clear the need for sea surface skin temperature and surface current. Computation of the surface saturation specific humidity brings in the need for surface salinity. The shape of the profiles of U(z), T(z), and q(z) depends on the stability of the atmosphere at the sea surface, with unstable conditions (e.g., warmer ocean than the air above) giving rise to a less rapid change with height compared to neutral and with stable conditions (e.g., cooler ocean than the atmosphere above) giving rise to more rapid change with height above the sea surface. Considerable work in the field has matured understanding of the stability dependence of the transfer coefficients, and the COARE 3.0 flux algorithm is a commonly used package of the bulk formulae for wind stress, latent heat flux, and sensible heat flux ([5]; available via ftp from ftp://ftp1.esrl.noaa.gov). The COARE 3.0 algorithm accepts the surface meteorological observations at their measured heights and returns air temperature and humidity at 2 m and wind velocity at 10 m. Another advantage of the COARE 3.0 code is that it includes a model of the diurnal surface layer in the ocean combined with a cool skin and uses that model to extrapolate ocean temperature from a measured, near-surface depth to a skin temperature estimate. Finally, the COARE 3.0 Flux algorithm, under the assumptions that rain falls with the wind velocity and cools to the dew point temperature, can provide the additional heat and momentum fluxes due to rain. This, of course, requires an observation of the rainfall.

Two remaining heat flux terms are the net shortwave radiation and the net longwave radiation. Incoming shortwave radiation (SW_{in}) is observed, and the reflected, outgoing shortwave radiation is parameterized in terms of an albedo, α, so that

$$Q_{SW} = \text{SW}_{\text{in}}(1-\propto) \quad (4)$$

The albedo, α, is close to 0.06; Payne [16], Taylor et al. [19], and Jin et al. [11] offer sun angle dependent models for the albedo. The net longwave requires an estimation of the outgoing longwave radiation, so that the net longwave is

$$Q_{LW} = LW_{\text{in}} - \left[\epsilon_s \sigma T_s^4 - (1-\epsilon_s) LW_{\text{in}}\right] \quad (5)$$

where the observed incoming longwave radiation is LW_{in}, ϵ_s is the longwave emissivity of the sea surface taken at 0.97, and T_s is the sea surface skin temperature.

The net heat flux is then

$$Q_{\text{net}} = Q_{\text{LW}} + Q_{\text{SW}} + Q_H + Q_B \quad (6)$$

where the sign convention used is for positive heat fluxes to be an oceanic gain. The freshwater flux is derived from the latent heat flux and the precipitation, where evaporation is

$$E = Q_H / \rho_w l_v \quad (7)$$

where ρ_w is seawater density and l_v is latent heat of vaporization, approximately 2380 kJ kg^{-1}. Freshwater flux is then $E-P$.

3 Sensors and Sensor Modules

Our approach toward making reliable and accurate surface meteorological and air-sea flux observations has been to develop a family of sensor modules, where the sensor, the required signal conditions, electronics, the digitization electronics, data logging, and communications interfaces are packaged together for each sensor in a sensor module [9]. Placing the signal condition and digitization electronics as close as possible to the sensor reduces the chance of noise and the challenges of small

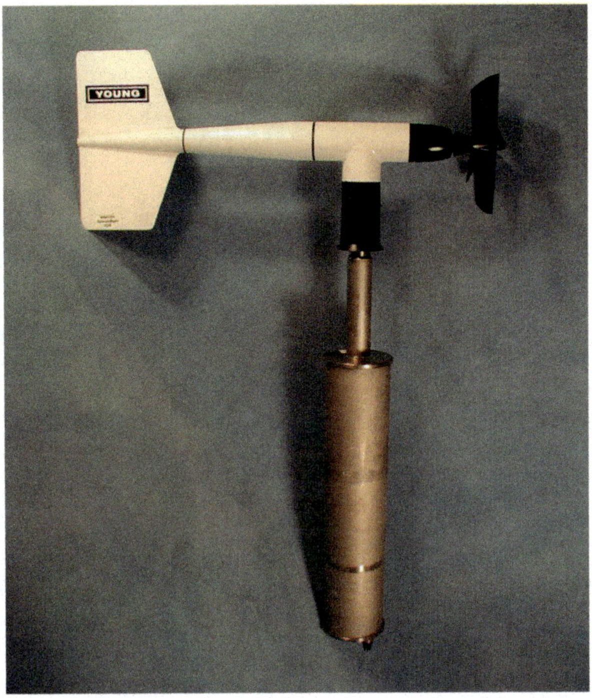

Fig. 1 R.M. Young Marine Wind Monitor Model 05106 fitted to an ASIMET module. Titanium tubing used for the housing has an outside diameter of 3.5 inches (8.89 cm)

signals carried along long cable runs. In addition, placing data storage in the module allows both for a redundant record to be recorded inside the module and for the module, when supplied with power, to be deployed as a stand-alone recording instrument. Normally, on a buoy installation, the full suite of sensor modules is wired together to a central battery and to a central logger. The central logger provides a complete record on a common time base in its data storage. The sensor modules are linked to the data logger via RS-485 communications and power cables. This system has been named the ASIMET system (Air-Sea Interaction Meteorological) and is in commercial production (http://www.starengineeringinc.com/markets/marine-navigation/).

Two types of anemometers have been in use with the ASIMET system, the R. M. Young propeller-vane anemometer (Marine Wind Monitor Model 05103, http://www.youngusa.com) (Fig. 1) and the Gill two-axis sonic anemometer (WindObserver II, http://gillinstruments.com/products/anemometer/wind_ob2.htm) (Fig. 2). In some locations, we have experienced wave damage to the propeller-vane anemometers. In locations with considerable marine bird activity and where birds attempt to land on the buoy, we have also seen damage to the propeller-vane anemometers. The two-axis sonic anemometer has the merit of no moving parts and is deployed to reduce the chance of damage from these sources. Both anemometers must be sited in a region of as undisturbed air flow as possible, on the windward side of the buoy tower where possible.

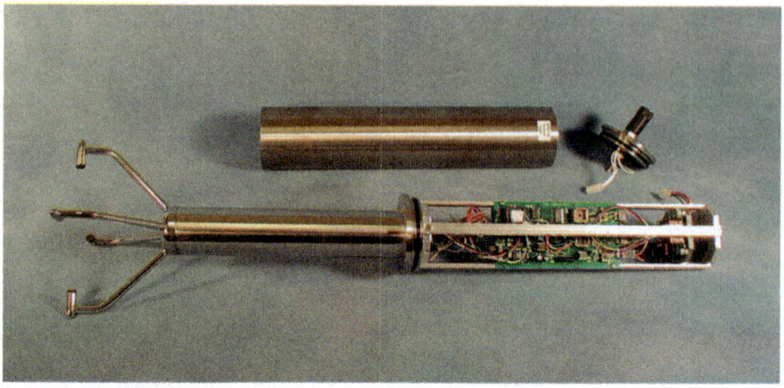

Fig. 2 Photograph of the ASIMET anemometer module equipped with the Gill Wind Observer II sensor, shown removed from the housing with the electronics with compass, signal conditioning, digitization, communications, and data storage electronics exposed. The bottom end cap has an underwater connector for the RS-485 data and power cable

The ASIMET relative humidity and air temperature sensor uses the Rotronic MP-101A sensor (http://www.rotronic-usa.com/products/meteorology/mp100a-mp400a/). This is a combination of a thin film humidity sensor and a thermistor (Fig. 3). The thin film capacitive sensor changes capacitance as water molecules migrate into the dielectric material, and the front end electronics provide a voltage proportional to relative humidity. Because the thermistor and humidity sensor are close together, this is a convenient arrangement for computation of specific humidity. There are two main challenges to use at sea: solar heating and the presence of saltwater. Sunlight can heat the sensor. To reduce this, the sensor is nested inside a stack of plate-like shields, arranged so that there is no direct path for the sunlight to reach the sensor and to minimize sunlight reflecting from the sea surface from reaching the sensor. This radiation shield is known as a Gill multiplate shield [7, 8]. In practice, these shields work well in sufficient wind, but when winds are low, below ~3 m s^{-1}, errors in air temperature can be as high as 3 °C in strong sunlight [1]. In low winds, an active ventilation system can be used, at the cost of power, to greatly reduce radiative heating error. The second major challenge is that of protecting the sensor from spray and salt, as the saltwater can damage the thin film sensor, and the close proximity of drops of saline water and of hygroscopic salt crystals could produce a local influence on the humidity. We protect the sensor by placing it inside a porous Teflon filter which allows air to pass but not liquid, and also is slick enough not to allow salt crystals to cling to its surface.

Use of either type of wind sensor, propeller-vane or sonic, requires that a compass be included in the electronics of the wind module. Present compasses are of the fluxgate type, which sense the horizontal component of the earth's magnetic field. Care is taken in the construction of the ASIMET modules and mounting brackets to not use any magnetic material that may influence the compass. In addition, as part of the predeployment testing of assembled buoys, the buoy is placed in an area that has been surveyed with a compass, aligned with a sighting reference, and then

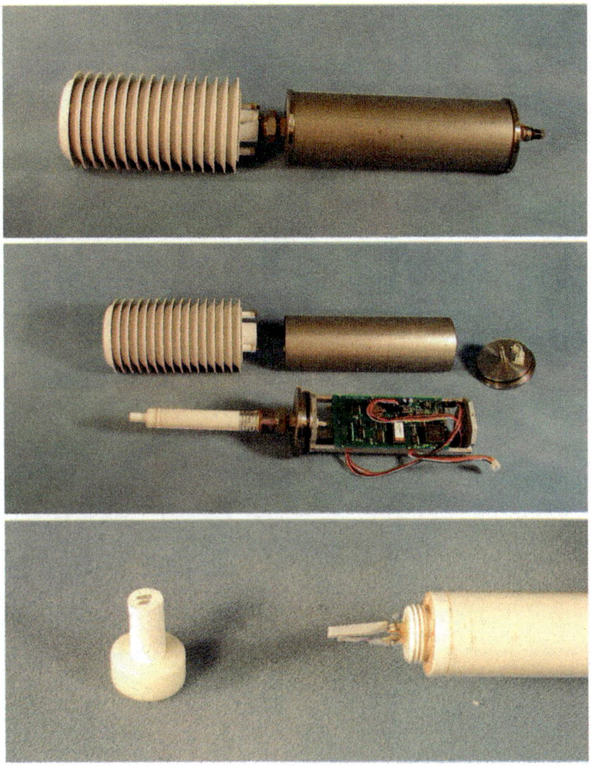

Fig. 3 Two pictures of the relative humidity/air temperature sensor module and a close-up of the sensor assembly. *Top*: assembled, with Gill multiplate radiation shield over sensor assembly; *Middle*: sensor module removed from tube and radiation shield; *Bottom*: Close-up of the sensor assembly showing the white, capacitive thin film humidity sensor, and, difficult to see, a thermistor temperature sensor

rotated, stopping every 45° to record data for ~5–15 min. The resulting comparison of recorded versus actual heading provides an estimate of the uncertainty in the compass, which is the major contributor to uncertainty in wind and wind stress direction.

For barometric pressure sensing, a Heise DXD digital pressure gauge (www.heise.com) has been used, as have other digital barometers. The sensor is fitted in a module, and the access to the outside is via a Gill pressure port [6], which is designed to reduce the impact of dynamic pressure fluctuations associated with the wind on the static atmospheric pressure (Fig. 4).

Incoming shortwave and longwave sensors have a common operating principle and design: radiation falls upon a surface that is painted with optical black paint, and a thermopile network below that surface outputs a voltage driven by the warming. For the incoming shortwave radiation measurement, the black surface lies inside one or more glass or quartz domes that allow shortwave radiation to pass and heat the black surface (Fig. 5). For the incoming longwave radiation measurement, the black surface lies inside an opaque silicon dome, which allows longwave but not shortwave radiation to pass (Fig. 5). However, because parts of the sensor close to the black surface also emit longwave radiation, the temperatures of the body and

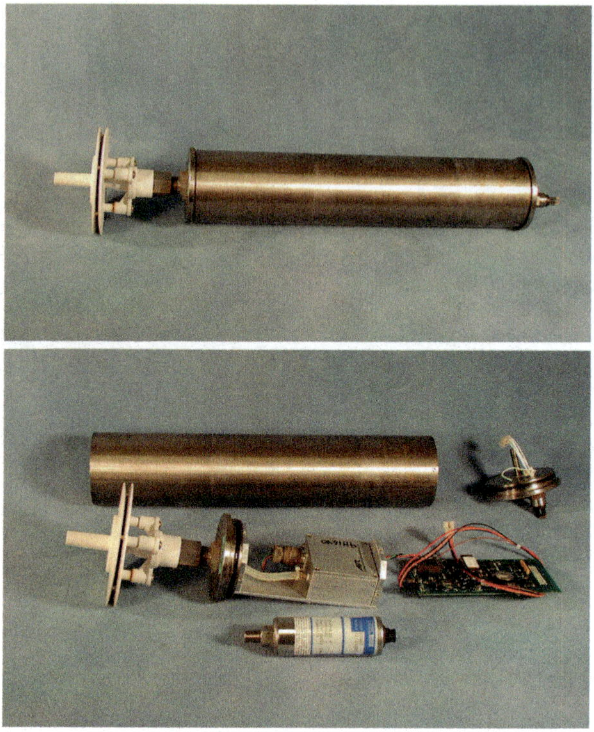

Fig. 4 The ASIMET barometric pressure module assembled (*top*) and apart (*below*). In the bottom image, two different choices of digital pressure sensor are shown as well as the recording and communications hardware on the circuit card and the Gill pressure port to the left, with a thin gap between two parallel plates and a central opening that communicates to the pressure sensor

dome must be measured and corrections made for their contributions to the voltage output from the thermopile.

For both shortwave and longwave radiometers, voltage outputs are in the range of 0–10 microvolts per W m^{-2} of incoming radiation; so, these small voltages must be amplified greatly prior to digitization. This requires stable electronics for signal conditioning and digitization, well shielded from radio frequency and other interference. For the longwave radiometer module, in particular, we found it essential to have amplifiers that were stable and would not change either gain or offset when powered off and powered on again. Some longwave radiometers have internal circuitry to perform the correction for the heating from the near-field radiation from the body and dome, but the ASIMET longwave module records body and dome temperatures as well as thermopile output, so that the computation of incoming longwave can be done outside the module and laboratory calibration information integrated.

The sought after incoming shortwave and longwave radiation are the values of the radiation that falls on a horizontal surface without obstruction and shadowing. Thus, radiometer modules (Fig. 6) should be mounted at the highest point, clear of shadows, and leveled with respect to the waterline of the buoy. At the same time, care may be needed to place them away from radio transmitters used for telemetry, as the radio frequency interference may be picked up by the thermopile networks and/or amplifiers. Buoys with low tension in the mooring line at their underwater bridle may tilt in response to winds and currents; these sensors have a cosine

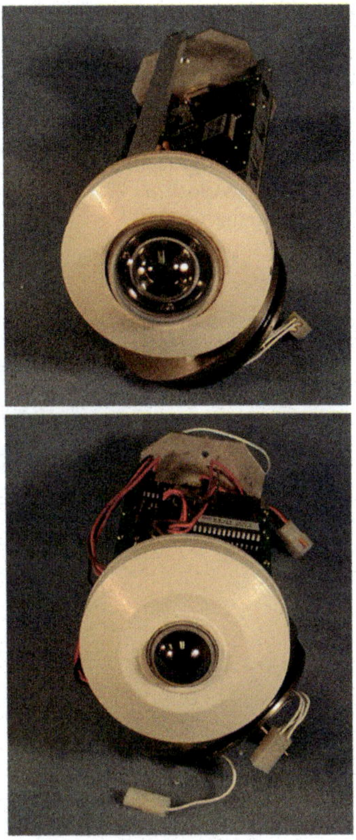

Fig. 5 (*Upper*) The top of an ASIMET shortwave radiation module, showing an Eppley Precision Spectral Pyranometer with clear domes fitted over the black sensor. (*Lower*) The top of an ASIMET longwave radiation module, showing the Eppley Precision Infrared Radiometer with an opaque silicon dome

response, so error at small angles will be proportional to the tilt angle. A buoy with large tension at the bridle tends to stay upright, with little mean tilt, and modest pitch and roll. MacWhorter and Weller [14] recorded buoy motion and then conducted a rooftop experiment in which they drove a two-axis pitch and roll table with a shortwave radiometer mounted on it and compared it to a stationary, leveled shortwave radiometer. Mean errors due to pitch and roll were less of a concern than the errors due to mean tilt, as a combination of the platform tilting toward and away from the sun together with the time response of the thermopile (~ 5 s) leads to rather small errors in the mean. Other challenges include dust, aerosols, or bird guano deposited on the radiometers as well as birds perching on the sensors. Salt spray, based on postcalibration of uncleansed sensors, has not proven to be a large source of error. We have found that some shortwave radiometers also show a change in their calibration with time associated with aging of the optical black paint on the disk under which the thermopile network is mounted. We have addressed this by testing shortwave radiation sensors from other vendors in search of greater calibration stability and also by carrying out annual laboratory calibrations against standards.

The standard rain gauge used with ASIMET is the R. M. Young Model 50,202 precipitation gauge (http://www.youngusa.com/Brochures/50202(0111).pdf). This has a cylinder which can fill with up to 50 mm of water. Once full, a siphon path is

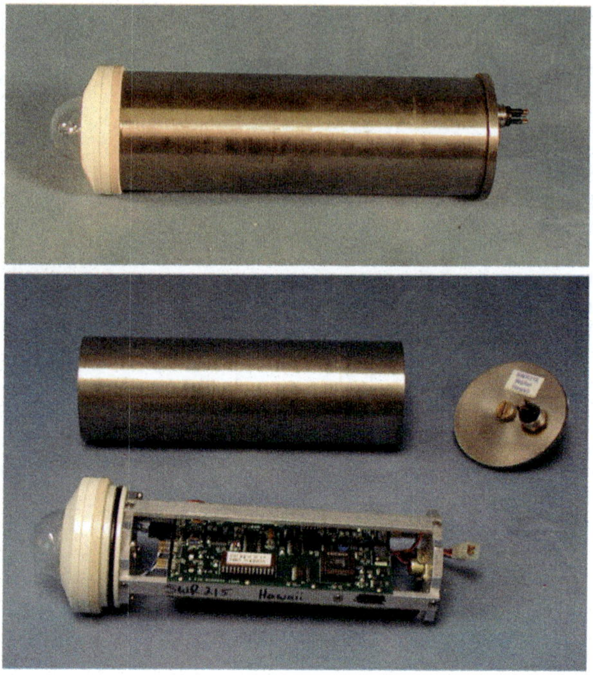

Fig. 6 An ASIMET shortwave radiation sensor module, in its case (*above*) and uncased (*below*). This configuration places the amplifier and digitization electronics as close as possible to the sensor

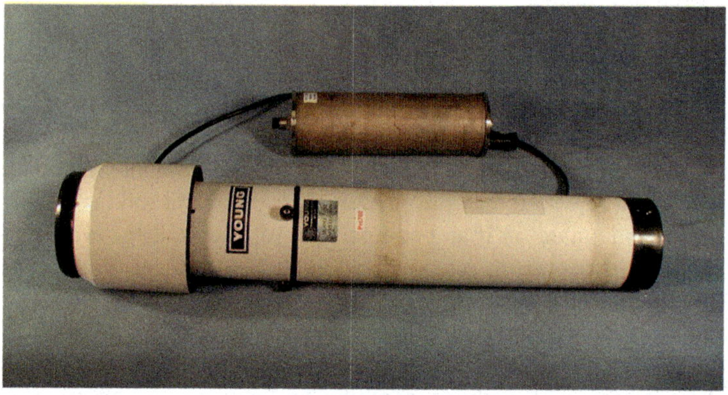

Fig. 7 An RM Young Model 50,202 self-siphoning rain gauge attached to an ASIMET module that digitizes the 0–5 volt DC output of the rain gauge can record data and can provide communication capability

completed and the gauge drains. The water acts as the dielectric inside a capacitor; a circuit outputs a 0–5.0 volt DC signal proportional to the water level inside. The ASIMET module (Fig. 7) records this voltage. The most challenging aspect of the precipitation module is to locate it in undisturbed air flow, doing this to ensure that the accelerated air flow is not blowing rain past the opening at the top of the gauge. Even with care, though, some rain is blown past and this error is wind speed dependent, growing to above 50% in wind speed above 10 m s^{-1} [17].

To obtain near-surface ocean temperature and salinity, the ASIMET system uses a Sea-Bird Electronics SBE 37-SM MicroCAT temperature and conductivity recorder mounted on the bridle legs of the buoy at about 1 m depth. This model has an RS-485 interface, and data are both recorded internally and acquired by the central ASIMET logger. Ocean currents are obtained by either a point current meter placed at ~10 m depth or by taking data from a shallow depth bin in the record from upward-looking acoustic Doppler current profiler.

Technical documentation on the ASIMET system can be found at http://frodo.whoi.edu/asimet/.

4 Buoy Installation, Data Logging, and Telemetry

A typical buoy installation is shown in Fig. 8. Typically, two complete and redundant ASIMET systems are deployed, each with separate battery supply, separate cabling, separate data logging, and separate telemetry. This is done for two reasons. First, it has ensured over all our deployments that a complete surface meteorological

Fig. 8 Two ASIMET systems installed on a 3 m diameter surface buoy. In this case, two propeller-vane anemometers are fitted (**a**), and two relative humidity and air temperature modules are fitted on the windward face (**b**) but mostly hidden from view. A barometric pressure module (**c**) is visible as is a rain gauge (**e**). A cluster of four radiometers, two each of incoming shortwave and incoming longwave, position their sensors at the highest point. A vane or fin (**f**) serves to orient the buoy with respect to the wind. In this case, a floating SST sensor is deployed (**g**) to supplement the SBE 37 SM on the buoy bridle

Fig. 9 A surface buoy equipped with two ASIMET systems lashed to the deck of a research vessel after recovery. The steel pipe bridle is visible under the hull; this is where the SBE 37-SM temperature and conductivity instruments are mounted. The buoy hull is closed-cell foam, with an aluminum box or well in the center to house data loggers and batteries. This buoy was also fitted with a Direct Covariance Flux System (DCFS), and the three-dimensional sonic anemometer from that system is visible between the propeller-vane anemometers

time series, with all variables, has been returned over 98% of the time. Second, when both systems are operational, sensor intercomparisons support examination of drifts and possible degradation. Figure 9 shows a recovered buoy lashed to the deck of a ship.

Within each ASIMET module, the sensors are sampled once per second and a 1-minute block average is formed for barometric pressure, incoming shortwave and longwave radiation, relative humidity, air temperature, rain level, and ocean temperature and conductivity. The wind velocity is vector-averaged over 1 min. Flash cards in each module record the 1-minute data. In addition, each ASIMET system has a data logger (Fig. 9). The data loggers query each sensor module once per minute and log that data. They also prepare 1-h averages which are available for satellite telemetry. Both Service Argos and Iridium have been used for satellite data telemetry. With an emphasis on low power consumption in the design of the ASIMET system, the buoy shown in Fig. 9 is capable of operating for 14 months or longer on alkaline battery packs housed in the buoy well together with the loggers.

The ASIMET systems have also been installed in ships. Research vessels such as the NOAA Ship Ronald H. Brown have ASIMET sensors cabled to their central data acquisition system. Wind, air temperature, and humidity sensors are mounted on a bow mast for clean air flow. Radiation sensors are mounted as high as possible in a shadow-free area upwind of the exhaust stack. Rain gauges are also mounted to obtain as undisturbed air flow as possible. Sea surface temperature comes from an intake pipe, a hull-mounted sensor, or from a sensor trailed in the water.

5 Accuracies Achieved in Surface Meteorology and Air-Sea Fluxes

The accuracies achieved while making surface meteorological measurements from surface buoys have been examined several ways. First, as mentioned above, at least two and sometimes three of the same sensors are mounted on the buoy, allowing the time series to be intercompared. Second, at the end of the typically 1-year long deployment, the new surface mooring is deployed before the existing surface mooring is recovered, allowing an overlap of the time series from the new buoy with freshly calibrated sensors with the time series from the sensors on the existing buoy. Third, for one or more days, the research vessel doing the deployment and recovery is parked bow into the wind within ~1/2 nm of the buoy so that shipboard data can be obtained for comparison with the data from the surface buoys (Fig. 10). Whenever possible, we have had the participation of another research group (Dr. Chris Fairall, NOAA Earth System Research Laboratory, Boulder, CO, USA), and they bring their own freshly calibrated sensors and, when necessary, mount their own bow tower on the vessel to obtain optimal mounting locations for their sensors. During the ship versus buoy comparisons, the telemetry from the buoy is monitored and the telemetered hourly surface meteorological data are compared to the shipboard data sets. Later, when the buoy is recovered and the 1-minute surface meteorological data available, these more frequently sampled data are brought into this intercomparison. Finally, all recovered sensors are removed from the buoy tower and preserved as recovered for postdeployment calibration.

Calibrations are an important component of the work to ensure sensor accuracy. During preparations for a deployment, each module is calibrated. A pressure test facility calibrates the barometric pressure sensors. A controlled temperature and humidity chamber is used to calibrate the humidity sensors. A water bath provides a temperature calibration capability. Combined with a black body cavity that is immersed in the water bath, it also provides the means to calibrate the longwave radiation sensors. Shortwave sensors are mounted on a rooftop facility and calibrated against higher quality standards that are returned to radiation calibration facilities on an annual basis. Before a deployment, each buoy is built up in the configuration in which it will be deployed, placed outside in an area of known magnetic field, and allowed to run for up to several weeks. During that time, the telemetered and recorded time series are examined, with particular attention to differences between the data from the same sensors on the redundant systems mounted on the buoy. As mentioned above, this predeployment test includes rotating the buoy through 45° steps to examine the performance of the compasses in the anemometers. The basic, 1-minute data from the modules are examined for any evidence of radio frequency interference by satellite data transmitters installed on the buoy.

Fig. 10 The Research Vessel Melville sitting next to a surface buoy deployed under the marine stratus clouds 800 nm west of northern Chile. The ship is holding bow into the wind near the buoy to support intercomparison of shipboard meteorological data with that from the buoy. Note the bow tower mounted on RV Melville just aft of the jackstaff to provide a place to mount the ship's sensors

Many of these practices evolved based on experience gained in the TOGA COARE [22], and that reference remains a useful document for those interested in buoy sensor performance and field intercomparisons. More recently, an examination of many years of data collection from ASIMET systems deployed on a surface buoy at 20°S, 85°W, some 800 nm west of northern Chile has been used to quantify the accuracies of the ASIMET sampled surface meteorology and derived air-sea fluxes in moderate conditions [4]. In this study, it was recognized that instantaneous uncertainty in a measurement from a sensor on a buoy stemmed from contributions from: (1) the laboratory calibration, (2) drift during the year deployed, and (3) errors unique to being in the field (e.g., from platform motion or orientation, from solar heating, from salt spray, from birds). Further, it was found useful to differentiate between instantaneous uncertainty and the uncertainties in averaged values, such as daily and annual. In developing understanding of these uncertainties, the laboratory calibrations, predeployment and postdeployment; the field intercomparisons with shipboard sensors; and the field intercomparisons of the same sensors on the same buoy and of the same sensors on fresh and deployed buoys were used.

Table 1 summarizes the findings of Colbo and Weller [4]. Some uncertainty arises in the laboratory calibration and can be quantified looking at the fit of calibration points to the resulting calibration equation(s). Drift can occur even with the sensor sitting on the shelf in the laboratory and is quantified by tracking calibration histories of undeployed sensors. In the field, a number of impacts are seen. As mentioned

Table 1 Summary of ASIMET sensor uncertainties stemming from laboratory calibration, sensor drift, and field impacts with estimates of total uncertainties in instantaneous, daily, and annual values

Sensor	Lab calibration	Drift	Field	Instantaneous	Daily	Annual
Longwave radiation (W m^{-2})	Coefficients of fit: 1.5 Noise: 0.5	2	Tilt: <2 Temperature Gradients: 4 Salt spray: <1 Solar: <1% SW_{in}	7.5	4	4
Shortwave radiation (W m^{-2})	2	<2	Tilt: <2 Temperature Gradients: 1–2 Salt spray: < 1	20 (more in broken cloud)	6	5
Humidity (%Rh)	Linear: 0.16 Cubic: 0.1	0.9	Under 95% RH: ±1 Heating in low winds: 3	1 3 (low winds)	1 3	1
Air temperature (°C)	<0.03	0.05	> 1, wind <1 m s^{-1} 0.7, wind = 2 m s^{-1} 0.4, wind = 3 m s^{-1} Radiation: 0.2	0.2 (more in low wind)	0.1	0.1
Barometric pressure (hPa)	0.06	1.5 (max) 0.2	Temperature: 0.1 Wind: <0.1 (wind <10 m s^{-1})	0.3	0.2	0.2
Sea surface temperature (°C)	0.001	0.05	Low wind: 0.1 Cool skin: <0.02	0.1	0.1	0.04
Wind speed (m s^{-1})	1%	+0.1	Tilt: <0.3% Sea state: Uncertain Very low wind: ±1	1.5%, 0.1 (more in low wind)	1%, 0.1 (max of these)	1%, 0.1 (max of these)
Wind direction (°)	Raw compass: 1° Buoy spin: 4°	2°	Low wind: 1 Flow distortion: <5	6 (more in low wind)	5	5

After Colbo and Weller [4]

above, the radiometers should be level, and any mean tilt introduces more significant error than pitch and roll. Postdeployment calibration shows a small effect due to salt spray. Uneven heating of the radiometer domes at low sun angles contributes to thermal gradients across the instrument body and thus to uncertainty. Comparison of pairs of radiometers revealed an unanticipated source of uncertainty that stems from the clocks in the modules. With clouds passing over, the incoming shortwave record can show high-amplitude fluctuations with short timescales. In this situation, when there is a clock drift and the time base of two shortwave radiometer modules become slightly shifted, instantaneous errors stem from real signal associated with incorrect times; hence, the value of 20 W m^{-2} stated by Colbo and Weller [4] in broken clouds.

Table 2 Accuracy in heat flux components, in net heat flux, in the magnitude of the wind stress, and in freshwater flux for an ASIMET system deployed in the subtropics

	Q_{LW}	Q_{SW}	Q_H	Q_B	Q_{net}	$\lvert\tau\rvert$	*E-P*
Percent error	10	2.5	5	15	20	20	10
Typical error	3.9 W m^{-2}	5 W m^{-2}	5 W m^{-2}	1.5 W m^{-2}	8 W m^{-2}	0.007 N m^{-2}	10 cm

After Colbo and Weller [4]

The use of the Teflon filter on the humidity sensors has proven effective. Absent active ventilation, error stems from radiative heating inside the Gill multiplate shield. The calibration curve derived from laboratory calibrations is based on calibration points at 95% RH and below, so greater uncertainty can be seen above 95% RH. Temperature sensors are stable and can be calibrated well. The challenges stem from radiative heating of the air temperature sensor in low wind and from sampling ocean temperature in the presence of a cool skin and a strong daytime surface temperature stratification in low wind. Barometric pressure sensors have proven reliable. Wind speed uncertainty, expected in low winds, is most appropriately expressed as a percent of the wind speed. As the buoy towers become more populated with sensors, we have increased the size of the wind vanes on the buoys to maintain their ability to orient into the wind and provide sensors such as the anemometers with good air flow. Before making this change, we have seen more recent buoy deployments where wind speeds do differ more than 1%. Wind direction uncertainty stems mainly from the characteristics of the fluxgate compass and our experience when spinning an assembled buoy, where we see the difference between observed and actual heading to exhibit a sinusoidal character with peak-to-peak amplitude approaching 5°.

Colbo and Weller [4] also considered how the sensor uncertainties impacted the fluxes computed using the formulae presented earlier. Table 2 summarizes their findings for the fluxes. The accuracies of the four individual heat flux components sum to more than the accuracy given for the net heat flux. This is because the largest errors in each component do not occur simultaneously. For example, in low winds and strong sunlight, heating errors may be seen in air temperature while at the same time the buoy would have little mean tilt-related error in the radiometric measurements. In the subtropics, then, the accuracy achieved in the net heat flux of 8 W m^{-2} meets the WOCE and TOGA COARE goals

In a much more challenging environment, we deployed a surface mooring for over a year in the Gulf Stream off the mid-Atlantic Bight of the east coast of USA [23]. Strong, cold winds, high waves, and strong currents all presented challenges, and this deployment to an extent explored a situation in which uncertainties in the air-sea fluxes computed by bulk formulae methods were expected to be large. On the other hand, signals were expected to be large. Weller et al. [23] reported the largest oceanic heat loss of the deployment was −1407.9 W m^{-2} during a cold air outbreak, and the largest oceanic heat gain was 832.8 W m^{-2} during the summer. Bigorre et al. [2] examined the pairs of sensors on this buoy to determine uncertainties in the observed surface meteorology and then looked at the propagation of these errors through the bulk formulae. The concern about possible flow distortion was realized in this heavily equipped buoy (fielded before the vanes were increased in size), and a low bias of 5% in wind speed was noted for winds less than 15 m s^{-1} and less than 10% for higher winds. Uncertainty resulting from the propagation of observational error through the bulk formulae was higher in low winds but asymptoted in higher wind speeds to approximately 12% in latent heat flux, to approximately 10% in sensible heat flux, and to approximately 15% in wind stress. Because the Gulf Stream regime was potentially one in which there would be uncertainties stemming from the use of the bulk formulae methods, Bigorre et al. [2] also compared the bulk formulae-based fluxes with fluxes from a Direct Covariance Flux System (DCFS) deployed on the same buoy. In wind speeds above 6 m s^{-1} and below 15 m s^{-1}, wind stresses were in good agreement. Below 15 m s^{-1}, buoyancy fluxes were in good agreement; above 15 m s^{-1}, bulk formulae (COARE 3.0) buoyancy fluxes were higher than DCFS fluxes. In general, though, Bigorre et al. [2] conclude that even in the Gulf Stream environment the uncertainties in the fluxes were close to or below 20%.

6 Conclusions and Future Work

Considerable progress has been made in both developing sensors suitable for unattended deployment at sea on surface buoys and in quantifying the uncertainties in the surface meteorological observations made with such sensors and in the air-sea fluxes of heat, freshwater, and momentum made using bulk formulae methods. Low-powered electronics form the basis of a system that can run for more than a year on alkaline batteries, thus not requiring power generation on the buoy and not involving the hazards of rechargeable batteries. Rugged sensor housings, the philosophy of placing signal-conditioning electronics close to each sensor in a sensor module with redundant data logging, and ongoing work at testing and calibrating sensors have proven their value. In moderate conditions, an accuracy of 8 W m^{-2} in net heat flux has been achieved.

Some challenges remain. In wind speeds below ~3 m s^{-1}, the passive radiation shields are not effective at preventing radiative heating, and active ventilation would be appropriate if the power were available. Buoy mean tilts should be monitored. Though not that significant an error source at small angles for anemometers, mean tilt is of concern for incoming shortwave and longwave radiation observations. The present rain gauge has errors, but the sampling of rain that is highly variable in space and time also presents challenges. It may be that use of a hydrophone at depth to record the acoustic signature of rain over an area is a better approach for the future [13]. Additionally, work should continue to improve the applicability of bulk formulae such as the COARE algorithm to the low wind speeds (<3 m $s^{-}1$) and the higher wind speeds (>20 m s^{-1}) in which most development of the algorithm has been done.

Acknowledgments Initial development of the ASIMET system was supported by the US National Science Foundation, Grant OCE-8709614. Subsequent support from the US National Oceanic and Atmospheric Administration's Climate Program Office and from the US Office of Naval Research has lead to significant improvements. The work reported here, including the laboratory calibrations, deployments, recoveries, and analyses, was done by the Upper Ocean Processes Group of the Woods Hole Oceanographic Institution; more about this group can be found at http://uop.whoi.edu. Over the years this work has benefited greatly from collaborations with Dr. Frank Bradley, CSIRO, Australia and Dr. Chris Fairall, NOAA ESRL, Boulder, CO.

References

1. Anderson SP, Baumgartner MF (1998) Radiative heating errors in naturally ventilated air temperature measurements made from buoys. J Atmos Ocean Technol 15:157–173
2. Bigorre SP, Weller RA, Edson JB, Ware JD (2013) A surface mooring for air-sea interaction research in the Gulf Stream. Part II: Analysis of the observations and their accuracies. J Atmos Ocean Technol 30:450–469
3. Bradley F, Fairall C (2007) A guide to making climate quality meteorological and air-sea flux measurements at sea. NOAA technical memorandum OAR PSD-311. Earth System Research Laboratory, Physical Sciences Division, Boulder
4. Colbo K, Weller RA (2009) Accuracy if the IMET sensor package in the subtropics. J Atmos Ocean Technol 26:1867–1890
5. Fairall CW, Bradley EF, Hare JE, Grachev AA, Edson JB (2003) Bulk parameterization of air-sea fluxes: updates and verification for the COARE algorithm. J Clim 16:571–591
6. Gill GC (1976) Development and testing of a no-moving parts static pressure inlet for use on ocean buoys. NOAA Data Buoy Office Progress Report 01–6–038-115, 43 pp
7. Gill GC (1979) Development of a small radiation shield for air temperature measurements on drifting buoys. NOAA Rep. Contract 01–7–038-827, National Data Buoy Center, Stennis Space Center, MS, 23 pp
8. Gill GC (1983) Comparison testing of selected naturally ventilated solar radiation shields. NOAA Rep. Contract NA-82-0A-A-266. Stennis Space Center, MS, 15 pp
9. Hosom D, Weller RA, Payne RE, Prada KE (1995) The IMET (improved meteorology) ship and buoy systems. J Atmos Ocean Technol 12:527–540
10. Jaimes B, Shay LK, Uhlhorn EW (2015) Enthalpy and momentum fluxes during Hurricane Earl relative to underlying ocean features. Mon Weather Rev 143:111–131
11. Jin Z, Charlock TP, Smith Jr WL, Rutledge K (2004) A parametrization of ocean surface albedo. Geophys Res Lett 31:L22301, doi:10.1029/2004GL021180
12. Large W, Yeager S (2004) Diurnal to decadal global forcing for ocean and sea-ice models: the data sets and flux climatologies. NCAR Technical Note: NCAR/TN-460+STR. CD Division of the National Center for Atmospheric Research
13. Ma BB, Nystuen JA (2005) Passive acoustic detection and measurement of rainfall at sea. J Atmos Oceanic Tech 22:1225–1248
14. MacWhorter MA, Weller RA (1991) Error in measurements of incoming shortwave radiation made from ships and buoys. J Atmos Oceanic Technol 8:108–117
15. McPhaden MJ et al (2010) The global tropical moored buoy array. In: Hall J, Harrison DE, Stammer D (eds) Proceedings of the "OceanObs'09: Sustained Ocean Observations and Information for Society" Conference, vol. 2, 21–25 Sep 2009, ESA PublicationWPP-306, Venice, Italy
16. Payne RE (1972) Albedo of the sea surface. J Atmos Sci 29:959–970
17. Serra YL, A'Hearn P, Freitag HP, McPhaden MJ (2001) ATLAS self-siphoning rain gauge error estimates. J Atmos Ocean Technol 18:1989–2002

18. Stephens GL, Li J, Wild M et al (2012) An update on Earth's energy balance in light of the latest global observations. Nat Geosci, 5, doi:10.138/NGEO1580
19. Taylor JP, Edwards M, Glew MD, Hignett P, Slingo A (1996) Studies with a new flexible radiation code. II: comparisons with aircraft shortwave observations. QJRMS 122:839–861
20. Webster PJ, Lukas R (1992) TOGA-COARE: the coupled ocean-atmosphere response experiment. Bull Am Meteorol Soc 73:1377–1416
21. Weller RA, Taylor PK (1993) Surface conditions and air-sea fluxes. CCCO-JSC Ocean observing system development panel, 131 pp, Texas A&M University, College Station, TX 77843-3146
22. Weller RA, Bradley F, Lukas R (2004) The interface or air-sea flux component of the TOGA coupled ocean-atmosphere response experiment and its impact on subsequent air-sea interaction studies. J Atmos Ocean Technol 21:223–257
23. Weller RA, Bigorre SP, Lord J, Ware JD (2012) A surface mooring for air-sea interaction research in the Gulf Stream. Part I: Mooring design and instrumentation. J Atmos Oceanic Technol 29:1363–1376
24. Zhu J, Shukla J (2013) The role of air-sea coupling in seasonal prediction of Asia-Pacific summer monsoon rainfall. J Clim 26:5689–5697

Drifter Technology and Impacts for Sea Surface Temperature, Sea-Level Pressure, and Ocean Circulation Studies

Luca R. Centurioni

Abstract The purpose of this chapter is twofold. First, we illustrate the technology used by the Lagrangian drifters deployed by Global Drifter Program (GDP), which is the principal component of the Global Drifter Array; second, we review and summarize the most recent studies on the impact of drifter data for calibration and validation of sea surface temperature (SST) satellite products, Numerical Weather Prediction (NWP) and climate studies, tropical cyclones (TCs)-ocean interaction and ocean circulation studies. Several types of drifters are described, starting from the simplest configuration that measures SST and sea-level atmospheric pressure (SLP), continuing with special drifters designed to measure sea surface salinity (SSS) and sea-level wind (SLW), and ending with air-deployable drifting thermistor chains that measure the temperature of the upper 150 m of the ocean, which are used to study the interaction of the ocean's mixed layer with TCs. We also discuss the implications of the satellite telecommunication technology on the accuracy of drifter's geolocation and on the timeliness of the near real-time data stream.

New opportunities for expanding the scope of the GDP are also discussed.

1 Lagrangian Drifter Technology

The use of Lagrangian drifters to measure near-surface ocean currents is more than a century old. As pointed out by Niiler [30] describes the early use of drifting buoys during the HMS Challenger expedition between 1872 and 1876 and presents the first complete and comprehensive historical and modern perspective of the Global Drifter Program (GDP). Here, we use the term Lagrangian, which in mechanics usually refers to the description of the trajectories of a system of particles from their position and time derivatives, interchangeably with water-following capabilities. Therefore, a good Lagrangian drifter is a device capable of following a volume of water at the surface of the ocean for a certain length of time with minimal relative motion, or slippage, with respect to the surrounding water. It should be emphasized

L.R. Centurioni (✉)
Scripps Institution of Oceanography, La Jolla, CA, USA
e-mail: lcenturioni@ucsd.edu

R. Venkatesan et al. (eds.), *Observing the Oceans in Real Time*, Springer Oceanography, DOI 10.1007/978-3-319-66493-4_3

that surface Lagrangian drifters are not deigned to follow the vertical motion of the water.

Starting in 1978, the Global Atmospheric Research Program (GARP), established under the framework of the World Meteorological Organization (WMO) and of the Intergovernmental Council of Scientific Unions (ICSU) with the objective to better understand the large-scale physical processes necessary to improve weather forecasting and the physical basis of climate, launched the First GARP Global Experiment (FGGE). Interestingly, FGGE had a substantial oceanographic component [United Nations Educational, Scientific and Cultural Organization (UNESCO) report SC-78/WS/91, available online at http://unesdoc.unesco.org/images/0003/000348/034822eb.pdf] and it supported the deployment of several hundred drifters designed to collect sea surface temperature (SST), air temperature, and sea-level atmospheric pressure (SLP). The FGGE program was one of the earliest examples of successful collaboration between oceanographers and meteorologists, which today is still a major thrust of the GDP. However, most FGGE drifters were expensive and did not have an efficient drogue (term used to define the sea anchor that ensures the water following capability of the drifter) attached. The inability of the FGGE drifters to accurately follow near-surface ocean currents was perhaps the main reason behind the decision to abandon this design [30], although some attempts were made to recover ocean current measurements from them [36].

A detailed account of the surface Lagrangian drifter designs in use before the GDP was established is given by Lumpkin and Pazos [23] and shall not be repeated here. It suffices to say that the drifter design that emerged in the early 1980s consisted of a surface buoy containing batteries, electronics such as a digital controller, analog signal digitizers, a satellite modem, and a suite of surface sensors. The surface buoy was attached to a drogue with a tether. In the early days of the GDP, considerable efforts were devoted to investigate the drifter and drogue design that would ensure optimal water-following capabilities, as reported by Niiler et al. [32]. Both holey-sock (Fig. 1) and TRISTAR (Fig. 2) drogues were tested extensively with vector-measuring current meters and dye to quantify the slip of the drogue through the surrounding water. The TRISTAR drogue was determined to be slightly more efficient than the holey-sock, although both types performed satisfactorily as long as the ratio of the drag area of the drogue to the sum of the drag area of all the other elements of the drifters (mainly the tether and the surface sphere) was larger than 40 [32]. The holey-sock drogue was chosen because it allows the drifter to be compacted in a smaller volume package than the TRISTAR, thus making storage, shipping, and deployment of the drifters easier and less expensive, thus opening the way to the implementation of a sustainable global array. The drag area ratio criterion is still used in modern drifters and it ensures that when the drogue is present the drifters can measure the ocean currents with an accuracy of about 0.9 cm s^{-1} for winds up to 10 m s^{-1} [32].

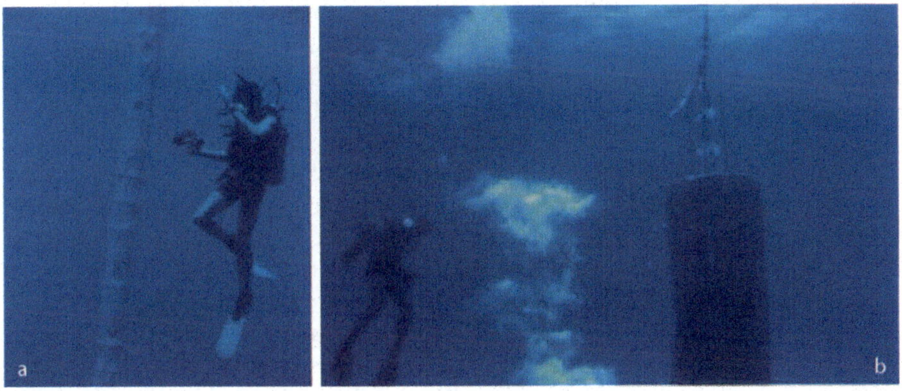

Fig. 1 (**a**) Holey-sock drogue. (**b**) Dye release and holey-sock drogues fitted with vector-measuring current meters to measure the slip of the drogue through the surrounding water

Fig. 2 (**a**) SVP drifter with TRISTAR drogue with Dr. Peter Niiler holding the surface buoy. (**b**) Diver checking the current meters used to measure the slip of the TRISTAR drogue through the surrounding water

1.1 The Velocity-Temperature (SVP) Drifter

The Surface Velocity Program (SVP) was the precursor of the GDP and was established in conjunction with the Tropical Ocean Global Atmosphere (TOGA) program in 1988 [15]. The simplest version of an oceanic drifter, commonly called the SVP drifter, was the system of choice used to implement the GDP. The GDP was the first component of the Global Ocean Observing System (GOOS) to reach completion with the ceremonial launch of drifter 1250 by Dr. Peter Niiler of the Scripps Institution of Oceanography and by Dr. Mike Johnson of NOAA from the Tall Ship Silva near Halifax, Nova Scotia, on September 18, 2005.

The SVP drifter used until 2002 consisted of a surface sphere with a diameter of 0.28 m, a subsurface float with a diameter of 0.2 m located at a depth of about 2 m, and a holey-sock drogue approximately 6 m long and with a diameter of 0.84 m [32]. The length of the tether was standardized so that the center of the drogue along its longest side was located at a depth of 15 m. Such depth was chosen to reduce the direct effect of the surface waves on the drogue (i.e., Stokes drift) and to measure ocean currents in the mixed layer over most of the oceans. In 2002, a smaller version of the SVP drifter was introduced, which consisted of a slightly larger surface buoy (0.38 m in diameter) and a shorter drogue of smaller circular cross section (about 0.6 m of diameter). The subsurface float was removed. The exact size of the surface buoy and of the drogue depends on several factors, including the diameter of the tether, but they are chosen so that the drag area ratio is larger than 40 [32], which, as of today, is a strict requirement for all the drifters operated by the GDP. The smaller version of the SVP drifter was introduced primarily to reduce the cost of the equipment and ease the logistics without affecting its Lagrangian properties.

The SVP drifter designed and produced today by the Lagrangian Drifter Laboratory (LDL) at the Scripps Institution of Oceanography (SIO) is presented here as an example of the modern technology.

In its simplest configuration (Fig. 3), the SVP drifter measures only SST and ocean currents at a depth of 15 m (i.e., the center depth of the drogue).

The drifter's hull has a diameter of 35 cm and is made of acrylonitrile butadiene styrene (ABS). The drogue consists of cylindrical tube of 600 den, or higher, ripstop fabric Cordura® nylon connected to the surface buoy with an impregnated steel wire-rope tether. A thermistor potted with a thermally conductive resin encapsulated into a stainless-steel enclosure is mounted in the bottom part of the surface buoy, 18 cm below the floatation line. The base of the thermistor enclosure that commu-

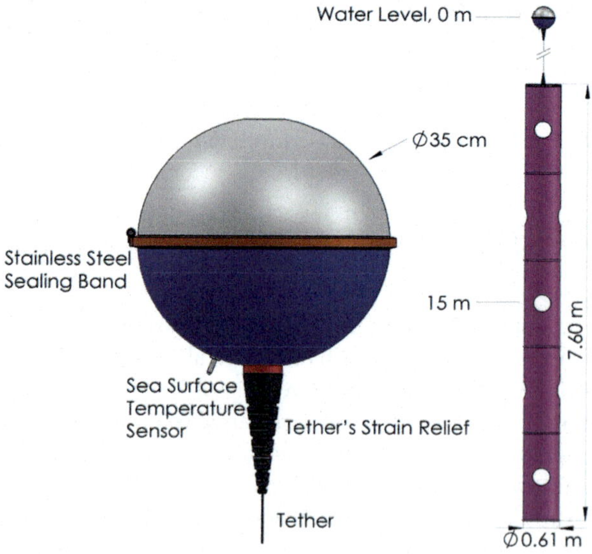

Fig. 3 Schematics of the SVP drifter designed and produced by the Lagrangian Drifter Laboratory at the Scripps Institution of Oceanography

nicates with the inside of the drifter is thermally insulated from the inside of the buoy with a special resin to avoid biased SST readings that may occur when the drifter's hull overheats under the sun. The bench accuracy of the SST measurements is ±0.05 °C after 5-point calibration across the sensing range of the thermistor (−5°C–40°C).

The presence of the drogue is detected with a strain gauge mounted inside the hull and near the tether attachment. The strain gauge senses the deformation of the hull due to the pull of the tether, that is enhanced when a drogue is attached and surface waves interact with the drifter's hull.

All drifters carry a satellite modem for data telemetry. The Argos satellite system has been used in the past to transmit the data to shore and to compute the drifter position using the Doppler shift of the transmitter carrier frequency, yielding a geolocation accuracy that ranges from ~200 m to ~2 km [22]. In 2016, the GDP has begun a transition to the Iridium satellite system. The main advantages of Iridium telemetry are that the data are received by the drifters' operators in a shorter time (typically 1 min compared to 90–120 min with Argos) and the positions are measured with a Global Positioning System (GPS) engine that yields typical geolocation accuracy of 2–50 m (rms). An LDL designed digital controller handles the duty cycle of the data collection (typically once per hour, but it can be changed underway using the two-way Iridium communication) and formats the data (SST and position) for satellite transmission. A comprehensive set of diagnostic data, that are crucial to monitor the health of the drifter array, such as the battery voltage, the hull's internal pressure, temperature, and humidity, are also formatted and appended to the data message.

The GDP drifters are fitted with diode-protected, alkaline D-Cell, 12 V, 56 Ah battery packs. Although alkaline D cells are not designed to withstand shocks and vibrations, they are preferred because they are less hazardous than primary lithium batteries. Since the target lifetime of the drifter is in excess of 400 days, it means that through its life the surface buoy undergoes several millions of impacts resulting from surface waves with mean periods ranging from ~3 s for wind waves to ~20 s for swell [20]. To prevent a quick degradation of the power source, expanding foam and a special resin are used to secure the cells forming the battery pack within a special bucket solidly connected to the drifter's hull. Other padding materials are added to the top of the potted battery pack to further minimize the adverse effects of impacts and vibrations, resulting in a life span in excess of 2 years. Other points of mechanical and electrical failure of drifters have been identified through years of examination of instruments recovered at sea or after running aground. These include the tether, the connections between the tether and the surface buoy, the drogue attachments, and the fabric used for the drogue.

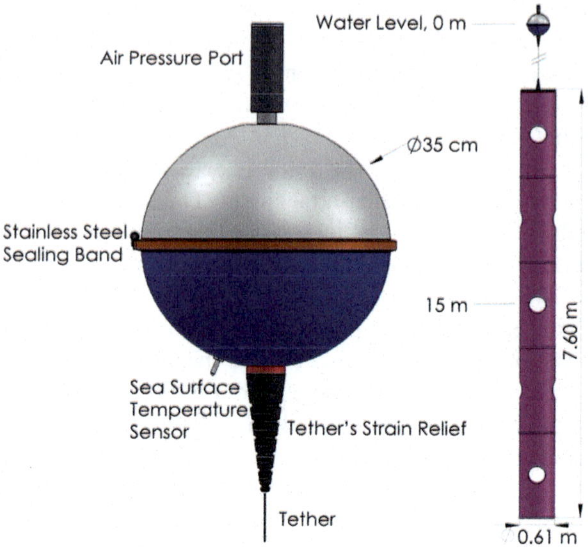

Fig. 4 Schematics of the SVPB drifter designed and produced by the Lagrangian Drifter Laboratory at the Scripps Institution of Oceanography

1.2 The Barometer Drifter (SVPB)

The SVP Barometer (SVPB) drifter (Fig. 4) has the same capability of the SVP drifter but it also carries a barometer to measure SLP [3, 16, 27, 30]. The air pressure sensor can either be a High Precision Barometer (HPB) by Honeywell, stable over the 2 years long nominal life span of the drifters with an accuracy of ±0.4 hPa, or an Integrated Pressure Transducer, also by Honeywell that has similar specifications and accuracy. Since ocean waves often pull the surface buoy underwater the air intake that communicates with the barometer is protected with a specially designed self-draining port and waterproofed with two GORE-TEX® membranes.

When the drifter goes underwater, erroneous SLP readings will occur, and these are filtered out by an on-board algorithm before the data message is formatted and queued for satellite transmission. Like the SVP, the life span of the SVPB drifter is in excess of 2 years.

1.3 The Salinity Drifter (SVPS)

The SVP Salinity (SVPS) drifter [7, 17] also stems from the SVP drifter design and it can be configured with a barometer. It carries Sea Bird Electronics SBE37 sensors and it is significantly more expensive than the SVP and SVPB versions. The salinity sensor is located at a depth of 0.5 m and a slightly bigger drogue is used to compensate for the larger drag area of the surface buoy and of the salinity sensor mount combined, as needed to keep the drag area ratio larger than 40 (Fig. 5). The life span

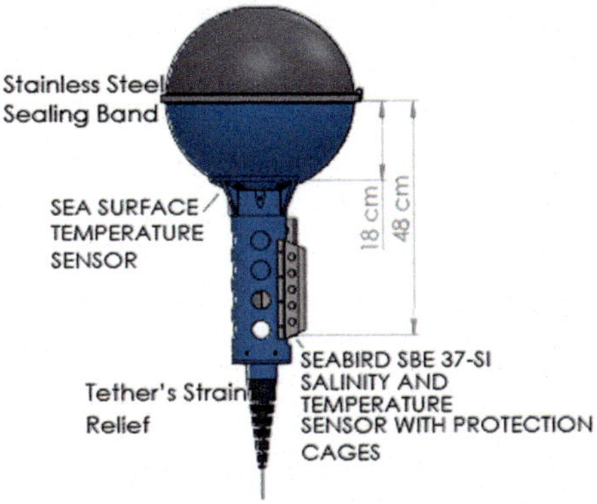

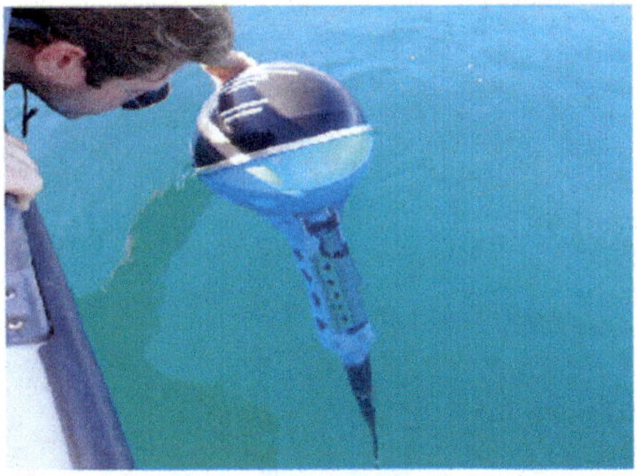

Fig. 5 *Left*, schematic of the SVPS drifter. The standard version has two SST sensors, one directly mounted on the drifter's hull and the second integrated in the SBE37-SI salinity sensor. The schematics of the drogues have been omitted for clarity. *Right*, one LDL SVPS being tested at the Scripps Institution of Oceanography in La Jolla, California

of the SVPS drifter is about one year and is significantly shorter than that for the SVP and SVPB drifters, which is about 1 year.

1.4 The Minimet Wind Drifter (SVPW) and the Autonomous Drifting Ocean Station (ADOS)

The SVPW or Minimet drifter (Fig. 6) was designed to be air-deployed in front of tropical cyclones. It carries the same air pressure sensors of the SVPB drifter and a sonic anemometer, for example, the Gill WindSonic that measures wind speeds up to 60 m s^{-1} with a 2% accuracy FS. The wind direction is measured with an internal compass integrated in the LDL controller and has direction accuracy of ±2° accuracy after calibration. The surface buoy of the Minimet is larger than the one used for the SVPB drifter and features a wind vane that has the purpose to stabilize the buoy in the presence of strong winds. An algorithm similar to the one installed on the SVPB drifter is used to discard the invalid anemometer data that occur, for example, when the drifter is submerged.

The ADOS (Fig. 6) can also be air-deployed. The surface buoy is identical to the one used for the SVPW drifter and a 150-m long tether replaces the drogue, thus making the ADOS a non-Lagrangian device. The device measures SST and 15 pods that measure water pressure and temperature are attached to the cable with 10 m spacing. The subsurface pods use inductive communication technology to send the data through the single-conductor impregnated steel wire rope combined with a sea water contact that closes the electrical circuit. The life span of the ADOS is approximately 3 months. Several air deployments have been made to date to measure the temperature structure of the upper ocean during the passage of tropical storms and

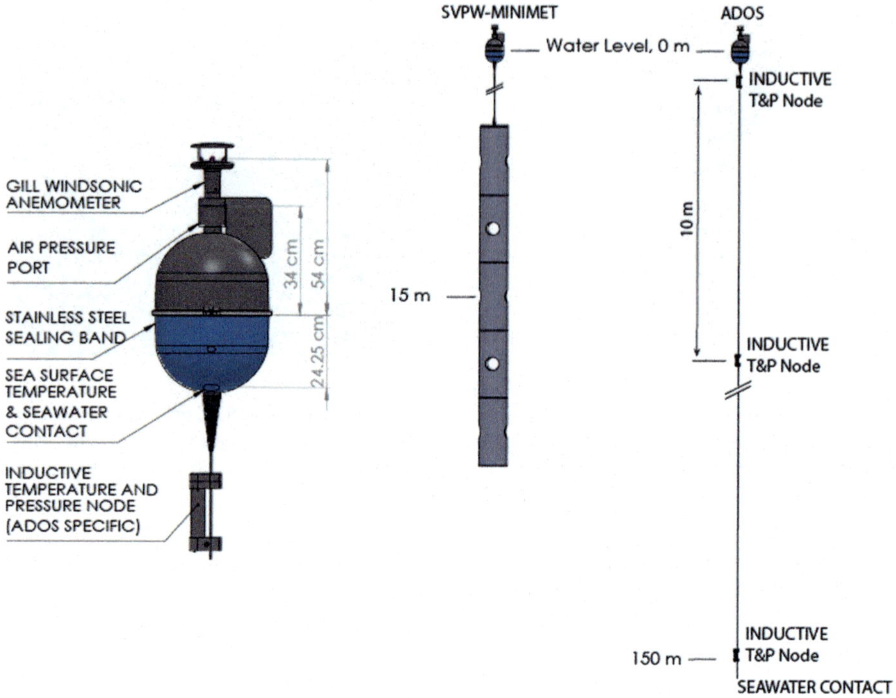

Fig. 6 *Left*, schematics of the surface buoy of the SVPW and ADOS drifters. A modified air pressure port is used to mount the wind sensor. The inductive temperature and pressure pod, specific to the ADOS, is also shown. *Center*: SVPW/Minimet configuration with drogue centered at 15 m depth. The size of the drogue is larger than that for the SVP and SVPB drifters to ensure that the the drag area ratio is larger than 40. *Right*: ADOS configuration with 150 m long tether, and temperature and pressure nodes spaced by 10 m

the evolution of the cold wake [9, 18, 29]. An enhanced version of the ADOS fitted with vertically profiling acoustic current meters was used to measure the properties of large-amplitude nonlinear internal waves in the northern South China Sea [4].

1.5 Coastal Ocean Dynamics Experiment (CODE) Drifter and River Drifter (RD)

The CODE drifter [10] was designed as an inexpensive water follower to track currents in the upper meter of the oceanic mixed layer. However, as of today, the cost of a CODE and SVP drifter is similar. The CODE drifter consists of a cylindrical hull, 10–15 cm in diameter that contains the batteries and electronics. The drag element consists of four sails made of plasticized nylon arranged in a cross-like shape (Fig. 7). The CODE drifter is slightly negative buoyant and four small floats

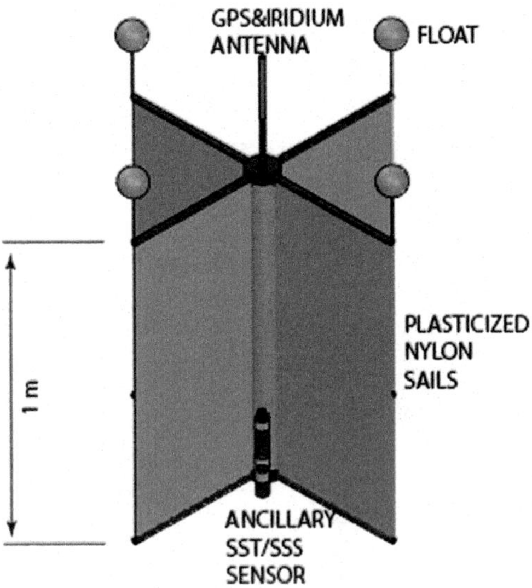

Fig. 7 Schematics of the CODE drifter configured with an external salinity and temperature sensor

connected to the end of the arms to which the sails are attached provide the extra buoyancy to ensure floatation. The CODE drifter carries the same SST sensors installed on the SVP drifter and can be configured with an external SSS sensor. The CODE drifter is considered an accurate water follower, and laboratory wave tank testing has shown that, if the drifter stays upright, the slip through the surrounding water is of the order of 1 cm/s for moderately sized waves [10]. However, it is likely that in presence of large breaking waves, the drag of the submerged part is affected by substantial tilting of the buoy, thus resulting in possible rectification of the relative motion of the drifter with respect to the wave orbital velocity. Poulain et al. [39] have found that the CODE drifters measure a wind-induced Ekman current that is approximately directed at 28–30° to the right of the wind (in the northern hemisphere). This is much less than the 45° angle predicted by the Ekman theory [13] and closer to the angle found for undrogued SVP drifters, which is generally more downwind, at about 17–20° to the right of the wind. This suggests that the CODE drifter may be subject to more wave-induced downwind leeway than the SVP drifter. The life span of the CODE drifter is about 1 year.

The River Drifter (RD) stems from the CODE drifter design and was conceived and designed by the LDL. This development originated from the need to have a Lagrangian drifter capable of operating in calm and shallow water bodies, such as lagoons and estuaries, or in the surf-zone, while also measuring the water depth and vertical shear of the three-dimensional currents underneath. A fleet of RDs can be used, for example, to study the dispersion of near-surface water and pollutants and their dependence on bed forms and on the vertical shear of the horizontal currents.

The central body of the RD is 0.6 m long, and the cross-sectional area of the sails and hull combined is 0.39 m^2 (Fig. 8). As for the CODE drifter, the hull contains the battery pack and the LDL controller to manage the sensors, the data acquisition, and

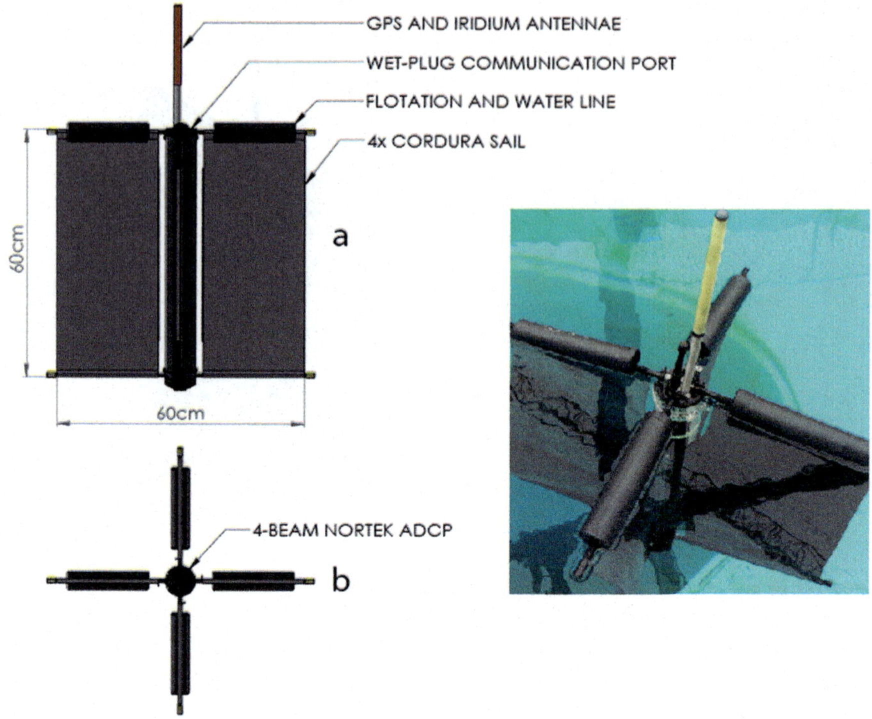

Fig. 8 Schematics of the side (**a**) and bottom (**b**) views of the river drifter adapted from [37] and flotation test of the river drifter (**c**)

telemetry. The sensor suite includes a GPS engine, data logger, as well as a specially designed acoustic Doppler current profiler (ADCP) by Nortek. The ADCP has four transducers, three to measure the three-dimensional water velocity and the fourth to measure the water depth. Two versions of the RD ADCP exist, 1 and 2 MHz, which allow operations at different water depths and with different vertical resolutions of the current. The data telemetry options include a direct RF link and the Iridium (Short-burst Data-SBD-modem) for global applications. A direct data link is also available through a wet-pluggable connector located next to the dual Iridium-GPS antenna which allows direct communications with the instrument for ADCP calibration, sensor configuration, data retrieval, and diagnostic output via RS-232 protocol.

Depending on the duty cycle and on the type of batteries, the endurance of the RD can vary from 1 to 7 days. A full description of the instrument and examples of applications can be found in Postacchini et al. [37].

2 The Global Drifter Program

The NOAA funded GDP [27, 30] is the principal component of the Global Surface Drifting Buoy Array, a branch of NOAA's GOOS and a scientific project of the DBCP. The DBCP is an international program coordinating the use of autonomous data buoys to observe atmospheric and oceanographic conditions over ocean areas where few other measurements are taken. The DBCP was created in 1985 as a joint body of WMO and of the Intergovernmental Oceanographic Commission (IOC) of UNESCO. The DBCP constitutes the data buoy component of the Joint WMO-IOC Technical Commission for Oceanography and Marine Meteorology (JCOMM).

The long-term objectives of the GDP are to:

- Maintain a global ocean observing network of at least 1250 Lagrangian drifters with a nominal resolution of 5° × 5° that, through the Argos and primarily the Iridium satellite systems, return data of Essential Climate Variables (ECV) including near-surface ocean currents (15 m depth), SST, SSS, SLP, SLW, and subsurface temperature in front of tropical cyclones;
- Innovate the drifter technology and expand the scope of the program by transitioning new sensors into operation; and
- Make the drifter data available in near-real time through the Global Telecommunication System (GTS) of the WMO Information System (WIS) in order to contribute to the World Weather Watch (WWW), and provide a data processing system and a delayed-mode, quality-controlled data archive for the scientific use of the drifter data.

The requirements of GOOS call for the GDP to maintain an array of 1250 drifting buoys, as needed to achieve a density of one drifter in 5° × 5° boxes. More details can be found in Lumpkin et al. [25] that also discusses the optimal implementation strategy to fulfill the goals of the GDP.

At the time of writing, the historical drifter dataset contains data starting from 1979, generated by over 20,000 surface Lagrangian drifters drogued at 15 m depth (Fig. 9), and it supports a detailed description of most oceanic mesoscale features and of the major persistent current system of the world's ocean (Fig. 10). Standardized quality control techniques and interpolating procedures onto regularly spaced 6-hourly time series are described in Niiler [30] and Maximenko et al. [27].

The GDP is managed by two groups that operate in close collaboration. The first, the drifter Operations and Data Assembly Center is located at the Atlantic Oceanographic and Meteorological Laboratory (AOML) in Miami, Florida, and oversees most of the global drifter deployments, performs the delayed-mode quality control of the drifter data, creates several gridded products, and archives the data. The second, the Lagrangian Drifter Laboratory at the Scripps Institution of Oceanography, La Jolla, California, provides most of the drifters for the global array either through fabrication and purchasing, oversees the drifter technology,

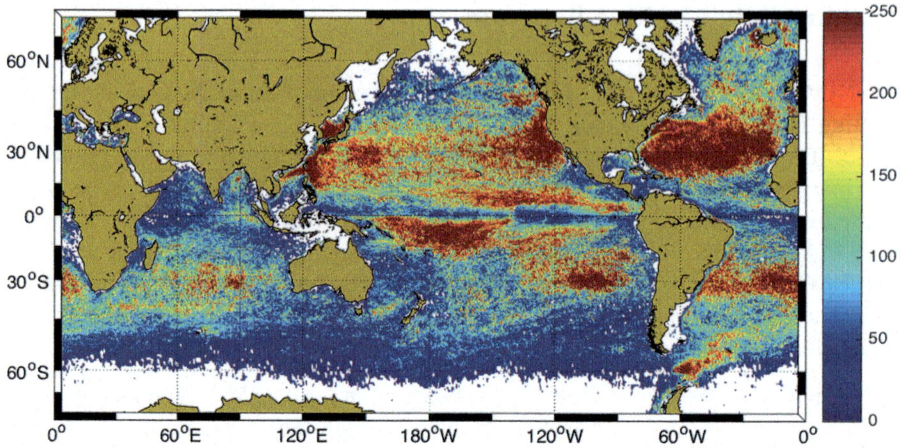

Fig. 9 Number of 6-hourly observations in 1/3° × 1/3° bins between February 15, 1979 and December 31, 2015. The total number of drifter observations in this time interval is 31,802,475

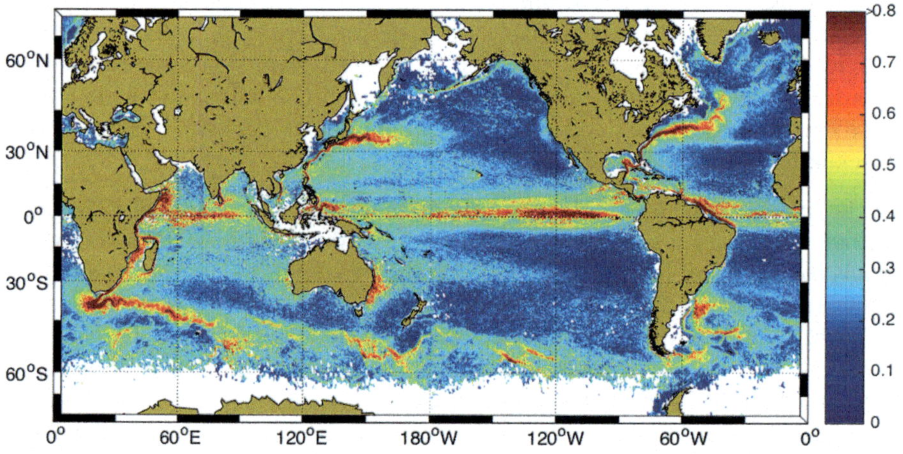

Fig. 10 Speed of the ocean currents obtained from bin-averaging the 6-hourly drifter observations in 1/3° × 1/3° bins

designs and fabricates new types of drifters, manages the real-time incoming data stream from the Iridium satellite system (Fig. 11), posts the data to the GTS, archives and distributes the data to its partners, maintains real-time statistics on the performance of the drifters, and provides project-specific web-based visualization interfaces. Multiple computer servers are operated by the LDL with considerable redundancies to ensure the continuity of the operations. Two server racks are operated in parallel, one located in the SIO colocation data center and one at the San Diego Supercomputer Center at UC San Diego. These two racks receive, process,

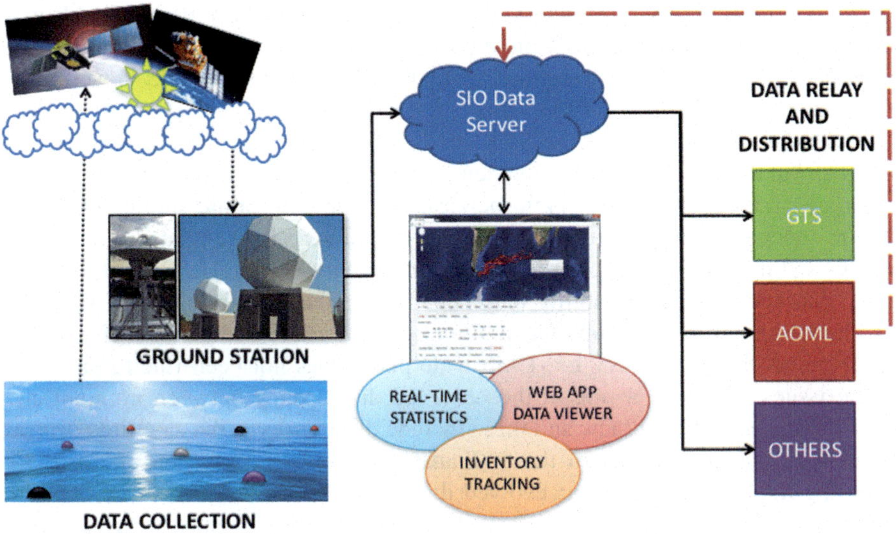

Fig. 11 Schematics of the GDP data route from the data collection in the field to the distribution of the converted data to the several partners and to the international community through the GTS

and relay real-time drifter data for collaborating partners. Further, a third remote system is operated using a commercial web service in the US Government Cloud data center for a real-time offsite archiving of all the incoming data.

3 Impacts of the Global Drifter Program Data

3.1 Sea Surface Temperature

Drifters measure and report SST at a depth of approximately 0.2 m every hour and are currently the largest source of global in-situ data [45] which are essential to validate and calibrate satellite SST products. Accurate global SST data are important for climate research and numerical weather prediction, among others. SST products can have small globally and annually averaged biases, but regional and seasonal biases can be much larger, often in excess of 0.5 °C [47]. For example, SST products derived from the advanced Along Track Scanning Radiometer are known to have differences between day and nighttime versions and have annually or globally averaged biases ranging from −0.04 °C to 0.2 °C [34]. Zhang et al. [47] introduced the Potential Satellite Bias Error (PSBE) metric, defined as the "residual satellite bias that cannot be further reduced at a given buoy density (sic)," to show that a global array of 1250 nearly uniformly distributed SVP drifters is needed, together with underway ship observations, to keep the regional and seasonal PSBE below 0.5 °C. The relationship between the PSBE and the number of drifters operating in the global array at any one time is almost linear and inversely proportional; it is assessed quarterly by NOAA's Climatic Data Center and it is used to evaluate the compliance of NOAA's contribution to GOOS, as required by the US Government Performance Results Act of 1993.

The nominal accuracy of the SST sensor of drifters deployed before 2014 is ±0.1 °C, but the use of more accurate temperature sensors, with smaller drift and an accuracy of ±0.05 °C is becoming increasingly common. Some studies have discussed the possibility that the accuracy of drifter's SST is actually worse than 0.1 °C, setting it at 0.23 °C after three-way error analysis [35]. However, the possibility that geophysical variability may negatively affect the analysis and bias, the accuracy estimate when pointwise in-situ observations are compared with satellite observations that sense SST is ~1 km wide pixels, has not been ruled out. Other subtle issues may play a role in the observed discrepancy between nominal and inferred drifter SST accuracy, and these include differences in the satellite retrieval algorithms and the stability of the SST analog electronic circuit and the drifter's sampling methodology that has evolved though the years. Nevertheless, it can be anticipated that more accurate SST drifter observations, together with measurements of the depth at which SST is sampled, of the sea state and improved geolocation resulting from the exclusive use of a GPS engine, will improve the quality of SST satellite products in the near future.

3.2 *Sea-Level Atmospheric Pressure: Climate Studies*

SLP observations constitute one of the longest data records of interest to climate science. Historically, atmospheric pressure sensors were well distributed over land and became progressively widespread over the ocean with the expansion of commercial shipping routes, the implementation of a Voluntary Observing Ship (VOS) program, and of the Global Drifter Array. In most areas, the drifter array represents the most accurate, and often unique, source of in-situ SLP data. The spatial distribution of SLP data over the ocean, if surface drifting buoys were removed, would be severely and negatively affected by the large gaps that exist in commercial shipping lanes. Notable examples of regions where there is very little ship traffic are the Southern Ocean and the Arctic region [45]. Even on trafficked shipping lanes, drifters are often the only source of most valuable SLP data when a storm develops and VOSs perform avoidance maneuvers.

The modern SLP drifter dataset contributes significantly to many re-analysis products used for climate assessment. Climate-related changes of the mean atmospheric load, or SLP, can induce long-term sea-level changes quantifiable in ~1 cm for a 1 mbar change, the so-called inverse barometer effect [46]. Although a formal assessment of the impact of SLP drifter data on the correction of satellite altimetry has never been performed (P.Y. Le Traon, Pers. Comm.), it is safe to assume that drifters are underpinning the sea-level correction in areas where they are the main source of oceanic SLP data, for example, in the Southern Ocean.

In-situ drifter SLP data are used in a variety of ways by climate scientists, including for trends computation to diagnose climate models and to construct climate indexes. The tropical Pacific SLP field is used to define the Multivariate El Niño–Southern Oscillation (ENSO) Index (MEI) [44]. The Southern Annular Mode (SAM) index is defined as the SLP difference between 40°S and 65°S [14]. As mentioned before, the drifters are probably the most reliable source of SLP data in this latitude band. Drifter data are widely used to compile global climate products such as the Hadley Centre's monthly historical mean sea-level atmospheric pressure dataset (HadSLP2) and the International Comprehensive Ocean-Atmosphere Data Set (ICOADS) v. 2.5 (ranging 1662–2007) and subsequent updates based exclusively on NOAA/NCEP GTS [45]. Substantial differences between several data products are observed in regions where data are sparse, particularly over the higher latitudes of the Southern Ocean and in the southeastern Pacific [1]. Improving the coverage of SLP observations from drifters is expected to improve the quality of several climate indexes. Other climate products that use the drifter (velocity, SST and SLP) data are the Climate Forecast System Reanalysis (CFRS) of the United States National Centers for Environmental Prediction [41] and the twentieth-century reanalysis project [8].

Reports of the weakening of the Pacific Walker circulation indicated by trends in the zonal SLP gradients across the equatorial Pacific exist [43]. Similar conclusions were reached by way of analysis of other climate variables, thus enhancing the confidence on such pattern of climate variability.

Other climatically relevant results based on SLP observations include changes in the occurrence and intensity of extratropical storms, monsoons variability, and trends in extreme winds events. The IPCC report, however, concludes that the confidence in some of these evaluations is low, primarily, because of inconsistencies between (and internal to) re-analysis products that must merge data from different years (some going back to early 1800s) and of different quality and "by the lack of long term data, particularly in the Southern Hemisphere (2013 Climate IPCC report pg. 2–62)." Therefore, it is important to ensure the continuity of the widely used in-situ SLP dataset coming from a variety of sources, including the Global Drifter Array.

3.3 *Sea-Level Atmospheric Pressure: Numerical Weather Prediction*

The impact of SLP drifter observations in Numerical Weather Prediction (NWP) is beneficial and very significant. Recent studies have addressed the issue of quantifying the implication of removing the drifter data from the GTS real-time data feed [3]. The vast majority of NWP systems run by many national meteorological agencies pull a large number of observations from the GTS in real time. The observations with in-situ platforms must be available with the shortest possible delay since the assimilation window of NWP models is typically 3–6 h long. The Iridium satellite telecommunication system is able to deliver the observations to the LDL's servers within 1 min from the moment the data are sent by the drifters. The LDL servers push the data to the GTS with a total delay shorter than 15 min from the time of data collection. Other GTS processing systems have very similar delays.

The Observing System Experiment (OSE), or data denial experiment, discussed by Centurioni et al. [3] and Horányi et al. [16], was performed at the European Center for Medium range Weather Forecasting (ECMWF) and with the objective to quantify the impact of the SVPB drifter data only. The principle of the OSE is that a data assimilation and forecast models, the ECMWF Integrated Forecast System (IFS) four-dimensional variational (4D-Var) system [19, 40], is used to produce a control run, in which all of the available data, including the SVPB data, are assimilated, and also a data denial run in which the SLP observations from the SVPB drifters are withheld. The main conclusions of the study are that the in-situ SVPB drifter data of SLP positively impact the forecast up to 5 days ahead and not only near the surface, but also higher in the troposphere, up to 250 hPa. The largest beneficial effects are observed in the mean sea-level atmospheric pressure field forecasts, but also in the predicted wind field and the study also suggests that the expansion of the SVPB drifter array to the tropics should be considered. Data denial studies are lengthy and expensive, but other approaches using adjoint-based Forecast Sensitivity Observation Impact (FSOI) can quantify the impact of any or all components of the observing system on a specific metric of forecast impact when the entire observational dataset is present in the assimilation system [2, 12, 21, 48], although the inherent linearization assumption restricts the range of the forecast assessment to about 2 days. Such assessments are often run operationally by ECMWF and by other agencies, such as the Global Modeling and Assimilation Office of the US National Aeronautics and Space Administration (NASA), and when the impact per observation or the fraction of beneficial observations is computed the in-situ SLP drifter data provide some of the largest values among the several components of the Global Observing System [3].

The demonstrated far-reaching role of the drifter barometer program in climate research and NWP would probably justify the installation of barometers on all of the GDP drifters, but, at the time of writing, only about 60% of the drifters carry a barometer.

Practical considerations that justify the barometer drifter program exist. One of the most difficult tasks for the GDP is to keep a nearly uniform distribution of drifters globally. A key to achieve this effort is the synergy with national meteorological services from, for example, the US National Weather Services, France, the United Kingdom, Canada, Australia, South Africa, New Zealand, India and many other partners, coordinated by the Data Buoy Cooperation Panel and implemented through the drifter barometer program upgrade and the direct purchase of barometers by the GDP. The collaboration with the meteorologists, who rely on a global network of marine barometers located at the interface between the ocean and the atmosphere, creates every year a large number of deployment opportunities (25%–30% of all deployments) of crucial importance for sustaining the Global Drifter Array.

3.4 *Subsurface Temperature, Air Pressure and Wind: The Seasonal Hurricane Array*

The track of a tropical cyclone (TC) is largely dictated by the large-scale atmospheric flow; therefore, the track forecasting depends crucially on the accuracy of the global atmospheric circulation models. Improvements in track forecasting have been achieved in the past few years but much less improvement has been obtained with respect to predicting the intensity [11]. The intensity of a TC is controlled by external processes that are regulated by the large-scale flow and by internal processes. These include TC-ocean interactions that affect the enthalpy and the momentum fluxes at the ocean-air interface. The numerical initialization of a TC is very challenging and the data to constrain the numerical models are seldom available. Furthermore, the ocean component of coupled models is often poorly represented and only recently the numerical ocean modules are being upgraded to provide more realistic representations of the ocean conditions.

The TC's enthalpy flux also depends on the thermal stratification of the upper ocean, and the amount of heat available can lead to TC intensification. The passage of a storm also creates some cooling, primarily caused by mixing and upwelling that can last long after the storm has passed [9, 18, 29].

The seasonal hurricane array of the GDP has been designed to provide the ocean and atmospheric models with the data needed for assimilation, especially on the scale of a TC inner core. In the case of a tropical cyclones threatening the US mainland, the 53rd Weather Reconnaissance Squadron (AFRC), also known as "Hurricane Hunters" usually deploys a transect of 10 ADOS and SPVW (Minimet) drifters. The transect can be about 300–500 km long and is designed to intersect the storm track about 24 h ahead of the forecasted intersection point. More ADOS drifters than Minimets are deployed on the right-hand side of the storm to capture the intensity of the developing cold wake [9, 29]. All the data (wind, SLP, SST, and subsurface temperature) are posted to the GTS in near-real time. The drifter data are also important for hurricane postseason assessments.

3.5 *Ocean Currents*

The quality-controlled ocean currents dataset that is released quarterly by AOML has supported more than 1000 peer-reviewed scientific publications (see http://www.aoml.noaa.gov/phod/dac/gdp_biblio.php for a partial list).

Several peer-reviewed publications have scrutinized the quality of the data collected by the drifters [26, 27, 30] and correction for biases have been discussed and attempted also for drifters that have lost their drogue [36].

At the time of writing, the ocean current data are only available in delayed time mode because the strain gauge data that indicate the drogue presence require a considerable amount of manual evaluation before a determination about the drogue presence can be made [26]. Undeniably, enabling a stream of real time, 15-m depth ocean currents from the global drifter array would be very useful for the evaluation and the future assimilation in ocean forecasting. The widespread use of the GPS technology to obtain the location of the drifters may offer a solution for the near real-time drogue presence detection problem. The GPS parameter time-to-first-fix is generally longer when the drogue is present because the drifters are frequently pulled underwater when waves are present and it takes longer for the GPS engine to acquire signal and navigation data.

A good synopsis of recent ocean circulations studies based on drifter observations can be found in Maximenko et al. [27] and a modern climatology is discussed by Lumpkin and Johnson [24]. A particularly powerful approach to remove the biases inherent to the irregular spatial and temporal sampling of the drifter array consists in combining satellite altimetry, wind reanalysis products, and sometimes the large-scale geoid to compute the near-surface geostrophic velocity field and the ageostrophic component of the surface velocity in the form of Ekman currents [13]. Such techniques provide a very detailed description of the near-surface flow and, in regions where the drifter velocity data are highly correlated with the altimetry-derived currents, can provide the temporal resolution that lacks in the drifter dataset [5, 6, 28, 31, 33]. A new exciting approach is the use of the high-frequency (super-inertial and super-tidal) drifter velocity data to measure, for example, the amplitude and phase of the baroclinic and barotropic tidal currents [38]. The introduction of Iridium-GPS drifters will provide the scientific community with a highly accurate, hourly velocity dataset that will improve the studies of global near-inertial and tidal currents.

4 Conclusions

The Global Drifter Array, supported by the US Global Drifter Program and by its international partners operating under the Data Buoy Cooperation Panel framework, is a unique infrastructure strategically located at the interface between the ocean and the atmosphere that addresses the pressing need for in-situ global observations needed to monitor the changing ocean-atmosphere system and to conduct and support basic research to understand the physical principles underlying these changes. It also addresses the need to understand how accurately weather prediction and coupled ocean-atmosphere models can represent the environment and produce forecasts, from weekly to decadal and inter-decadal scales. Global observations of essential climate variables, such as SST and sea-level atmospheric pressure are critical for data assimilation and for the evaluation of forecasts of the state of the ocean and of the atmosphere on a variety of temporal scales.

More than 1000 peer-reviewed scientific articles published using GDP data have demonstrated the tremendous impact of the Global Drifter Array and of its data in basic research, weather forecasting, climate change studies, and many other fields and in operational efforts. Existing measurements of SST, SLP, and ocean currents at regional scales are already capable of supporting studies of regional climate change, indicated also by changes of the coupled atmospheric and oceanic circulations. However, considerably large uncertainty is still associated with our understanding of the physics of air-sea interaction and of the oceanic mixed layer. A proper physical representation and effective parameterizations of how the momentum imparted by the wind to the ocean surface sets the ocean in motion is still missing, as demonstrated by the fact that in-situ velocity observations from drifters are very difficult to assimilate in ocean circulation models. Improving such physical description is critically important and the Global Drifter Array is well placed to make a very significant impact in the next decade because technical developments and new sensors will allow the collection of key global observations. For example, global measurements of the velocity shear in the upper ocean will not only provide a wealth of data for assimilation in numerical models, model skill evaluations, and impact studies, but will help us gain a much deeper understanding of the physics of the mixed layer, where the wind stress, turbulent momentum fluxes, surface waves and internal waves, heat and salt fluxes all interact in a nonlinear fashion to couple the ocean and the atmosphere.

Ultimately, understanding the relative role of geostrophic and ageostrophic currents, the sum of which is at present exclusively measured globally with SVP drifters in the upper ocean, that is, where the transfer of momentum and vorticity from the wind occur, will improve the physical descriptions of heat and freshwater exchanges with the atmosphere and of their lateral transport within the mixed layer.

The departure of upper ocean currents from geostrophy is known to be significant. Recent publications [5, 28] have shown that the geostrophic circulation maps that are computed from dynamic height or from dynamically balanced ocean topography are very different from maps obtained with direct measurements of near-surface ocean velocity. While surface drifters measure globally the total near-surface velocity, the vertical structure of the ageostrophic circulation in the upper ocean is virtually unknown. Understanding how heat and freshwater are forced, advected [7, 42], and mixed in the upper ocean requires a much better knowledge of the three-dimensional upper ocean circulation than we currently have. A better knowledge of the Ekman currents and of the convergence patterns of the ageostrophic currents that drive the vertical circulation of the upper ocean will be crucial to constrain ocean circulation and climate models toward a more realistic representation of upper ocean physics and will ultimately lead to improved ocean state and climate forecasts.

Acknowledgments This work was supported by the NOAA grant # NA10OAR4320156 "The Global Drifter Program." The US Office of Naval Research and the Climate Program Office of NOAA are gratefully acknowledged for their substantial support for the development of Lagrangian drifter technology.

References

1. Allan R, Ansell T (2006) A new globally complete monthly historical gridded mean sea level pressure dataset (HadSLP2): 1850–2004. J Clim 19:5816–5842
2. Cardinali C (2009) Monitoring the observation impact on the short-range forecast. Q J R Meteorol Soc 135:239–250
3. Centurioni L, Horányi A, Cardinali C, Charpentier E, Lumpkin R (2016) A global ocean observing system for measuring sea level atmospheric pressure: effects and impacts on numerical weather prediction. Bull Am Meteorol Soc (In Press)
4. Centurioni LR (2010) Observations of large-amplitude nonlinear internal waves from a drifting Array: instruments and methods. J Atmos Ocean Tech 27:1711–1731
5. Centurioni LR, Ohlmann JC, Niiler PP (2008) Permanent meanders in the California Current System. J Phys Oceanogr 38:1690–1710
6. Centurioni LR, Niiler PN, Lee DK (2009) Near-surface circulation in the South China Sea during the winter monsoon. Geophys Res Lett, 36, L06605, doi:06610.01029/02008GL037076
7. Centurioni LR, Hörmann V, Chao Y, Reverdin G, Font J, Lee D-K (2015) Sea surface salinity observations with Lagrangian drifters in the tropical North Atlantic during SPURS: circulation, fluxes, and comparisons with remotely sensed salinity from Aquarius. Oceanography 28:96–105
8. Compo GP et al (2011) The twentieth century reanalysis project. Q J R Meteorol Soc 137:1–28
9. D'Asaro EA et al (2013) Impact of typhoons on the ocean in the Pacific. Bull Am Meteorol Soc 95:1405–1418
10. Davis RE (1985) Drifter observations of coastal surface currents during Code – the method and descriptive view. J Geophys Res (Oceans) 90:4741–4755
11. DeMaria M, Sampson CR, Knaff JA, Musgrave KD (2014) Is tropical cyclone intensity guidance improving? Bull Am Meteorol Soc 95:387–398
12. Gelaro R, Zhu Y, Errico RM (2007) Examination of various-order adjoint-based approximations of observation impact. Meteorol Z 16:685–692
13. Gill A (1982) Atmosphere-ocean dynamics, vol 30. Elsevier, 662 pp
14. Gong D, Wang S (1999) Definition of Antarctic oscillation index. Geophys Res Lett 26:459–462
15. Hansen DV, Poulain PM (1996) Quality control and interpolations of WOCE-TOGA drifter data. J Atmos Ocean Tech 13:900–909
16. Horányi, András, Carla Cardinali, and Luca Centurioni (2017) The global numerical weather prediction impact of mean-sea-level pressure observations from drifting buoys', Quarterly Journal of the Royal Meteorological Society, 143:974–85.
17. Hormann V, Centurioni LR, Reverdin G (2014a) Evaluation of drifter salinities in the subtropical North Atlantic. J Atmos Ocean Tech 32:185–192
18. Hormann V, Centurioni LR, Rainville L, Lee CM, Braasch LJ (2014b) Response of upper ocean currents to typhoon Fanapi. Geophys Res Lett 41:2014GL060317
19. Janisková M, Lopez P (2013) Linearized physics for data assimilation at ECMWF. In: Park SK, Xu L (eds) Data assimilation for atmospheric, oceanic and hydrologic applications, vol Vol. II. Springer, Berlin Heidelberg, pp 251–286
20. Kinsman B (1984) Wind waves: their generation and propagation on the ocean surface. Dover Publications, xxiii, 676 pp
21. Langland RH, Baker NL (2004) Estimation of observation impact using the NRL atmospheric variational data assimilation adjoint system. Tellus A 56:189–201
22. Lopez R, Malarde JP, Royer F, Gaspar P (2014) Improving Argos Doppler location using multiple-model Kalman filtering. IEEE Trans Geosci Remote Sens 52:4744–4755
23. Lumpkin R, Pazos M (2007) Measuring surface currents with surface velocity program drifters: the instrument, its data, and some recent results. Cambridge University Press, New York
24. Lumpkin R, Johnson GC (2013) Global ocean surface velocities from drifters: mean, variance, El Niño–Southern Oscillation response, and seasonal cycle. J Geophys Res Oceans 118:2992–3006

25. Lumpkin R, Centurioni L, Perez RC (2016) Fulfilling observing system implementation requirements with the global drifter array. J Atmos Ocean Tech 33:685–695
26. Lumpkin R, Grodsky SA, Centurioni L, Rio M-H, Carton JA, Lee D (2012) Removing spurious low-frequency variability in drifter velocities. J Atmos Ocean Tech 30:353–360
27. Maximenko N, Lumpkin R, Centurioni L (2013) Chapter 12 – ocean surface circulation. In Gerold Siedler SMGJG, John AC (eds) International geophysics, Academic Press, pp 283–304
28. Maximenko N et al (2009) Mean dynamic topography of the ocean derived from satellite and drifting buoy data using three different techniques*. J Atmos Ocean Tech 26:1910–1919
29. Mrvaljevic RK et al (2013) Observations of the cold wake of Typhoon Fanapi (2010). Geophys Res Lett 40:316–321
30. Niiler PP (2001) The world ocean surface circulation. In Siedler G, Church J, Gould J (eds) Ocean circulation and climate. Academic Press, pp 193–204
31. Niiler PP, Maximenko NA, McWilliams JC (2003a) Dynamically balanced absolute sea level of the global ocean derived from near-surface velocity observations. Geophys Res Lett 30
32. Niiler PP, Sybrandy A, Bi K, Poulain PM, Bitterman D (1995) Measurements of the water-following capability of holey-sock and TRISTAR drifters. Deep-Sea Res 1(42):1951–1964
33. Niiler PP, Maximenko NA, Panteleev GG, Yamagata T, Olson DB (2003b) Near-surface dynamical structure of the Kuroshio extension. J Geophys Res (Oceans) 108:3193
34. O'Carroll AG, Watts JG, Horrocks LA, Saunders RW, Rayner NA (2006) Validation of the AATSR Meteo product sea surface temperature. J Atmos Ocean Tech 23:711–726
35. O'Carroll AG, Eyre JR, Saunders RW (2008) Three-way error analysis between AATSR, AMSR-E, and in situ sea surface temperature observations. J Atmos Ocean Tech 25:1197–1207
36. Pazan SE, Niiler PP (2001) Recovery of near-surface velocity from undrogued drifters. J Atmos Ocean Tech 18:476–489
37. Postacchini M, Centurioni LR, Braasch L, Brocchini M, Vicinanza D (2016) Lagrangian observations of waves and currents from the river drifter. IEEE J Ocean Eng 41:94–104
38. Poulain P-M, Centurioni L (2015) Direct measurements of World Ocean tidal currents with surface drifters. J Geophys Res Oceans 120:6986–7003
39. Poulain P-M, Gerin R, Mauri E, Pennel R (2009) Wind effects on drogued and undrogued drifters in the Eastern Mediterranean. J Atmos Ocean Tech 26:1144–1156
40. Rabier F, Järvinen H, Klinker E, Mahfouf JF, Simmons A (2000) The ECMWF operational implementation of four-dimensional variational assimilation I: experimental results with simplified physics. Quart J Roy Meteorol Soc 126:1143–1170
41. Saha S et al (2010) The NCEP climate forecast system reanalysis. Bull Am Meteorol Soc 91:1015–1057
42. Schmitt RW et al (2015) From salty to fresh-salinity processes in the upper-ocean Regional Study-2 (SPURS-2): diagnosing the physics of a rainfall-dominated salinity minimum. Oceanography 28:150–159
43. Vecchi GA, Soden BJ, Wittenberg AT, Held IM, Leetmaa A, Harrison MJ (2006) Weakening of tropical Pacific atmospheric circulation due to anthropogenic forcing. Nature 441:73–76
44. Wolter K, Timlin MS (2011) El Niño/Southern Oscillation behaviour since 1871 as diagnosed in an extended multivariate ENSO index (MEI.ext). Int J Climatol 31:1074–1087
45. Woodruff SD et al (2011) ICOADS Release 2.5: extensions and enhancements to the surface marine meteorological archive. Int J Climatol 31:951–967
46. Wunsch C, Stammer D (1997) Atmospheric loading and the oceanic "inverted barometer" effect. Rev Geophys 35:79–107
47. Zhang H-M, Reynolds RW, Lumpkin R, Molinari R, Arzayus K, Johnson M, Smith TM (2009) An integrated global observing system for sea surface temperature using satellites and in situ data: research to operations. Bull Am Meteorol Soc 90:31–38
48. Zhu Y, Gelaro R (2008) Observation sensitivity calculations using the Adjoint of the Gridpoint Statistical Interpolation (GSI) analysis system. Mon Weather Rev 136:335–351

Origin, Transformation and Measurement of Waves in Ocean

P. Chandramohan, M.V. Ramanamurthy, K. Jossia Joseph, Satya Kiran Raju Alluri, D. Shyamala Varthini, and K.N. Navaneeth

Abstract The Wave characteristics play a major role in the upper ocean dynamics, planning port operations, issuing sailing notifications and design of coastal/offshore structures. Waves in open sea exhibit very random and complex behavior with the superposition of many sinusoidal waves, each having its own height, period and direction. The design waves of varied return period based on extreme wave statistics is important for designing the coastal structures. This necessitates the accurate measurement of wave characteristics for various applications. The present chapter explains the state-of-the-art instruments used for measuring waves along with the advantages and disadvantages of each technique. Besides, a case study of estimation of wave climate using measurements and numerical modelling and the general wave characteristics in north Indian Ocean are also presented.

1 Introduction

Waves play a major role in the sea to keep it hydrodynamically active, increase the circulation process, churn and clean the ocean and keep the shoreline beautiful. They influence considerably the air-sea interaction processes of the coupled ocean-atmosphere system. The morphology of the coastline and associated shoreline features are determined by the wave-induced currents, which are responsible for sediment transport. Knowledge of the surface gravity waves is essential for engineering applications as these influence the engineering structures [1]. The period of oceanic waves ranges from few seconds (capillary waves) to several hours (tides). Wind-induced surface gravity waves, generated through the friction or drag between air and water, are most common and have maximum impact on human activity. The normal band for the wind-generated gravity waves is between 1 s and 30 s.

P. Chandramohan (✉)
Indomer Coastal Hydraulics, Chennai, India
e-mail: ponpcm52@gmail.com

M.V. Ramanamurthy
ICMAM PD Chennai, Chennai, India

K.J. Joseph • S.K.R. Alluri • D. Shyamala Varthini • K.N. Navaneeth
NIOT, Chennai, India

R. Venkatesan et al. (eds.), *Observing the Oceans in Real Time*, Springer Oceanography, DOI 10.1007/978-3-319-66493-4_4

Wind-driven surface gravity waves are distinguished in two states: '*sea*', when the waves are being generated by the wind that raised them, and '*swell*', when they reach out of the area of influence of the wind. The *sea* usually is of shorter period (higher frequency) than *swell*. In case of seas around India, as a rough rule of thumb, 7 s is considered separating sea from swell, although there is some overlap between them. While the sea is shorter in length, steeper, rugged and more confused, the swell is relatively high, nearly uniform along the crest and longer in length.

2 Generation of Waves

Wind blowing over the water surface generates waves. In deep oceans, the wind can blow unobstructed with various magnitudes, in different directions and for various durations. The turbulence set by the wind gives rise to varying energy inputs such as a vertical pulse as pressure on the surface and the horizontal energy transfer as shear varying with durations and energy level at the same place as the wind blows. These different types of inputs would have both vertical and horizontal components. The relative importance of vertical and horizontal forces depends on the wind force and stage of sea state. These processes would have different levels of effectiveness at different stages of wave generation and growth. At the very initial phases, the normal pressure fluctuations are most significant. Afterwards, the tangential stress becomes more effective. When the waves are of considerable dimensions, 'sheltering' would have significant effects. During faster stages (non-linear) of growth, wind shear (a net effect of wind velocity changes with height over the sea surface) plays a dominant role. There are many wave generation theories currently available in literature. Each one stresses some particular features and fails in some other aspects.

The relationship between wind speed and wave height is obvious. Wave height is, however, not solely dependent on the wind speed. Other quantities such as the duration or period for which the wind has blown and the distance or fetch of water over which the wind blows also influence the resulting wave height. The simplest possible case of wave generation by the wind is termed fetch-limited growth. In fetch-limited growth, a wind of constant speed and direction blows perpendicular to a long and straight coastline. In addition, the wind is assumed to blow for a long enough period such that the duration of the wind has no influence on the resulting waves.

The wind waves while propagating from deep water to the shore can be classified by the water depth in which they travel, such as deep water, transitional depth and shallow water waves (Table 1). The classification criterion is made according to the magnitude of (d/L) at a given point (ratio of water depth to wave length) and the

Table 1 Classification of waves based on relative depth (d/L)

Classification	d/L	kd	tanh (kd)
Deep water	1/2 to ∞	π to ∞	1
Transitional	1/20 to 1/2	$\pi/10$ to π	Tanh (kd)
Shallow water	0 to 1/20	0 to $\pi/10$	kd

resulting limiting values taken by the function tanh ($2\pi d/L$). When the hyperbolic tangent ($kd = 2\pi d/L$) is large, then tanh (kd) tends to 1, and when kd is small tanh (kd) tends to kd.

As waves propagate from deep water to shallow water, there will be significant change in the characteristics due to processes such as wave shoaling, refraction, diffraction, wave breaking, etc. Understanding these complex processes is essential for the structural and functional design of any coastal system. The wave characteristics play a major role in the upper ocean dynamics, planning port operations, issuing sailing notifications and design of coastal/offshore structures. The information on wave height and direction is significant input for optimizing the breakwater design, planning the breakwater layout and for the alignment of navigational channels. The wave direction near a breakwater opening is essential to estimate the wave tranquility inside the port basin and alongside the berths. The breaking wave characteristics close to the shore determine the volume and direction of littoral drift, in turn the shoreline erosion and the shoreline management. Design for waves of various return periods based on extreme wave statistics are important for designing the coastal structures and to decide the berm height and allowable overtopping section.

Hence, to understand the wave transformation phenomena, extensive field measurements or physical model studies are required which are time-consuming and expensive. So, numerical models in conjunction with available field measurements are commonly used for understanding the nearshore dynamics.

3 Wave Measurement Systems

Planning a wave measurement programme depends on the wave characteristics in the region of interest, and the intended application which determines the period of measurement. Measurement for a period of typically 1 year is required for deriving typical wave characteristics, whereas the information on the extreme conditions demands measurements over several years. Moreover, the wave measurement system should be designed to survive the extreme conditions and exposure for long periods in a very unfriendly environment. The design of coastal structures requires long-term wave measurement for estimating the design wave height for various return periods, but leads to higher cost and delay in the execution of a project. In general, the wave measurements combined with satellite-derived or simulated wave data at the region of interest for sufficiently long period are utilized for preliminary planning after validating with measured wave data.

The earliest records of wave characteristics can be found in ship reports based on visual observations, which were widely used for various applications. The sailing ships routinely recorded wave height, period and direction with geographical coordinates. The compilation of ship records provided a rich source of wave variability, but is limited to major ship routes, and is prone to human errors. Waves acts as a major driving force behind many coastal processes such as sediment transport, shoreline changes, etc. The present modelling capabilities are inadequate to accurately simulate these processes and cannot be estimated using satellite products or ship records, which indicate the vital role of wave measurement.

The complexity of wave characteristics necessitates complex measuring devices to accurately measure and estimate various wave parameters. With the advancement in technology, sophisticated electronic instruments are used for wave measurements at selected locations. Waves are primarily measured based on (i) principle of resistance, (ii) pressure variation (iii) acoustic principle, (iv) accelerometer, (v) GPS-based measurement and (vi) remote sensing/ radar techniques.

3.1 Principle of Resistance

A Wave staff, which is used to measure wave height, works on the principle of resistance, that is, the resistance of a sensor changes with a change in sea surface elevation. A wave staff is attached to a fixed structure and penetrates the water surface. When the staff is immersed in the column of water, the resistance would change according to the variation in the depth of immersion. This variation in resistance can be related to the wave height. A Wave staff is made of nickel chromium wire that is spirally wound about an insulating staff (e.g. PVC pipe). The spiral winding provides more a resistance per unit of staff length. Alternating currents must be used with resistance wire staff to prevent electrolytic effects due to chemical ions in sea water. Wave direction can also be estimated with the help of three sensors mounted in a triangular configuration. The measurement accuracy is normally affected by biofouling and fouling caused by flotsam and oil. Water splash gives rise to problem in wave height and it needs a platform or structure for mounting the wave gauge.

3.2 Pressure Variation

When the wave propagates in the sea, it exerts dynamic pressure, in addition to the static pressure. The magnitude of the dynamic pressure would change with the wave height and wave period. The dynamic pressure is given by the function of

$$\rho g \eta \,(\cosh[2\pi\left(\left(z+d\right)\right)/\left(L\right]))\,/\,(\cosh\left(2\pi d/\left(L\right)\right)$$

where,
ρ: density of sea water
g: acceleration due to gravity
η: sea surface height
d: bottom depth
z: depth of the sensor below the surface
L: wavelength

This pressure fluctuation can be sensed by a pressure sensor placed at the bottom of the seabed or mounted on a platform or attached to a structure below the sea surface, which can be converted to wave height and period. Strain gauges and quartz crystal pressure sensors are the most often used. The dynamic pressure reduces hyperbolically from the surface to bottom, and its value becomes very low below 10 m water depth. These gauges are operated in shallow depths from 5 to 10 m depending upon the wave period range expected, and the accuracy is improved by applying a correction factor for depth attenuation.

A standard procedure for analysis of pressure data is computation of a wave pressure spectrum $E_p(f)$ and correction of this spectrum to provide normal wave spectrum $E_p(f)$ for wave heights. The appropriate equation is

$$E(f) = \frac{E_p(f)}{K^2(f)}$$

where depth attenuation factor is given by

$$K(f) = \frac{\cosh\left[k(z+D)\right]}{\cosh\left[\mathrm{kD}\right]}$$

Here, k is the wave number, z is the depth of the sensor below the surface and D is the bottom depth. The main advantages of such pressure gauges are that they do not require any surface-penetrating structure, have a high signal voltage output and do not require frequent maintenance. The main disadvantages however are that they require signal correction, depth limitation, limited frequency response subjected to fouling and are difficult to install in some areas.

3.3 *Acoustic Principle*

This method utilizes the principle of reflection of sound waves. The sound pulses, which are transmitted towards the surface by a transducer, get reflected from the surface and they rebound to the transducer. The time interval between transmission and reception determines the vertical distance of the sea surface above. The use of an acoustic sensor is not simple, and sensors are expensive. Variations in temperature and salinity affect the speed of sound in sea water and the calibration. Foam or bubbles at the sea surface adversely affect acoustic measurement. An advantage of the sensor is that it can be placed at the bottom in deep water. Figure 1 shows Acoustic Doppler Current Profiler (ADCP) which utilizes the acoustic principle.

3.4 *Principle of Acceleration*

The acceleration of a water particle below the wave profile changes continuously. The characteristic of the change in acceleration is measured using the accelerometer, and wave height is obtained from acceleration.

Although an accelerometer buoy is conceptually simple, there are several technical considerations required for proper design and operation. Measurement of vertical

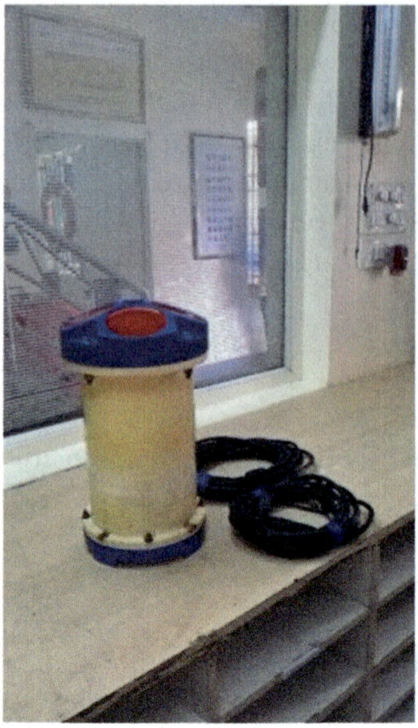

Fig. 1 Acoustic Doppler Current Profiler (ADCP)

accelerations should not be contaminated by horizontal accelerations. This problem is often treated by use of a specially gimbaled accelerometer, supplemental use of a gyroscope to maintain vertical reference or use of mechanical devices such as a pendulum to approximately maintain an accelerometer vertical. Double integration is either performed electronically or numerically by a board microprocessor. The mooring obviously must allow the buoy to move freely with the waves, and finally the buoy must be designed to follow the wave without significantly amplifying the wave-induced accelerations.

A Wave rider buoy is the most widely used instrument for measuring surface waves. It consists of a spherical buoy designed to float on the sea surface and is moored to the seabed with special arrangement such as a rubber cord, to maintain the free-floating characteristics of the buoy (Fig. 2). The waverider follows the sea surface by rising and falling along with it, resulting in the vertical acceleration of the buoy. The wave elevation can then be obtained by integrating the acceleration twice. The waverider is designed to transmit the recorded information to the shore station by wireless means, or it can transmit through an ARGOS satellite to any desired destination. While the measurement of acceleration leads to understanding the wave heights, the horizontal tilts of the buoy provide measurement of wave slopes, which provide information on the wave direction.

Wave measurements are also carried out with moored data buoys equipped with a motion reference unit (MRU), which incorporates solid state accelerometers, angular rate sensors and servo flux gate compasses for all three axes. There are no moving parts inside the sensor, and advanced digital signal processing algorithms are used for calculating of the wave motions. The MRU is highly configurable, and its dynamic response can be programmed, typically for a cut-off frequency of 0.025 Hz.

Fig. 2 Deployment of Waverider Buoy

The wave measurement is based on rapid sampling of heave, pitch and roll, together with the heading, which is then transformed to heave and slopes in a geographical coordinate system. The analysis method to derive directional spectra is the traditional method used for the discus buoy and is carried out onboard the buoy. The advantages are the highly accurate wave measurements, and they can be deployed in deep ocean.

3.5 GPS-Based Measurement

A GPS motion sensor consists of a GPS antenna and a GPS receiver. The GPS system calculates the velocity of the buoy from changes in the frequency of GPS signals according to the Doppler principle (Fig. 3). For example, if the buoy is moving towards the satellite, the frequency of the signal is increased, and vice-versa. The velocities are integrated through time to determine buoy displacement. The measurement principle is illustrated in Figure 3 which shows a satellite directly overhead and a satellite at the horizon. In practice, the GPS system uses signals from multiple satellites to determine three-dimensional buoy motion.

3.6 Remote Sensing/Radar Techniques

Ocean application satellites have been provided with sensors such as radar altimeter and synthetic aperture radar to monitor the wave climate on their path using sea surface imageries. A radar altimeter transmits a short pulse, and the shape of the forward edge of the return pulse that is reflected from the sea surface is dependent on a measure, such as significant wave height. Shore-based radar also holds promise for wave measurements. Synthetic aperture radar (SAR) can give information on wavelength and direction. While instantaneous information on wave

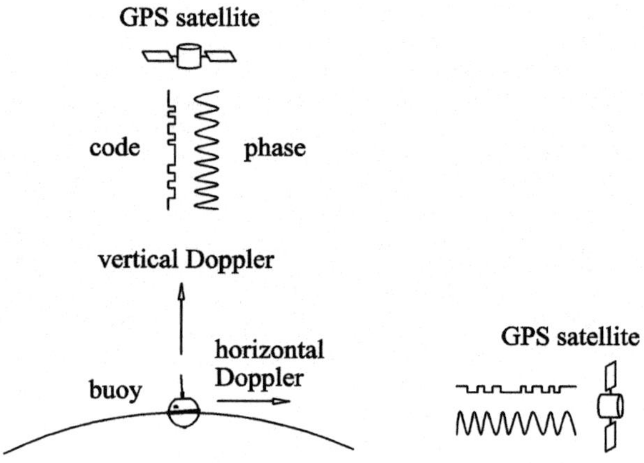

Fig. 3 GPS motion sensor

parameters is obtained to some extent, continuous data on the consecutive waves for longer periods are still a problem with radar revolving with satellites around the orbit. On the other hand, a microwave remote sensor mounted above sea surface can measure waves with height, period and direction in real time. They are designed for remote sensing of the ocean surface from stationary and floating platforms. The sensor head is mounted at about 50–100 m above the mean sea level. During a typical measurement sequence, observations are taken from six directions with an angular increment of 30°. Based on the processing circuitry, the wave information is displayed as the wave height spectra, one for each look direction of the antenna.

3.7 *Pilot Project on Wave Measurement Evaluation and Test (PP-WET)*

Wave measurements are carried out globally by many organizations utilizing many instruments/sensors. The quality and reliability of these measurements have not always been quantified. To resolve the issue, standardizing the wave measurement against a reference measurement is proposed by experts in wave measurement and buoy networks at the JCOMM (Joint WMO-IOC technical Commission for Oceanography and Marine Meteorology) Technical Workshop on Wave Measurements from Buoys (October 2008, New York, USA). The participants initiated a pilot project, PP-WET to develop procedures and protocols for the continuous testing and evaluation of wave measurements. The PP-WET is supported by JCOMM and leading agencies such as Environment Canada, European Services, Marine Observation Program, NOAA. Many institutes participated in PP-WET with an objective to establish the validity of wave measurements against the reference standard Wave Rider buoy to build confidence in the global community.

4 Estimation of Wave Climate Using Measurements and Numerical Modelling

The estimation of the nearshore wave climate using numerical modelling and measurement is explained through a case study. Puducherry, known for tourism/recreation, has lost its natural beach due to natural causes or man-made activities. To restore the lost beach using suitable coastal protection measures, it is essential to understand nearshore wave transformation. So, a spectral based wind-wave model was constructed to simulate the growth, decay and transformation of wind-generated waves from offshore to coastal areas.

A regional scale wind-wave model (Fig. 4) was set up primarily for the North Indian Ocean, and the results are used to force the local wave-to-wave model to arrive at an understanding of the nearshore wave transformation along Puducherry coast. The reliability of the model depends on the validation with offshore, coastal and nearshore measurements. The deep-water buoy data and coastal buoy data were collected under the Ocean Observation Network programme of ESSO-MoES.

The waves in the sea exhibit very random and complex behaviour with the superposition of many sinusoidal waves, each having its own height, period and direction.

The offshore buoy BD08 (18N, 90E) deployed at water depth of 2000 m (Fig. 5) and the coastal buoy deployed at water depth of 30 m (Fig. 6) near Puducherry are used for validation of the regional wave model. The local wave model is validated with the nearshore data measurements at 5 m water depth (Fig. 7).

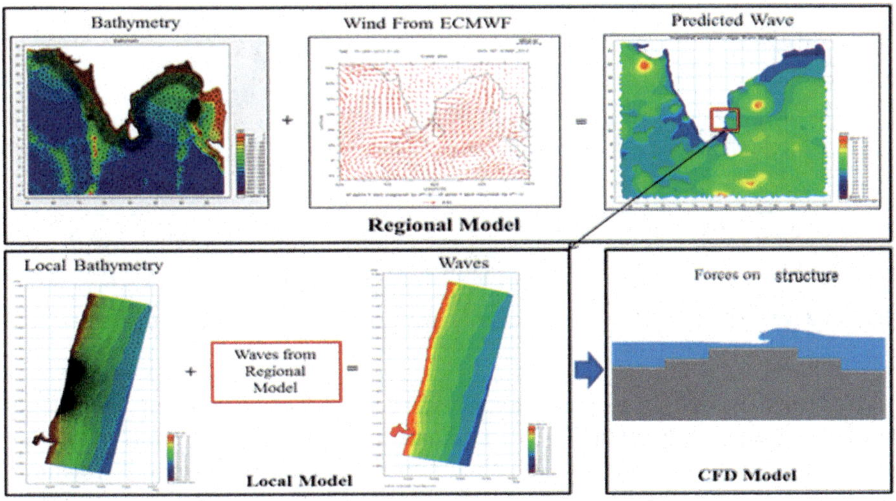

Fig. 4 Prediction of nearshore wave climate

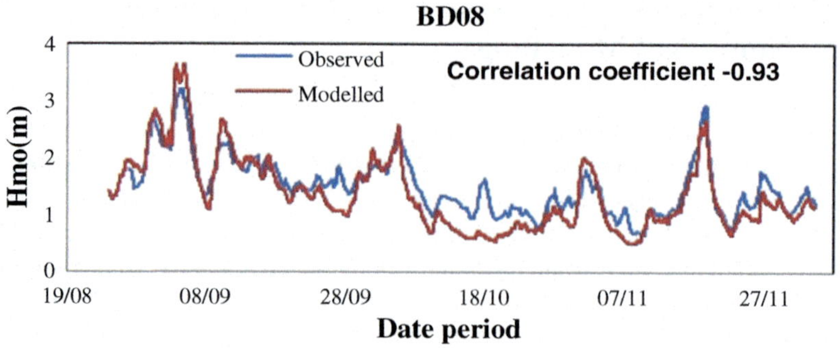

Fig. 5 Validation with deep-water buoy data

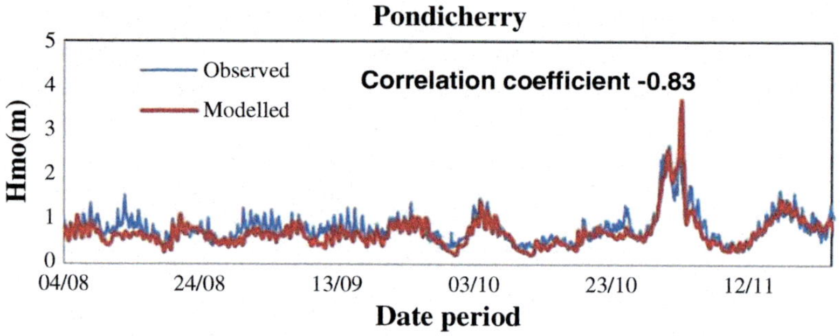

Fig. 6 Validation with coastal buoy data

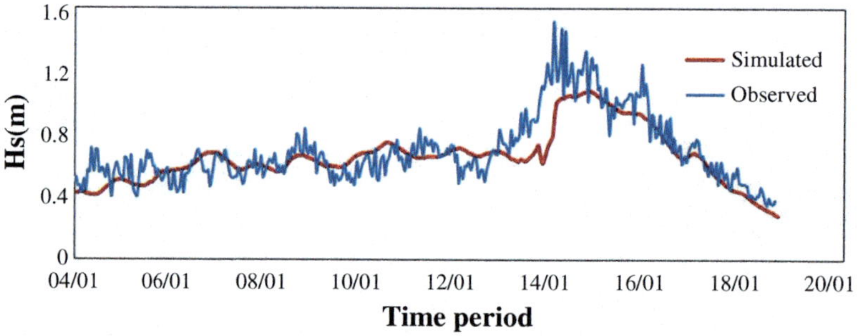

Fig. 7 Validation with nearshore wave measurements (DWR)

Numerical modelling in conjunction with coastal and offshore measurements will help in arriving at suitable structural and functional configuration of various marine structures such as port and harbour structures, intake and outfall structures and coastal protection schemes. However, considering the complexity of the project, physical modelling and extensive field measurements may be required.

5 Wave Data Analysis

The wave measurement systems generally record a wave trace (Fig. 8) for a specific period (typically 17 min), sampled at specific frequency (1 Hz) and recorded at fixed intervals (normally every 3 h). The parameters measured are heave or vertical displacement, which is processed by applying Fourier analysis to derive the non-directional wave spectra that provide the wave height and wave period parameters. To measure the directional parameters, additional information such as roll and pitch or horizontal slope measurements is required. The two-dimensional directional wave spectra generated by applying FFT provides the various wave parameters. The critical wave parameters are significant wave height, maximum wave height, one-tenth wave height, peak wave direction, zero-crossing wave period, wave period corresponding to peak energy and wave spectra defining the sea and swell wave energy, which provide a reliable information on the wave climate for various applications.

Normally, the wave climate of the proposed project area is arrived based on the direct approach of measuring the directional waves for at least 1 year and is represented in terms of measures of wave height, period and direction. Single parameter distribution (SPD), joint distribution and rose vector plots of wave parameters are utilized to estimate the wave characteristics and its variability at the selected location (Table 2 and Fig. 9). The detailed analysis of SPD tables represents the relative distribution of wave parameters under various categories. The joint variability of wave parameters is an important design criterion, provided by joint distribution tables and rose vector plots, which helps to identify the characteristics and extreme conditions.

The joint distribution (Table 2 and Fig. 9) indicates that 90% of the wave height is less than 1.5 m. The dominant wave directions are southerly waves representative of the southwest monsoon and east-northeasterly waves indicating northeast monsoon. The severe sea conditions are observed in east-northeasterly waves during northeast monsoon but limited to ~20% of occurrence.

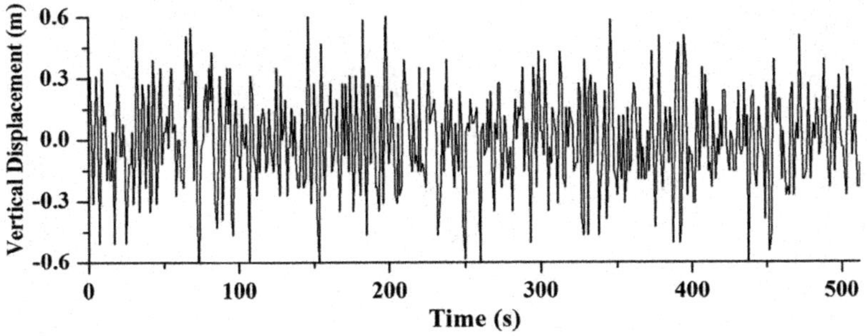

Fig. 8 Recorded wave data

Table 2 Joint distribution of wave height and wave direction (%)

Wave height Direction	0–0.5	0.5–1.	1.–1.5	1.5–2	2–2.5	2.5–3	3–3.5	3.5–4	%age	Avg.	Maxi
0–15											
15–30									0.01	0.68	1.36
30–45		0.04	0.12	0.07	0.02				0.25	0.77	2.36
45–60		0.64	1.8	0.5	0.07				3.01	0.83	2.41
60–75	0.01	2.99	4.84	1.15	0.09	0.02	0.01	0.02	9.12	0.71	3.67
75–90	0.14	5.85	4.45	0.54	0.16	0.08	0.02		11.23	0.6	3.42
90–105	0.38	5.93	2.35	0.17	0.08	0.03			8.93	0.63	2.84
105–120	0.54	5.24	0.84	0.05					6.66	0.63	1.79
120–135	0.33	4.95	0.31	0.01					5.6	0.66	1.63
135–150	0.81	7.12	0.84						8.77	0.63	1.65
150–165	0.35	8.03	2.47	0.08					10.93	0.66	1.94
165–180	0.08	7.24	6.68	0.64	0.08				14.72	0.59	2.27
180–195	0.02	4.19	8.46	2.92	0.17				15.75	0.75	2.28
195–210	0.01	1.24	2.41	0.45					4.1	0.77	2.03
210–225	0.04	0.28	0.26	0.11					0.69	0.56	2.02
225–240		0.11	0.05						0.16	0.66	1.36
240–255		0.01	0.02	0.01					0.04	0.69	1.68
255–270		0.01	0.01	0.02					0.04	0.99	1.87
270–285											
Percentage	2.69	53.86	35.90	6.71	0.66	0.12	0.03	0.02			

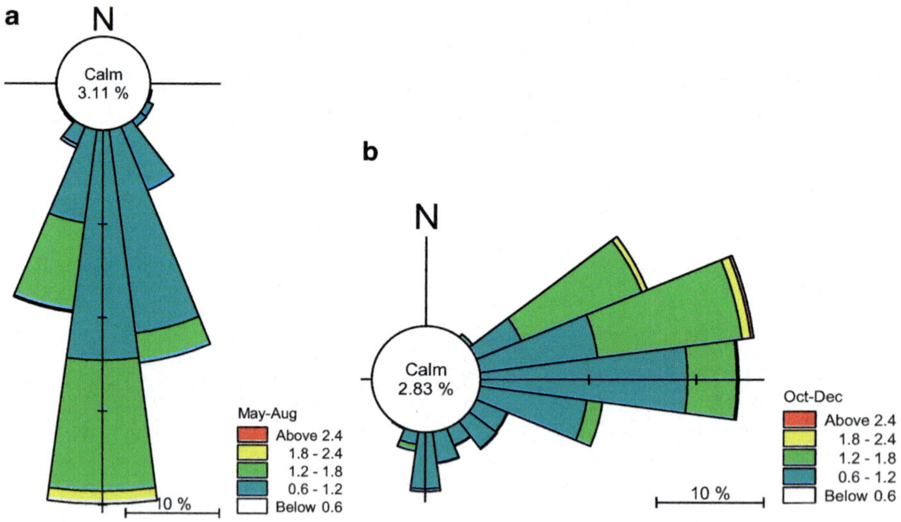

Fig. 9 Rose diagram of wave during (**a**) SW monsoon and (**b**) NE monsoon in coastal waters off Chennai, India

The joint distribution analysis also indicates the most predominant significant wave height for each month or each season that can be selected to carry out near-shore wave transformation studies to get the wave climate at different water depths and also at the breaker zone to use as input for wave tranquility studies, seabed sediment transport, littoral drift, navigational simulation study and other aspects of planning.

5.1 *Wave Characteristics in North Indian Ocean*

The wave information and its variability have significant impact on various spectra of science, technology and society. The waves in the Arabian sea (AS) and Bay of Bengal (BoB) in general exhibit significant seasonal variability in line with changing wind pattern. The major wave direction in open ocean remains southwesterly for most part of the year. However, the mean wave direction changes remarkably during the northeast monsoon season. The limited northward extent restricts the propagation of waves further north, and the southern Indian Ocean remains the major source of swell waves in AS and BoB.

The wave spectra which indicate the distribution of wave energy exhibit single-peaked, double-peaked and multipeaked spectra in north Indian Ocean (Fig. 10). The wave spectra in AS and BoB exhibit a combination of all the three and vary with season. The single-peaked spectra mostly exhibit the dominance of wave energy from strong local winds (Fig. 10a). The double-peaked spectra (Fig. 10b) are observed mostly during non-monsoon seasons and distant storms, indicating the significant contribution of swell waves along with locally generated sea waves. The multipeaked spectra (Fig. 10c) indicate a complex wave pattern with superposition of waves from multiple sources and in general lesser wave energy.

The time series plot of wave characteristics in AS and BoB reveals the striking variability in wave climate (Fig. 11). The dominance of southwest monsoon is well reflected with strong waves during the southwest monsoon season in both AS and BoB, with comparatively calm sea state for the rest of the year, except during cyclone passage. The significant wave height reaches more than 5 m during southwest monsoon in AS, whereas that of BoB hardly crosses 4 m. The average wave period in AS varies between 6 and 8 s during southwest monsoon. The wave period lies mostly between 4 and 6 s for the remaining period, with intermittent higher wave period indicating the significant contribution from swells in AS. The average wave period in BoB exhibits comparatively lesser variability with an average of 6 s with intermittent peaks during non-monsoon season.

The AS exhibits nearly steady southwesterly waves during June–September, whereas that of northern BoB remains steady southwesterly for most of the period except during November, December and January. The mean wave direction in AS is dominantly northeasterly during October–February, followed by a slow transition to steady southwesterly. The mean wave direction in BoB during northeast monsoon is scattered and exhibits a quick transition to southwesterly waves during February

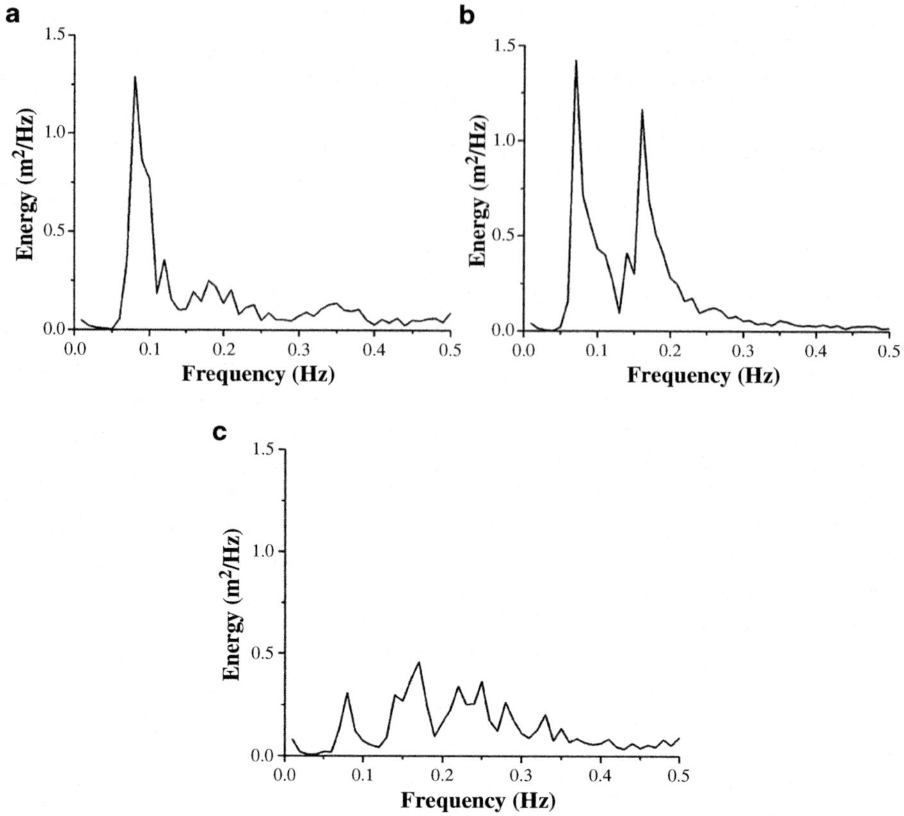

Fig. 10 (**a**) Single-peaked, (**b**) double-peaked and (**c**) multipeaked wave spectra

In general, AS exhibits higher variability in wave characteristics compared to that of BoB, with stronger waves during southwest monsoon and remarkable variability in mean wave direction and average wave period.

5.2 Design Waves

As waves are generally random, being driven by the near surface winds, a statistical approach is often taken to define the design conditions. The measurement of waves collected through various instruments are subjected to long-term wave statistical analysis for the estimation of design waves for different return periods of 1, 5, 10, 25, 50 and 100 years. The most commonly used methods for estimating design wave height are Gumbel distribution, Weibull distribution, Generalized Extreme Value Analysis and Generalized Pareto distribution.

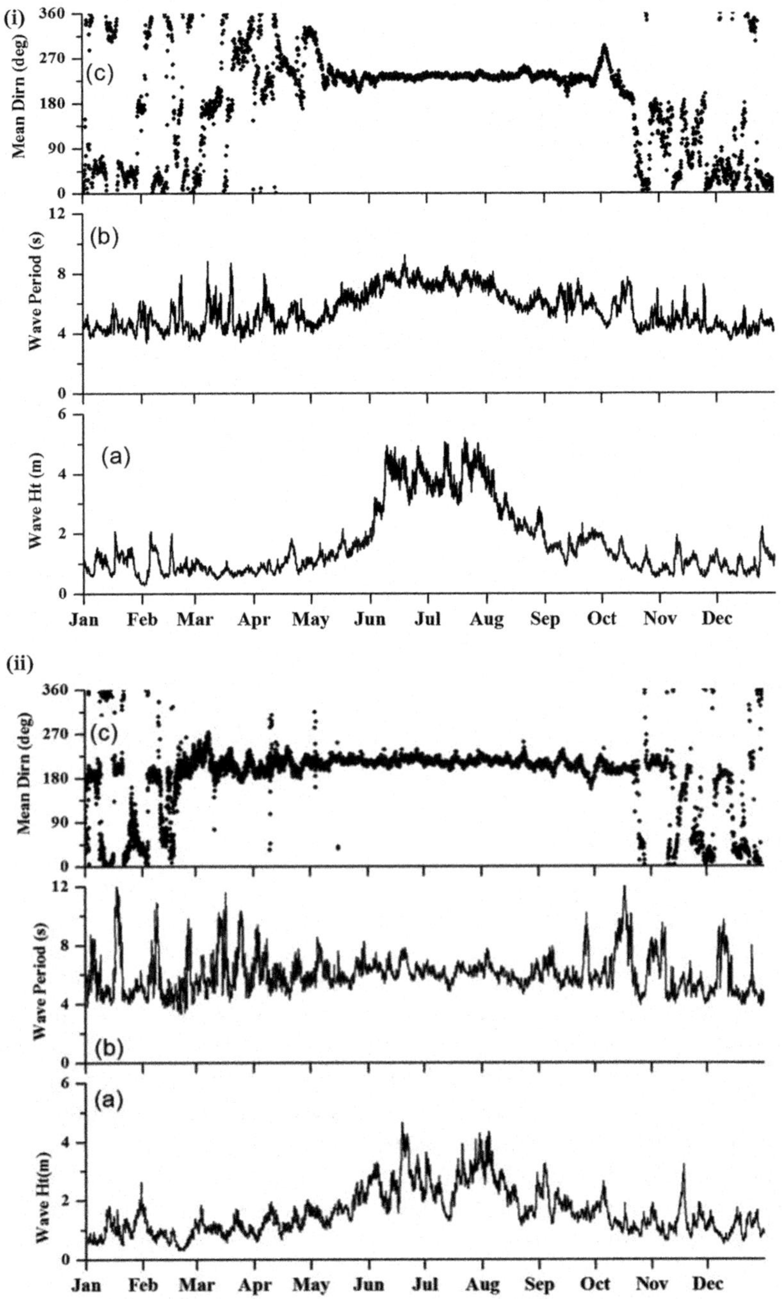

Fig. 11 Time series of wave parameters in northern AS (i) and northern BoB (ii). (**a**) Significant wave height, (**b**) average wave period and (**c**) mean wave direction

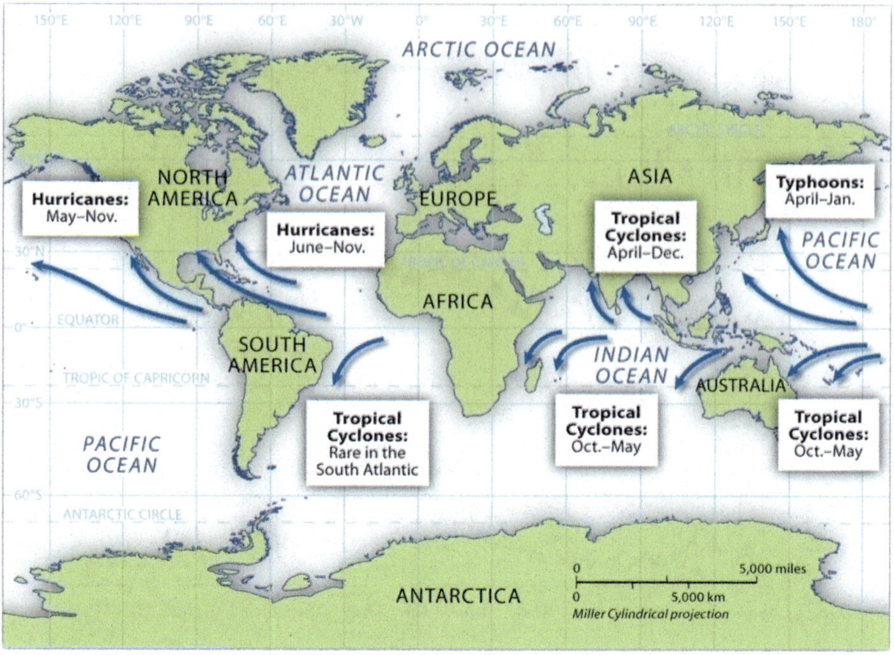

Fig. 12 Extreme wave conditions experienced around the world (Source: http://2012books.lardbucket.org/books/regional-geography-of-the-world-globalization-people-and-places/s08-05-tropical-cyclones-hurricanes.html)

5.3 *Extreme Waves*

While the normal wave climate varies and repeats in annual cycle with seasonal wind, these are waves which arise out in ocean due to the enormous amount of energy released during a catastrophic event. The potential for extreme cyclonic wind along the coastal region around the project area was first assessed by examining the available records of the parameters of cyclonic storms, which have crossed in a particular region. Hurricane or cyclone or typhoon can cause waves to rise more than 15 m in Atlantic Ocean; up to 10 m was recorded along the Indian coast. The duration of storm-induced wind waves generally prevail for a day or two if the eye of the storm crosses the vicinity. A wide variety of storms are encountered over the oceans of the world. These storms are grouped into the three main families experienced in maritime regions: tropical cyclones, midlatitude cyclones and hybrid systems (Fig. 12).

Reference

1. Engineers, US Army Corps Of "Coastal engineering manual." Engineer Manual 1110 (2002): 2–1100

Part III
Subsurface Observations

Oceanographic Floats: Principles of Operation

Eric D'Asaro

Abstract Oceanographic floats have become one of the most widely used oceanographic tools. As part of the international ARGO program, over 3000 floats monitor the temperature and salinity of the global ocean. Specialized floats carrying a broader instrument suite measure the ocean's biological and chemical properties. Dense arrays of floats make more detailed measurements of regional variations and physical processes. This chapter reviews the physical bases of float operation, the design bases that allow for different float behaviors, and describes an operational float control system that generates these behaviors.

1 Introduction

Oceanographic floats have become one of the most widely used oceanographic tools. As part of the international ARGO program, over 3000 floats monitor the temperature and salinity of the global ocean. Specialized floats carrying a broader instrument suite measure the ocean's biological and chemical properties. Dense arrays of floats make more detailed measurements of regional variations and physical processes. This chapter reviews the physical bases of float operation, the design bases that allow for different float behaviors, and describes an operational float control system that generates these behaviors.

An oceanographic "float" is a device that floats at mid-depth in the ocean. It is distinct from a "drifter" (Chap. 6), which floats at the surface. Its depth is determined by the subtle differences between the variations of its density with depth and the variations in the density of seawater. A float comes to the rest at the depth where these match. The float then follows the water surrounding it thereby measuring the water's motion. In this mode, a float is a Lagrangian current sensor. A float

For Observing the Oceans in Real Time – Instruments, Measurements and Experience
May 30, 2017

E. D'Asaro (✉)
Applied Physics Laboratory and School of Oceanography, University of Washington,
Seattle 98105, WA, USA
e-mail: dasaro@apl.washington.edu

R. Venkatesan et al. (eds.), *Observing the Oceans in Real Time*, Springer
Oceanography, DOI 10.1007/978-3-319-66493-4_5

at rest in the ocean requires no energy to maintain its depth and can thus be an exceedingly quiet and stable platform with a lifetime of many years. Since the density of the ocean almost always increases with depth, a float can change its depth by varying its density, moving downward by making itself denser and upward by making itself less dense. In this mode, a float is a profiler. Equipped with sensors, for example temperature, and salinity, it can make vertical profiles of these quantities. Such profiling requires energy, so that the useful float lifetime is determined by the number of profiles, the energy required per profile, and the energy available to the float.

Gould [11] and Rossby [28] review the history of floats. Henry Stommel [29] and John Swallow [30] collaborated to develop the first "Swallow floats" to measure deep oceanographic currents. These floats emitted acoustical signals and were tracked using receivers on a ship. Much longer ranges (1000 km) and durations (1 year) were achieved by using bottom-fixed receivers and exploiting the mid-depth sound channel. These "SOFAR floats" played a crucial part in the landmark 1973 MODE experiment [22]. Starting in the 1980s, improved electronics and ARGOS satellite tracking and telemetry [2] allowed these roles to be reversed. RAFOS floats (SOFAR spelled backward) use bottom-fixed transmitters and acoustic receivers on the floats [27]. At the end of the mission, the floats surface and transmit their data. RAFOS floats are small and relatively inexpensive and continue to be deployed to explore regional circulation and transport, e.g. Bower et al. [3].

The profiling float technology that led to the current ARGO array was jointly developed by Russ Davis and Doug Webb [9]. These floats profile by changing their volume, pumping oil in and out of an external bladder. They surface after each upward profile, determine their position using GPS and send data using the ARGOS, or more recently, Iridium, satellite communications systems. Unlike the acoustically tracked floats described above, they can operate autonomously anywhere in the world ocean. Although they do measure the average current between fixes, their primary mission is profiling. Originally called ALACE floats, and with a variety of manufacturer brand names, the generic name "ARGO float" will be used here.

The future of float technology lies in several areas. First, advances in small, low-power sensors, and improvements in satellite telemetry are allowing variables other than temperature and salinity to be measured by ARGO float technology. Appropriate sensors for oxygen, nitrate, pH, and various bio-optical properties [4] are commercially available and are being deployed in significant numbers. Second, although most floats either drift at a nearly constant depth or profile, other operational modes can enable unique measurements. For example, a float can continuously match its density to that of an evolving mixed layer, thereby measuring the mixed layer motions [8], tracking the evolution of a bloom within the mixed layer [1] or measuring air-sea fluxes beneath a hurricane [7]. A float can follow a surface of constant density to trace its evolution and the evolution of properties along it [26]. A float can photographically survey the bottom by remaining a few meters above the topography [25] as it is swept across these features by the currents. These applications rely on fine control of the float's buoyancy to move it vertically in response to its environment. This chapter will describe the physics and fluid dynamics of float motion that enable fine float control and a control system based on this physics and used by the author to implement such control.

2 Float Density and Behavior

Equations of State A key parameter controlling float behavior is its density $\rho_f = M/V_f$, the ratio of its mass M to its volume V. The mass varies only slowly during a deployment due to corrosion, biofouling, and, sometimes, float pieces falling off. The volume, however, changes due to temperature, pressure, and intentional control. D'Asaro [6] models this as

$$V_f = V_0 + B - \gamma_f V_0 P + \alpha_f V_0 \left(T - T_0\right) + V_{\text{air}} \frac{P_{\text{atm}}}{P + P_{\text{atm}}} \tag{1}$$

where V_0 is the reference volume, B is the active change in volume by the float, γ_f is the compressibility of the float, that is, its fractional decrease in volume due to pressure P, α_f is the thermal expansion coefficient of the float, that is, its fractional increase in volume due to temperature T referenced to T_0, and V_{air} is the volume of 'air-like' material at atmospheric pressure P_{atm}. The compressibility and thermal expansion terms are set by the mechanical properties of the float, with the compressibility term usually being more important. Both laboratory and field measurements clearly indicate the presence of high compressibility near the surface represented by the "air" term [5, 6]. Although some of this is due to transient bubbles on the outside of a float, stable non-zero values persist throughout the life of some floats suggesting that pockets of air can persist within the rubber seals or other float components.

Equation (1) is an "Equation of State" for a float: $\rho_f(T, P, B)$ gives the density of the float as a function of T, P, and B, just as the equation of state of seawater $\rho_w(S, T, P)$ [20] gives its density as a function of S (salinity), T, and P. Any object placed in the water at a particular oceanic location will float at the surface, if its surface density is less than the surface density of the water, that is, $\rho_f(T_s, 0, B) < \rho_w(S_s, T_s, 0)$, sink to the bottom if its bottom density is greater than that of the bottom water, i.e. $\rho_f(T_b, 0, B) > \rho_w(S_b, T_b, 0)$, or float within the water column at a depth where $\rho_f = \rho_w$.

Response to Vertical Displacements Properly designed floats have a unique ability to measure the very small vertical velocities typical in the ocean. Vertical velocities are much smaller than horizontal velocities, up to perhaps 0.1 m s^{-1} in strong turbulence, but 10^{-5} m s^{-1} or less in evolving mesoscale eddies. Although small, these velocities are important both dynamically and because they move properties vertically in the ocean.

Understanding floats' ability to measure vertical velocity requires an understanding of the changes in seawater properties in response to pressure changes. Under many conditions, the mixing of heat and matter between a parcel of water and neighboring parcels is sufficiently small to be negligible, that is, changes in the

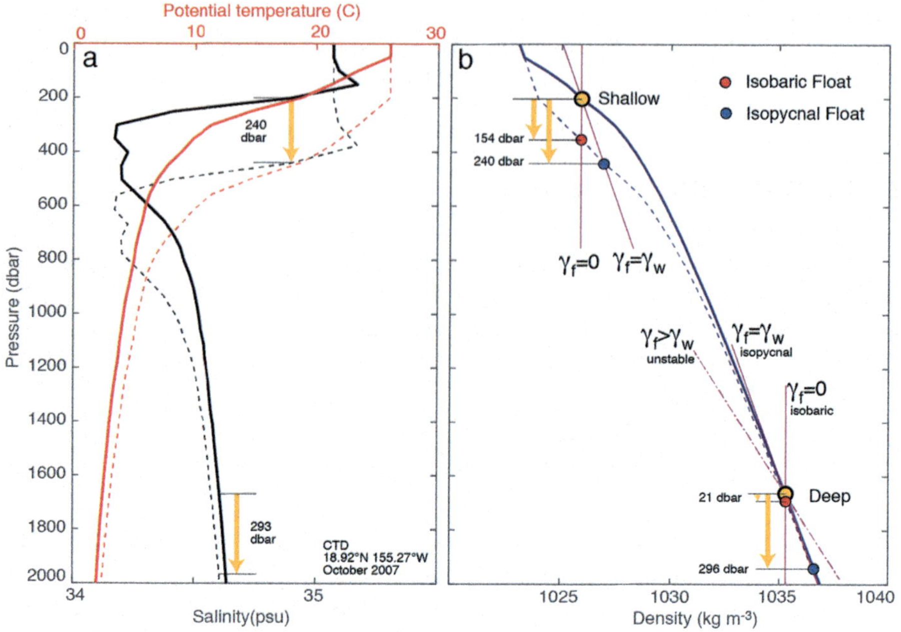

Fig. 1 (**a**) Profiles of potential temperature (*red line*, top axis) and salinity (*black line*, bottom axis) for a typical subtropical ocean. Profile is displaced downward (*orange arrows*) to form a displaced profile (*dashed lines*) simulating passage of a strong mesoscale eddy. (**b**) Profile of in-situ water density for these T and S profiles. Response of oceanographic floats to the vertical displacement. Floats are initially at 201 dbar ("shallow") and at 1666 dbar ("deep") as shown by the *yellow circles*. Behavior of these floats depends on their compressibility γ_f compared to that of water γ_w as shown by the *purple lines* and the *red* and *blue circles*. Effects of thermal expansion are ignored here. See text for details

properties of this parcel are adiabatic. If such a parcel is moved to higher pressure, it compresses; its volume decreases and its temperature increases. If the pressure is returned to its original value, these changes are reversed. Oceanographers thus define "potential temperature" θ and "potential density" σ as the temperature and density that a water parcel would have if moved adiabatically from its current temperature, salinity, and pressure to a reference pressure. The quantities θ, σ and S are thus conserved under adiabatic pressure changes. Modern formulae for these and other seawater properties can be found in the study by Millero et al. [21], IOC [16] and by searches for "TEOS-10," the Thermodynamic Equation of Seawater-2010. Note that TEOS-10 uses the name "conservative temperature" for θ.

Figure 1 shows an example of water property changes in response to vertical motion for a typical profile from the subtropics. The profiles of θ and S (Fig. 1a, heavy curves) define ρ_w (Fig. 1b, heavy blue). Vertical velocity is simulated by moving the θ and S profiles downward corresponding to a pressure change ΔP to produce the dashed lines in Fig. 1a, b; a simple displacement is appropriate because neither θ or S changes due to such displacements. The displacement is shown by two orange arrows in Fig. 1a: at 200 dbar, the "shallow" depth, and at 1666 dbar, the "deep" depth. At the deep depth, water density is primarily a function of pressure; it increases proportionally to $\gamma_w \Delta P$, where γ_w is the compressibility of seawater. At the shallow depth, changes in θ and S are also important and the density increases more rapidly with depth.

Isobaric Behavior The behavior of floats at each depth is diagnosed in Fig. 1b. A critical parameter is the ratio of the compressibilities of the float and the water γ_f/γ_w. For most floats this is less than 1, typically 0.3–0.5. This is approximated by assuming $\gamma_f = 0$, that is, ρ_f is constant, and shown by the vertical purple lines in Fig. 1b. Such a float initially rests at the orange circles and moves along the purple line to the red circles in response to vertical displacement.

At the deep depth, the density remains a function of pressure and thus changes very little as the temperature and salinity change due to vertical displacement. The deep float therefore remains at nearly the same pressure, moving downward by only 7% of the water's displacement. Thus, a float with $\gamma_f/\gamma_w \ll 1$ is called *isobaric* because it stays at nearly a constant depth in the deep ocean. This design is stable and simple. Most floats, particularly those whose mission is not primarily Lagrangian, are designed to be close to isobaric in the deep ocean.

Isopycnal Behavior At the shallow depth, even an isobaric float partially responds to the vertical displacement of the profile, 64% of the profile's displacement in our example. The float's equilibrium point will move with the water if the compressibility of the float is increased so that $\gamma_f = \gamma_w$. In Fig. 1b, the float then moves along the slanted purple line from the orange circles to the blue circles. A float designed in this way is called *isopycnal*, because it follows a surface of potential density. Rossby et al. [26] first designed a float in this way by adding a compressible component to a glass-hulled RAFOS float. D'Asaro [6] produced a similar effect in an aluminum hull by machining a series of stiff rings spanned by thin compressible arched panels.

The above example ignores the effect of thermal expansion. Thermal expansion of seawater α_w is a strong increasing function of temperature, being nearly zero at the freezing point, matching that of an aluminum float at about 8 °C, and increasing to many times that value at tropical temperatures. A more complete analysis of float behavior (e.g. [26]) shows an ideal isopycnal float should have coefficient $\alpha_f = \alpha_w$., so that both the pressure and temperature responses of the float match that of seawater. Since α_w is not constant, and most materials have a nearly constant α_f, this is difficult to do passively.

Stability The stability of a float to vertical displacements is also an important consideration. For an isobaric float and a fixed ocean profile, a float moved upward (along the solid purple line) from its equilibrium position (Fig. 1b, orange dot) does not change its density. It is thus heavier than the surrounding, shallower water and is pushed back toward the equilibrium position. As γ_f increases, this stabilizing force decreases. For the deep float (Fig. 1b) an example with $\gamma_f > \gamma_w$ (dashed-dot purple line) is unstable; the float becomes lighter than the surrounding water as it is moved upward and is thus pushed away from its equilibrium point. Instability can also be caused by the thermal expansion coefficient. A float expanding faster than water that is, $\alpha_f > \alpha_w$, is destabilized if the temperature increases upward. Similarly, if as in Fig. 1a, a stabilizing temperature occurs with an unstable salinity gradient, a float with $\alpha_f = \alpha_w$ is unstable. This can be counteracted by adjusting γ_f as was done by Rossby et al. [26]. The main point is that a nearly isopycnal float can be on the edge of instability. Practically, this limits the degree to which a float can be made isopycnal.

The above discussion considers only the effect of the float's equation of state on its behavior. All of the effects are passive and are set by the material properties and mechanical design of the float. They thus require no energy input. These effects can be overcome and more complex behaviors realized by actively controlling the float's buoyancy, that is, by varying *B*. This requires a more complex mechanical design, a software control system, and energy input. However, all of these features exist in modern floats. The remainder of this chapter focuses on active control of such floats.

3 Float Drag in a Stratified Ocean

Float Drag A float is moved upward in the water by increasing its volume, that is, B in (1) and downward by decreasing its volume. Each new B sets a new equilibrium depth, as described above. When the float is not at its equilibrium depth, a buoyancy force acts to move the float toward equilibrium. The rate at which it moves and the time it takes to get to the new equilibrium depth is set by the hydrodynamic drag, that is, the speed at which the float moves for a given force. Active control of the float's motion thus requires an understanding of the drag properties of the float.

Homogeneous Ocean In quiet homogeneous water, the drag of a moving object is well studied. The key parameter is the Reynold's number Re = $W\, l/\nu$, where W is the velocity, l is a characteristic size of the object, and ν is the kinematic fluid viscosity, typically about 10^{-6} m^2 s^{-1} for water. For *Re* larger than a few hundreds, the drag is characterized by a drag coefficient C_D that usually varies only slowly with *Re*. The drag force is given by

$$F_{\text{Hdrag}} = \frac{1}{2} C_D \rho_w A W^2, \tag{2}$$

where A is the area of the object perpendicular to the flow direction. For flow perpendicular to a flat plate, $C_D = 1.1$; for a smooth sphere, $C_D \sim 0.5$; and for a highly streamlined object, $C_D \sim 0.04$. For a float with W of 0.01–0.1 m s^{-1} and l of about 1 m, *Re* is 10^5–10^6. In this range, the flow around a sphere or cylinder is transitional between laminar and turbulent [15].

Stratified Ocean Most of the ocean is stably stratified, with the water's density increasing downward. As detailed below, the flow around an object sinking or rising through a stratified fluid can be quite different from that in a homogeneous fluid and increases the drag. The importance of stratification for the flow around an object, and thus for the drag, is measured by the Froude number $F = W/Nl$ where N is the buoyancy frequency approximately given by $N^2 = -\frac{g}{\rho}\frac{d\sigma}{dz}$ and g is the acceleration of gravity (see [16] for details on N^2 computation). The value of N increases with increasing vertical density gradient. A parcel of water moved away from its equilibrium depth is moved back toward equilibrium on a timescale of about N^{-1}; the buoyancy period is $2\pi/N$.

In the absence of stratification, that is, $F \gg 1$, and for $Re > 10$, the drag is controlled by eddies that form behind the object and become increasingly turbulent with increasing *Re*. Figure 2b shows a numerical simulation of the wake behind a sphere at $Re = 200$ and large F [31]. Eddies are apparent behind the sphere. These trap water behind the object and carry it downward with the sinking object. In a stratified fluid, the trapped water is lighter than the water surrounding the object. Thus, in Fig. 2a, the water in the wake (colored pink) is carried downward from a position far above the top of the figure. This lighter water is buoyant, pulling the sphere upward and thus increasing the drag. This buoyancy also suppresses the wake eddies so that for sufficiently low F, they disappear. This is shown in Fig. 2cd, with $F = 2$. For a sphere, the eddies are suppressed at about $F = 20$ [12]. Without the eddies, the flow converges strongly behind the sphere, and is accelerated upward by its buoyancy to form a narrow column of light water above the sphere. Figure 3 graphically shows the formation of this column. A sinking body also radiates internal gravity waves with an intrinsic frequency very near N [23]. The wave drag becomes an increasingly large part of the drag at very low F. The flow behind a sinking object in a stratified fluid is thus very different from that in a homogeneous fluid. Turbulence is suppressed, and the drag results from a combination of internal wave radiation and a buoyant wake, which can extend far behind the object.

Stable vertical density stratification increases the drag of floats over that given by (2) and causes floats to move more slowly than would be predicted by (2). For $F \gg 1$, water moves past the object so fast that the stratification is irrelevant while for $Fr \ll 1$, stratification dominates the flow. Similarly, stratification dominates the drag for $W \gg F\,N\,l$ and is negligible for $W \ll F\,N\,l$. For typical float sizes ($l \sim 1m$), the transition speed ($F \sim 1$) is about $W \sim 0.01$ m s^{-1} in the upper ocean ($N \sim 0.01$ s^{-1}) and about $W \sim 0.001$ m s^{-1} in the deep ocean ($N \sim 0.001$ s^{-1}). The strongest effects occur in highly stratified density interfaces where $N \sim 0.1\ s^{-1}$ and $F < 1$ for $W < 0.1$ m s^{-1}.

Despite the complex flow, the excess drag due to stratification is approximately linear in velocity and given by

$$F_{\mathrm{Sdrag}} = \rho_w A N L W \tag{3}$$

where L is the length scale. [13, 14, 31]. The total drag is then given by

$$F_{\mathrm{drag}} = F_{\mathrm{Hdrag}} + F_{\mathrm{Sdrag}} \tag{4}$$

the sum of the quadratic component (2), dominant at high speeds, that is, $F \gg 1$, and a linear component (3) dominant at low speeds, that is, $F \ll 1$. For a sphere, the data by Hanazaki et al. [13]‘s data is fit well using (4) with $C_D = 0.75$, close to the accepted drag coefficient for a sphere at $Re = 200$ and $L = 1.25$ a, close to the

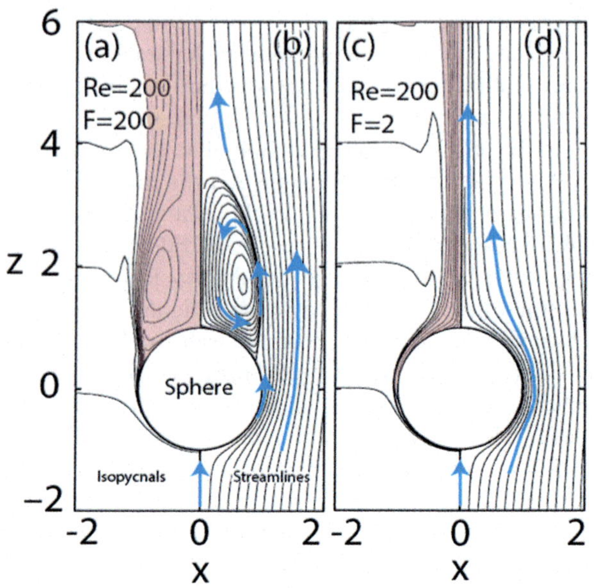

Fig. 2 Steady-state flow around a sphere sinking through a stratified fluid. Each panel is split with isopycnals plotted on the left side of the sphere and streamlines plotted on the right side. (**a**) Isopycnals at F=200, Re=200. Fluid lighter than the top isopycnal is shaded. (**b**) Streamlines for same. *Arrows* show flow direction. (**c**) Isopycnals at F=2, Re=200. (**d**) Streamlines for same. (Modified from [31], their Fig. 2)

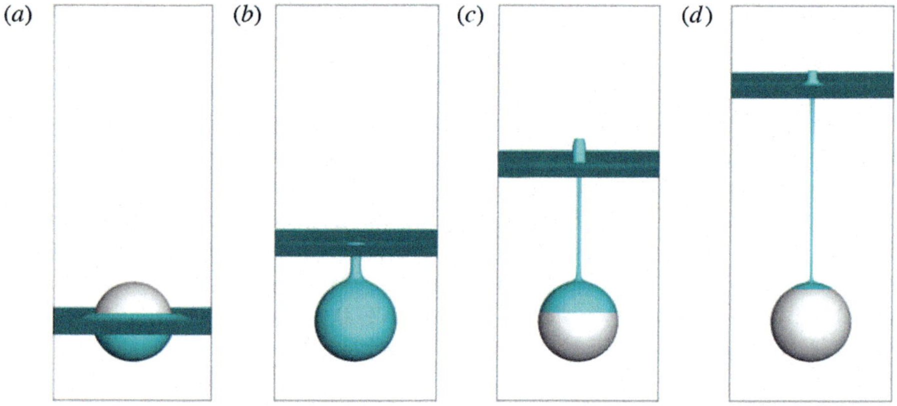

Fig. 3 Deformation of an isopycnal surface behind a sphere sinking through a stratified fluid with Fr=0.3 and Re=200 (From [13], their Fig. 4)

sphere's radius (Fig. 4). These data, scaled using typical float dimensions yield a curve defining the float's vertical velocity as a function of its drag, or equivalently its buoyancy, which covers the range of typical values. As expected, the range of F (colors in Fig. 4) spans 1, indicating that buoyancy is important, but the float Re (text in Fig. 4) is very much bigger than that in the study by Hanazaki et al. [13]. Furthermore, floats are far from spherical. Additional hydrodynamical effects not captured in Fig. 4 are thus possible.

The major prediction of the above discussion is that the drag on a float can be strongly affected by the stratification, especially at low speeds. In particular, the

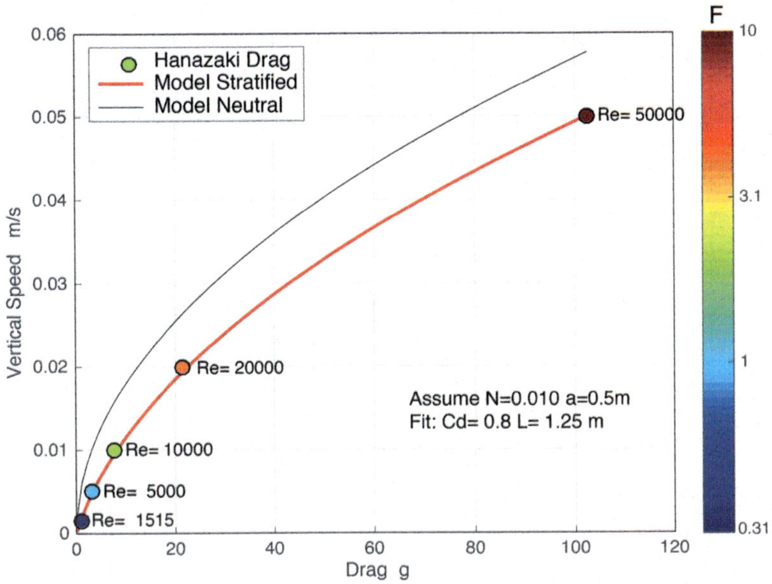

Fig. 4 Fit of Eq. (4) to Hanazaki et al. [14]'s data (dots digitized from their Fig. 23) for the drag of a sphere in a stratified fluid. *Dot color* indicates Froude number. Reynolds numbers for each data point is annotated. *Red line* is model fit. *Gray line* is model with N=0

drag is linear, rather than quadratic at low speeds. D'Asaro [6] (his Fig. 11) shows such a linear drag, but the data were insufficient to demonstrate an explicit dependence on stratification N. In this case, the float repeatedly oscillated vertically at speeds of a few mm s^{-1} due to a control system designed for drag law (2) rather than (4). Its vertical motion was slow due to a large horizontal drogue with area πa^2, a = 0.7 m. The float buoyancy was evaluated using (1) and was used to create a buoyancy/drag curve similar to that in Fig. 4. A best fit of (4) to this data yields L = 1.1 m. This is about 1.6 a, close to that found in Fig. 4.

Figure 5 shows similar drag curves from a float profiling and settling on isopycnals in the upper 500 m in a region with large variations in stratification. Data are fit to (4) to yield the colored lines. For weak stratification (red), the points nearly follow the quadratic drag relationship (2, gray line), while for strong stratification (black squares) and small speeds they follow the linear relationship (3, black line). The transition from linear to quadratic is seen most clearly in the insert figure. The best fit value of L = 1.7 m, is similar to that found by D'Asaro [6], even though the drogue for this float is very much smaller. The relationship between float geometry and the value of L is thus complex and needs further study. Nevertheless, Fig. 5 clearly shows the dependence of drag on stratification and the ability of (4) to fit this dependence for real floats. Most importantly, it shows that in the stratified ocean, float drag is linear rather than quadratic at small speeds with the slope of the linear relationship dependent on the stratification.

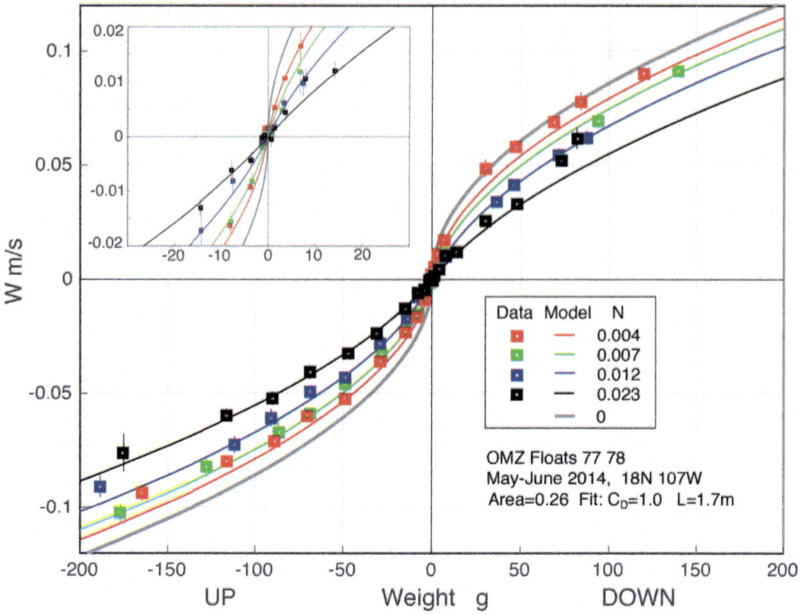

Fig. 5 Drag curve for 2014 OMZ floats. The float weight in water is computed from Eqs. (1) and (8). The float vertical velocity W is computed from the rate of change of pressure. An average profile of stratification N is computed from the float CTD during fast downcasts and interpolated to each time. Data is averaged in bins of weight and N. The mean weight and mean vertical velocity are plotted as colored squares. Vertical bars are two standard deviations of the mean. Colors indicate average N in each category as listed in the legend. Lines are fit of model (5) to the data. *Gray line* is the model curve for N=0. Inset in upper left shows details for small weights and speeds

4 Float Maneuvers

A float is moved vertically by varying B (Eq. 1). Several types of maneuvers are possible as described in this section. Section 5 then develops a model for float behavior based on the drag and buoyancy effects described above and a control algorithm for the float. Combined, these reduce the problem to the well-known equation for a damped, simple harmonic oscillator as described in Sect. 6. The use of this approach to conduct various maneuvers is described in Sect. 7. Examples are shown from deployments of various models of the APL/UW float "MLFII" described by D'Asaro [6].

The simplest and most common float maneuver is a profile. For the simplest down profile (Fig. 6a), a float at the surface decreases B until it sinks past a target depth. It then increases B until it comes to rest. Similarly in an up profile (Fig. 6b), the float starts at some depth and moves upward to the target depth. An ARGO float mission typically consists of a down profile to 1000 m at which the float passively

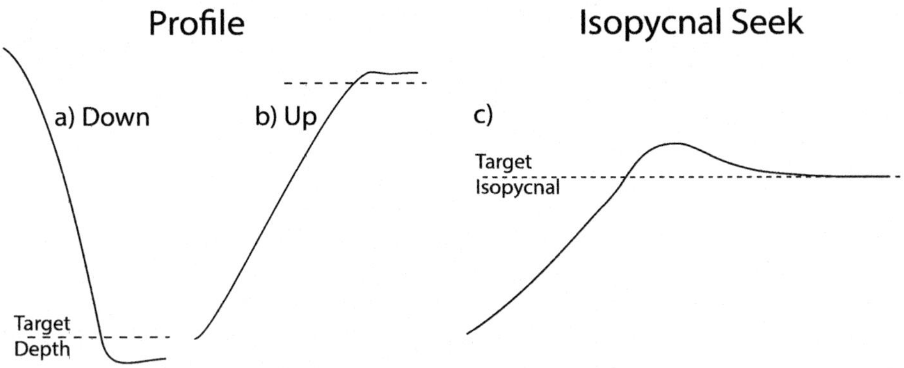

Fig. 6 Simple float maneuvers. (**a**) A down profile ending near a target depth. (**b**) An up profile ending near a target depth. (**c**) An isopycnal drift ending exactly on a target isopycnal

drifts for many days, a second down profile to 2000 m, followed by an up profile to the surface during which the sensors are turned on and data are taken.

A more difficult maneuver is isopycnal seeking. The goal is to bring a float to a specified isopycnal, that is, a surface of constant potential density, and remain on this isopycnal (Fig. 6c). For Lagrangian floats [6] with CTD measurements of potential density on both the top and bottom of the float, a typical mission goal is to maintain the target isopycnal between the two CTDs by actively varying B. This approach avoids the danger of instability, for example, Fig. 1b, of the purely passive approach used in isopycnal RAFOS floats. If all of the constants in Eq. (1) are known, and Eq. (1) is accurate, then isopycnal seeking may seem easy: the T and S values of the target isopycnal are used in (1) and it is solved for B. In practice, the constants in (1) are not always known well, particularly at the start of a mission. Oceanographic profiles are often not as simple as in Fig. 1; T and S can vary in complicated ways so that their values at the target isopycnal may not be known. Under these conditions, a dynamic approach is needed in which B is continuously varied based on the measured T and S values so as to move the float toward the target.

Even more difficult is the goal of making a float Lagrangian, that is, having it follow the water surrounding it exactly. As discussed in detail by D'Asaro [6], this will occur first, if the float has the same density as the water and second, if the changes in velocity across the float are small. The first can be done by actively changing B so that the float's density matches a measured seawater density. The accuracy to which this can be done is primarily controlled by how well the float's equation of state is known. The second is difficult; the resulting errors need to be assessed for any particular application.

5 A Dynamic Float Control Algorithm

Buoyancy and Drag The equation for vertical float position *Z*, positive upward, is

$$\frac{dZ}{dt} = W. \tag{5}$$

Vertical velocity *W* is set by the float's buoyancy *b*, here defined in weight units, that is, kg not Newtons. The analysis is simplest if the linear drag (3) is assumed rather than the more complete (4); the following analyses will thus be only strictly be accurate when the float is moving very slowly, that is, in the final stages of an isopycnal seek. The water density ρ_w is further approximated by a constant ρ_0 whenever it is not used to compute buoyancy (Boussinesq approximation). The vertical force balance in equilibrium is

$$g\,b = \rho_0 A\,N\,L\,W \tag{6}$$

so that *W* computed from (6) is the terminal velocity of the float and ignores initial transients. Combining (5) and (6)

$$\frac{dZ}{dt} = \frac{g}{\rho_0 ALN} b \tag{7}$$

The equation for float buoyancy is

$$b = M - \rho_w\left(S,T,P\right)\left[V\left(T,P\right) + B\right] \tag{8}$$

where ρ_w, the water density, is a known function of *S*, *T*, and *P* and *V*(*T*,*P*) is the float volume, a known function of *T* and *P* as in (1).

Behavior of an Ideal Isopycnal Float As a first reference point, the equations for an exactly isopycnal float are derived. Such a float should have a constant potential density σ_{goal}. If it is displaced a distance z from σ_{goal}, the difference $\Delta\sigma$ between its density and that of the surrounding water is

$$\Delta\sigma = \sigma_{\text{goal}} - \sigma = \rho_0 \frac{N^2}{g}\left(z - z_0\right). \tag{9}$$

So, the float's density is greater than that of the surrounding water when it is displaced upward. Setting $z_0 = 0$ for simplicity and approximating the float's volume by V_0 in (1) the float's buoyancy is

$$gb = g\,V_0 \Delta\sigma = -V_0\,\rho_0\,N^2\,z. \tag{10}$$

Combining (7) and (10)

$$\frac{dz}{dt} = -\omega_I z \tag{11}$$

with the solution $z = Z_{t=0}e^{-\omega_I t}$. The float moves toward the target depth with an inverse decay time

$$\omega_I = \frac{NV_0}{AL} \equiv NC. \tag{12}$$

The constant $C = V_0/AL$ is the ratio of the float's true volume to the volume AL in the drag expression (3). For the float in Fig. 5, $C = 0.15 = 6.7^{-1}$, so the float decays toward the target isopycnal in several buoyancy periods. For a float with a larger drag, AL is larger, C is smaller and the decay is slower. For the standard MLFII float with the drogue deployed, for example in the study by D'Asaro et al. [8], $C \approx 28^{-1}$; it decays toward an isopycnal about four times slower than the float in Fig. 5.

Mixing changes the density of water parcels and thus moves them to a different isopycnal. If the mixing occurs sufficiently rapidly, a float will follow the water rather than staying on its original isopycnal σ_{goal}. The float thus behaves as a Lagrangian float rather than isopycnal float on the timescale of mixing. However, once off its isopycnal, the mismatch in density, that is, (9), will cause a float to return to σ_{goal} on a time scale of $(CN)^{-1}$. The float is thus isopycnal, rather than Lagrangian at these longer timescales. The balance between these two effects can be controlled by varying the drag of the float.

More formally, we write a forced version of (11) by imposing a diapycnal velocity $E(t)$, due to mixing, that forces the float off its isopycnal,

$$\frac{dz}{dt} = -\omega_I z + E(t). \tag{13}$$

Substituting $E(t) = \tilde{E}e^{i\omega t}, z(t) = \tilde{z}e^{i\omega t}$, the response function is

$$|\tilde{z}|^2 = \frac{|\tilde{E}|^2}{\omega^2 + \omega_I^2} \tag{14}$$

At high frequencies, $\omega \gg \omega_I$, the restoring force, that is, ω_I, has no effect and the float is Lagrangian; it follows the water with a displacement $\tilde{E}(i\omega)^{-1}$. At low frequencies, $\omega \ll \omega_I$, there is a steady balance between the diapycnal velocity, the restoring force and the drag; the float is displaced from its isopycnal by $\tilde{E}(\omega_I)^{-1}$, or about 3 cm for a standard MLFII with an open drogue, $E = 1$ m/day and $N = 0.01$ s^{-1}.

Control for Lagrangian Behavior As described above, a float will be maximally Lagrangian if its density matches that of the surrounding water. This can be done by actively varying B in response to the measured values of T, S, and P using

$$B = \frac{M}{\rho_w(S,T,P)} - V(T,P) + B'. \tag{15}$$

Substituting (15) into (8) yields

$$b = \rho_0 B'. \tag{16}$$

The first two terms in (15) cancel out buoyancy effects due to variations in the difference between the volume of seawater displaced by the float, $M/\rho_w(S, T, P)$, and the variations of the actual float volume $V(T, P)$. If there are no errors in M, $V(T, P)$ or $\rho_w(S, T, P)$, or in the measurements of appropriate values of S,T, and P, then for $B'=0$ the float will remain neutrally buoyant with its density always matching that of the surrounding water. It will thus be Lagrangian except for the effects of its finite size. Although there are some subtle stability errors with this approach [6], it can yield useful measurements of Lagrangian trajectories [8].

Control for Isopycnal Behavior A float can be given isopycnal behavior by adding an effective compressibility to active control of B,

$$B' = V_{\text{off}} - \hat{\gamma}\frac{\Delta\sigma}{\rho_0}V_0 \tag{17}$$

where $\hat{\gamma}$ is a non-dimensional compressibility. Consider the case with $V_{\text{off}} = 0$. Combining (17), (16), (10), and (7) yields

$$\frac{dz}{dt} = -\hat{\gamma}\omega_I z. \tag{18}$$

Notice that this is the same as (11) for $\hat{\gamma} = 1$; with this value the float is isopycnal. Similarly, for $\hat{\gamma}$ 1 the float is isobaric; for $\hat{\gamma} > 1$ the float seeks isopycnals faster than an isopycnal float; for $\hat{\gamma} = 0$ the float is Lagrangian and for $\hat{\gamma} < 0$ the float is unstable. Thus the entire range of behaviors shown in Fig. 1 can be imposed on a float by varying $\hat{\gamma}$.

Isopycnal Seeking If the coefficients of (1) are not accurately known or change with time, control using (15) and (17) is insufficient to accurately move a float to an isopycnal. For example, to locate a float to within 1 m with $N = 0.01\ s^{-1}$ requires knowing its mass and volume to an accuracy of 10^{-6}. This is nearly achievable for mass using an accurate scale, but is more difficult for volume. One solution is to "ballast" the float in a laboratory tank of water, adjusting its weight until it is neutral and then calculating the necessary mass changes for the expected oceanic conditions. With care this can work well, but often requires weighing the float under pressure and controlled temperature conditions. Another solution is to ballast the float only roughly in the laboratory and then conduct ballasting maneuvers in the ocean. In this case, the parameters of (1) are not well known initially and are determined during the mission, both over the first few days of deployment and later when necessary. The ballasting maneuvers require a control system that moves the float to the target isopycnal without accurately knowing the parameters of (1), in particular V_0.

Such isopycnal seeking is implemented by adding an additional control term

$$\frac{dV_{\text{off}}}{dt} = -\frac{\Delta\sigma V_0}{\rho_0}\frac{1}{\tau} \tag{19}$$

where τ is an equilibration time over which V_{off} is adjusted to move the float to σ_{goal}. This is equivalent to adjusting the base float volume V_0 in (1) until the float is neutral at σ_{goal}. Differentiating (17) and using (9), (16), (19), (7), and (8) yields

$$\frac{d^2Z}{dt^2} + \hat{\gamma}\omega_I\frac{dZ}{dt} + \omega_I\tau^{-1}Z = 0. \tag{20}$$

Substituting exponential solutions $Z(t) = \tilde{Z}e^{-\lambda t}$, with λ having units of frequency, so that a real, positive value of λ corresponds to stability, into (20) yields

$$\lambda = \frac{1}{2}\hat{\gamma}\omega_I \pm \frac{1}{2}\sqrt{(\hat{\gamma}\omega_I)^2 - 4\omega_I\tau^{-1}}. \tag{21}$$

Alternative parameters are

$$\tau_c = \frac{4}{\hat{\gamma}^2 CN} = \frac{4}{\hat{\gamma}^2\omega_I} \tag{22}$$

and

$$\lambda_c = \frac{2}{\hat{\gamma}\tau_c} = \frac{1}{2}\hat{\gamma}\omega_I. \tag{23}$$

Using these parameters (20) becomes

$$\frac{d^2Z}{dt^2} + 2\lambda_c\frac{dZ}{dt} + \frac{\tau}{\tau_c}\lambda_c^2 Z = 0, \tag{24}$$

yielding solutions characterized by

$$\lambda = \lambda_c\left[1 \pm \left(1 - \frac{\tau_c}{\tau}\right)^{1/2}\right] \tag{25}$$

6 Control Regimes

Note that (20) is the equation for a damped simple harmonic oscillator and thus has well-defined oscillatory and exponential regimes. These are governed by two adjustable parameters. The parameter $\hat{\gamma}$ sets the rate at which a float moves to its isopycnale, e.g. (18). The parameter, τ sets the rate at which this isopycnal moves toward σ_{goal}, e.g. (19). Two different types of behavior are set by the value of τ/τ_c:

Oscillatory For $\tau < \tau_c$, damped oscillations occur with an oscillation frequency $\sqrt{\tau_c/\tau - 1}$ and an exponential decay rate $\lambda_c = 0.5\,\hat{\gamma}\,N\,C$. Note that stronger stratification leads to more rapid decay. Examples are shown by the red and magenta curves in Fig. 7.

Critical The transition to purely damped solutions occurs for $\tau = \tau_c$. The solution is purely exponential with $\lambda = \lambda_c$. For a given value of $\hat{\gamma}$, this results in the fastest possible damping with no oscillations. The black curve in Fig. 7 is critically damped.

Exponential If $\tau > \tau_c$, the solutions are the sum of two decaying exponentials. For $\tau \gg \tau_c$ the fast rate, with $\lambda = 2\lambda_c = \hat{\gamma}\omega_I$, corresponds to the motion of the float toward its rest isopycnal as in (18). The slow rate, with

$$\lambda = \frac{1}{\tau\hat{\gamma}} = \frac{\lambda_c}{2}\frac{\tau_c}{\tau}, \tag{26}$$

corresponds to the motion of the rest isopycnal toward the target isopycnal and is always slower than the critical rate. Note that the time to reach the target isopycnal is proportional to $\hat{\gamma}\tau$ and is thus independent of N. Examples of decay in this regime are shown by the blue and cyan curves in Fig. 7.

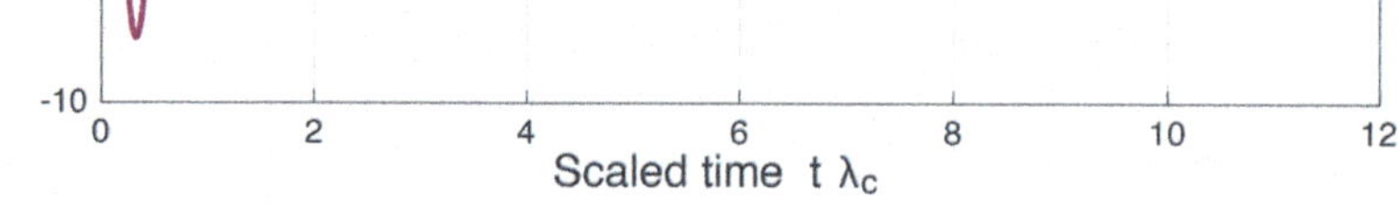

Fig. 7 Solutions to Eq. (24). Vertical coordinate Z is distance from target isopycnal, initialized at an arbitrary unit of 10 and zero velocity. Time is scaled by λ_c. *Colored curves* show solutions varying from damped oscillations for $\tau < \tau_c$ (*red, purple*), to critical for $\tau = \tau_c$ (*black*), to exponential for $\tau > \tau_c$ (*blue, cyan*)

7 Usage and Performance

The above algorithm has been used in operational APL/UW floats since 2001 and proven to be a robust tool, with some issues. It is used in several different ways.

Float Ballasting For proper operation, a float's mass must be adjusted so that its density can match that of the water in which it is operating. This is done first roughly in a water tank or off the back of a boat. A more accurate measurement is then made by deploying the float in local waters and bringing it to an isopycnal at about 60 m depth using the control algorithm. Since this is usually done on multiple floats simultaneously using a small boat, it is important that the float equilibrates as quickly as possible and with high reliability. Accordingly, the algorithm is tuned to operate close to the critical point with $\hat{\gamma} > 1$ to speed up the convergence. Figure 8 shows two examples with τ/τ_c, having estimated values of 0.7 (blue) and 2.4 (red). In each, the float starts at the surface, profiles to 60 m and then begins an isopycnal seek to the isopycnal at 60 m. Both floats undergo large oscillations in the vertical, with the red float both reaching the surface and bringing the volume controller (B in Eq. 1) to its smallest value. Nevertheless, they both equilibrate to the target isopycnal within 10^4 s. The final equilibrium has variations in B less than 1 cc, which is sufficient for setting the final float mass and thereby ballasting the float. Note that during the final equilibration, the float's pressure continues to vary as the target isopycnal moves up and down, but the float remains on this isopycnal.

Stairstep Settles A more complex sampling pattern is created with a set of isopycnal seeks each with a different target isopycnal. Figure 9 shows an example from the oxygen minimum zone off Mexico. The steps are flatter when viewed in isopycnal coordinates (red) rather than in pressure coordinates (black) because the isopycnals move vertically due to internal wave displacements.
Such an "isopycnal staircase" has two uses. First, as in Fig. 8, the equilibrium at the end of each step defines a point where T, S, P, and thus ρ_w are known. Since the float is neutrally buoyant, $\rho_w = \rho_f = M/V_f$. Since the float mass M is known, V_f can be computed. Given a set of such points, the constants in the float's equation of state (1) can be estimated. For float operations where this is important, such as Lagrangian measurements in the mixed layer (e.g. [8]), multiple staircase settles are typically run and V_0 , γ_f , and V_{Air} are computed. Accurate estimates of both γ_f and V_{Air} require equilibrium values to be obtained both deeper than a few 10's of meters and shallower than 20 m. The shallow measurements are particularly difficult to obtain because the ocean is often weakly stratified there. Rarely are there sufficient variations in temperature to distinguish the effects of γ_f from those of α_f. Accordingly, α_f is usually estimated from the physical properties of the float materials. However, with care, all of the coefficients of (1) can be computed with sufficient accuracy to estimate the float buoyancy to a few tenths of a gram, an accuracy of a few parts per million relative to a typical MLFII mass of 50 kg. Under most conditions, all of the coefficients appear to be stable over many days; the exponential decay in V_{Air}

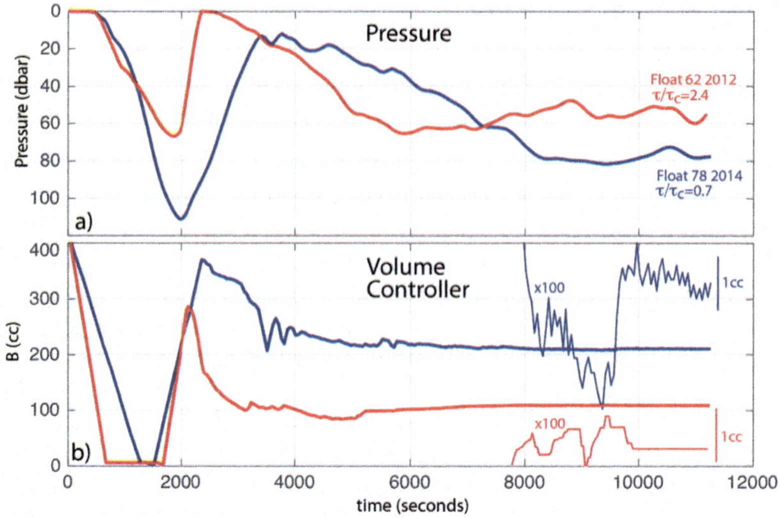

Fig. 8 Examples of isopycnal settles for two floats deployed on short-term ballasting missions in Puget Sound. (**a**) Float pressure. (**b**) The volume of the float volume controller, B in Eq. (1). *Thin lines* in show the same data magnified by a factor of 100 and offset to show the final settling. *Thin labeled lines* show the scale for the *thin lines*. Both floats start at the surface, sink to 60 m setting the target isopycnal at this depth, and then start the isopycnal settle. The final settling is typically not at exactly 60 m, because the target isopycnal has moved to different depth

described by D'Asaro [6] does not appear to actually occur. Typically, the value of V_0 (or equivalently M) changes at a rate of a few grams per month due to corrosion and fouling with much larger changes occurring if float pieces fall off. Fouling can be controlled by using antifouling paint, at the cost of a large, but steady change in float weight due to paint ablation. More difficult control problems occur when fish push on the float [18].

Second, during the isopycnal segments of the stairsteps the float remains accurately on an isopycnal so that there are minimal variations in oceanographic properties. This is an ideal environment for measurements using sensors with a slow response time. Thus the stairsteps in Fig. 9 were specifically designed to accommodate measurements by gas tension sensors (GTD) similar to those described by McNeil et al. [19] as well as special high accuracy oxygen sensors. Isopycnal sampling of about 9000 s allowed these sensors to fully equilibrate and thus produce an accurate measurement.

Isopycnal-Lagrangian Drifts The same control algorithm can be used to make a float optimally Lagrangian while still remaining close to a target isopycnal. For this application, $0 < \hat{\gamma} \le 1$, so that the float is between isopycnal and Lagrangian behavior, and $\tau \gg \tau_c$ so as to keep the float in the exponential decay regime. The float's motion will thus be described by (18) and its frequency response by

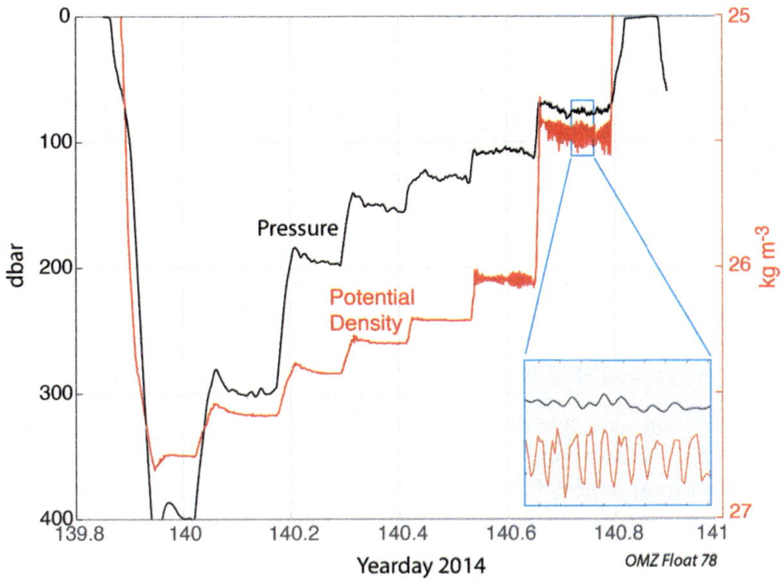

Fig. 9 A stairstep profile consisting of a set of isopycnal settles, each targeting a successively shallower isopycnal. The time series of pressure (*black*, left axis) is noisier than that of potential density (*red*, right axis). The blue inset details high frequency oscillations during the shallowest profile

$$|\tilde{z}|^2 = \frac{|\tilde{E}|^2}{\omega^2 + \tilde{\gamma}^2 \omega_I^2}. \tag{27}$$

Since $\tilde{z}$ is the distance from the isopycnal, it measures diapycnal water motions. Since these can only occur due to mixing or float imperfections, $\tilde{z}$ is a measure of mixing.

By (27), the float will follow such diapycnal displacements with frequencies $\omega \gg (\tilde{\gamma}\omega_I)^{-1}$ and will thus be Lagrangian at these frequencies. The float will not measure diapycnal displacements with frequencies $\omega \ll (\tilde{\gamma}\omega_I)^{-1}$, but instead will remain on σ_{goal}. This behavior allows the float to be Lagrangian at high frequencies, without being subject to potential instability at low frequencies, that is, the unstable case in Fig. 1. The Lagrangian frequency band can be made wider by either decreasing $\hat{\gamma}$, or decreasing ω_I by increasing the drogue area. These behaviors have not been well explored with real floats in the ocean, although Lien et al. [17] show some evidence for diapycnal float motions in strong mixing events.

8 Issues and Future Progress

Low Stratification The control approach outlined here requires the ocean to be stably stratified. As the stratification becomes weaker, control becomes increasingly ineffective. For example, for a standard MLFII with a large drogue and $\tilde{\gamma} = 1$, the equilibration times are $\tau_c = 2\lambda_c^{-1} \approx 110 N^{-1}$ or about 11,000 s for a typical upper ocean $N = 0.01\ s^{-1}$. However, for a deep ocean or mixed layer stratification $N = 0.001\ s^{-1}$, these times increase to longer than a day. Furthermore, for a fixed τ, the decreasing N will move the float into the oscillatory regime. The resulting weakly damped oscillatory motions will resemble the first parts of Fig. 8 in which the float is poorly controlled. Practically, this system works best for values of $N \approx 0.01\ s^{-1}$. Increasing care is needed as N drops below $0.003\ s^{-1}$.

High Stratification At high stratification, it is difficult to accurately estimate the density of the water surrounding the float. For $N = 0.01\ s^{-1}$, the density difference across a 1.4 m long float is 0.015 kg m^{-3}. For a float with 50 L displacement, this corresponds to a 0.7 g difference between the calculated buoyancy at the top and bottom of the float. This is acceptably small and with CTDs on both the top and bottom of the float, and some averaging, can be reduced further. For $N = 0.06\ s^{-1}$, these increase to 0.015 kg m^{-3} and 25 g, which are too large to compute a useful ballast. However, with such a strong stratification, precise ballasting is not needed to stabilize the float on the stratification.

A subtler problem is seen in the last two settles of Fig. 9. The control system oscillates, with rapid changes in B of a few 10's of cc and in pressure of a few meters. These oscillations are more common at higher N and have a frequency close to the local N. Their cause is not well understood. One contributor may be the time delay in the control system between when S, T, and P are measured and when B is changed. This can be up to 20 s and thus becomes a significant fraction of the buoyancy period for large N. However, including such a delay using the steady-state drag formulation (4) does not lead to accurate modeling of the oscillations. Studies of the unsteady motion of floating objects in a stratified fluid [24] show complex behaviors with oscillations near N that are not well described by (4). It seems likely that the observed float oscillations are due to the control system resonating with and thus exciting such oscillations. Practically, these can be controlled by reducing the strength of the control system feedback during each settle. The difference between the measured and target potential density $\Delta\sigma$ is artificially reduced from its full value at the start of the settle to a much smaller value by the end. This allows the control system to bring the float to the target isopycnal during the start of the settle, while suppressing the oscillations once it has reached the isopycnal.

Turbulent Layers In the weakly stratified and turbulent upper mixed layer, the drag on a float is poorly understood. D'Asaro et al. [8] report direct measurements of the motion of a float relative to the water and find these to be very much less than those predicted by (2). This is consistent with some previous work [10], but in general turbulence has been found to both increase and decrease the settling speed of sinking particles. Further work is needed to understand the effect of turbulence on float drag.

Finite Float Size The current control algorithm does not accurately model the finite size of the float or account for the details of its shape. Thus, the details of the flow around the float are not well understood and there is little guidance on how to predict the float drag parameter L in a stratified fluid. Even with two CTDs, the variations in stratification across the float lead to large uncertainties in its computed buoyancy for large stratifications.

Improved Algorithms The current control algorithm has proven to be a useful and robust tool. Nevertheless, in its current configuration, optimal performance can only be obtained by manually adjusting the two control parameters and the constants in the equation of state. In an ocean with variable N, the control parameters should optimally be varied with N. Ideally, information from the entire float trajectory could be combined with the control model and the full equation of state to obtain globally optimal values of the equation of state, drag model, ocean state and control model for any given mission. Better control could also be obtained with an improved understanding of float dynamics, particularly behavior on timescales comparable to N and in the presence of turbulence where fundamental understanding limits our ability to model and control float motion.

References

1. Alkire MB, D'Asaro E, Lee C, Perry MJ, Gray A, Cetinić I, Briggs N, Rehm E, Kallin E, Kaiser J, González-Posada A (2012) Estimates of net community production and export using high-resolution, Lagrangian measurements of O2, NO3-, and POC through the evolution of a spring diatom bloom in the North Atlantic. Deep Sea Res Part I Oceanogr Res Pap 64, June 2012:157–174, ISSN 0967–0637, doi:https://doi.org/10.1016/j.dsr.2012.01.012
2. Benson E (2012) One infrastructure, many global visions: the commercialization and diversification of Argos, a satellite-based environmental surveillance system. Soc Stud Sci 42(6):843–868. http://www.jstor.org/stable/41721363
3. Bower AS, Lozier MS, Gary SF, Böning CW (2009) Interior pathways of the North Atlantic meridional overturning circulation. Nature 459(7244):243. https://doi.org/10.1038/nature07979
4. Claustre H et al (2010) Bio-optical profiling floats as new observational tools for biogeochemical and ecosystem studies: potential synergies with ocean color remote sensing. In: Hall J, Harrison DE, Stammer D (eds) Proceedings of OceanObs'09: Sustained Ocean Observations and Information for Society (Vol. 2), Venice, Italy, 21–25 September 2009, ESA Publication WPP-306, doi:https://doi.org/10.5270/OceanObs09.cwp.17
5. D'Asaro EA, Farmer DM, Osse JT, Dairiki GT (1996) A Lagrangian float. J Atmos Oceanic Tech 13(6):1230–1246
6. D'Asaro EA (2003) Performance of autonomous Lagrangian floats. J Atmos Ocean Technol 20(6):896–911
7. D'Asaro E, McNeil C (2007) Air-Sea gas exchange at extreme wind speeds. J Marine Sys 66:92–109

8. D'Asaro EA, Thomson J, Shcherbina AY, Harcourt RR, Cronin MF, Hemer MA, Fox-Kemper B (2014) Quantifying upper ocean turbulence driven by surface waves. Geophys Res Lett 41:102–107
9. Davis RE, Webb DC, Regier LA, Dufour J (1992) The Autonomous Lagrangian Circulation Explorer (ALACE). J Atmos Ocean Technol 9:264–285
10. Fornari W, Picano F, Brandt L (2016) Sedimentation of finite-size spheres in quiescent and turbulent environments. J Fluid Mech 788:640–669
11. Gould J (2005) From Swallow floats to Argo: the development of neutrally buoyant floats. Deep-Sea Res II 52(3–4):529–543
12. Hanazaki H, Kashimoto K, Okamura T (2009) Jets generated by a sphere moving vertically in a stratified fluid. J Fluid Mech 638:173–197
13. Hanazaki H, Nakamura S, Yoshikawa H (2015) Numerical simulation of jets generated by a sphere moving vertically in a stratified fluid. J Fluid Mech 765:424–451
14. Higginson RC, Dalziel SB, Linden PF (2003) The drag on a vertically moving grid of bars in a linearly stratified fluid. Exp Fluids 34(6):678–686
15. Hoerner SF (1965) Fluid dynamic drag: practical information on aerodynamic drag and hydrodynamic resistance [Jun 01, 1965], ISBN-10: 9991194444
16. IOC, SCOR and IAPSO (2010) The international thermodynamic equation of seawater – 2010: calculation and use of thermodynamic properties. Intergovernmental Oceanographic Commission, Manuals and Guides No. 56, UNESCO (English), 196 pp
17. Lien RC, D'Asaro EA, McPhaden MJ (2002) Internal waves and turbulence in the upper central equatorial Pacific: Lagrangian and Eulerian observations. J Phys Oceanogr 32(9):2619–2639
18. Lien RC, D'Asaro EA, Menkes CE (2008) Modulation of equatorial turbulence by tropical instability waves. Geophys Res Lett 35(24)
19. McNeil C, D'Asaro E, Johnson B, Horn M (2006) A gas tension device with response times of minutes. J Atmos Ocean Technol 23(11):1539–1558
20. Millero FJ (2010) History of the equation of state of seawater. Oceanography 23(3):18–33. https://doi.org/10.5670/oceanog.2010.21
21. Millero FJ, Feistel R, Wright DG, McDougall TJ (2008) The composition of standard seawater and the definition of the reference-composition salinity scale. Deep-Sea Res I Oceanogr Res Pap 55(1):50–72
22. MODE Group (1978) The mid-ocean dynamics experiment. Deep Sea Research 25(10):859–910
23. Mowbray DE, Rarity BSH (1967) The internal wave pattern produced by a sphere moving vertically in a density stratified liquid. J Fluid Mech 30(03):489–495
24. Prikhod'ko YV, Chashechkin YD (2006) Hydrodynamics of natural oscillations of neutrally buoyant bodies in a layer of continuously stratified fluid. Fluid Dyn 41(4):545–554
25. Roman C, Inglis G, McGilvray B (2011) Lagrangian floats as sea floor imaging platforms. Cont Shelf Res 31(15):1592–1598, ISSN 0278–4343, https://doi.org/10.1016/j.csr.2011.06.019
26. Rossby T, Levine ER, Connors DN (1985) The isopycnal Swallow float – a simple device for tracking water particles in the ocean. Prog Oceangr 14:511–525
27. Rossby T, Dorson D, Fontaine J (1986) The RAFOS system. J Atmos Ocean Technol 3:672–679
28. Rossby HT (2005) The evolution of Lagrangian methods in oceanography. In: Griffa A, Kirwan AD, Mariano AJ, Özgökmen T, Rossby T (eds) Lagrangian analysis and prediction of coastal and ocean dynamics. Academic Press, London
29. Stommel H (1955) Direct measurements of sub-surface currents. Deep-Sea Res 2:284–285
30. Swallow JC (1955) A neutral-buoyancy float for measuring deep currents. Deep-Sea Res 3:74–81
31. Torres CR, Hanazaki H, Ochoa J, Castillo J, Van Woert M (2000) Flow past a sphere moving vertically in a stratified diffusive fluid. J Fluid Mech 417:211–236

Measuring Ocean Turbulence

Emily L. Shroyer, Jonathan D. Nash, Amy F. Waterhouse, and James N. Moum

Abstract Ocean turbulence (and turbulence in general) tends to be tremendously intermittent, events often dominating average values. Or, put another way, the distribution of turbulence tends to be highly skewed, requiring significant systematic observations to capture the important dynamics that control time and space averages. It is thus imperative to link large-scale processes (macroscale) to turbulence energetics (microscale) to characterize the dynamics of a particular regime and to develop a quantitative understanding of the role of turbulence in ocean momentum and scalar budgets.

1 Introduction

The ocean contains a tremendous range of dynamics, as exemplified by a number of phenomena from breaking surface gravity waves to convective entrainment at the base of a cooling mixed layer to shear-driven mixing in internal waves. Ocean turbulence provides a mechanism for energy dissipation and momentum transfer across time and space scales, and it also serves as the means by which tracers such as heat, salt, and nutrients are irreversibly mixed. Consequently, turbulence in the ocean affects both the local ecosystem and global climate. Locally, turbulent mixing controls mixed layer nutrient budgets, heat content, and sea surface temperature. Globally, turbulent fluxes affect the sequestration of carbon. And the ocean's overturning circulation is modulated by turbulence in a variety of ways, e.g., through deep convection, thermocline mixing, and abyssal boundary mixing.

Turbulence converts kinetic energy to internal energy through viscous heating, thereby dissipating fluid motions. For this reason, turbulence requires a constant source of energy or it will decay. In the ocean, the energy source for turbulence is provided through shear or buoyancy of the larger-scale mean flow. Turbulent mixing

E.L. Shroyer (✉) • J.D. Nash • J.N. Moum
Oregon State University, Corvallis, OR, USA
e-mail: eshroyer@coas.oregonstate.edu

A.F. Waterhouse
Scripps Institution of Oceanography, La Jolla, CA, USA

R. Venkatesan et al. (eds.), *Observing the Oceans in Real Time*, Springer Oceanography, DOI 10.1007/978-3-319-66493-4_6

also acts to dissipate the variance of both active (temperature and salinity) and passive (oxygen, CFCs, neutrally buoyant dye) tracers at an enhanced rate relative to laminar flows. Diffusive mixing in a turbulent fluid is typically several orders of magnitude greater than molecular values. This enhanced diffusivity is arguably one of the most important characteristics of a turbulent fluid, and provides much of the practical motivation (e.g., quantification of turbulent heat or nutrient transports) in the study of turbulence.

In this chapter, we briefly summarize the most pertinent concepts needed to describe turbulent motions, provide a short history of early developments in measurement of ocean turbulence, and review methods by which we quantify ocean mixing. This work builds off of many reviews of turbulence theory relevant to studies in the ocean. These resources include [4, 5, 18, 21, 63]. The reader is also referred to the introductory text by [69, 71], which cover many of these topics in more detail.

1.1 Turbulence in the Ocean

As traditionally applied in geophysical flows, the term "turbulence" encompasses motions as diverse as 100-km scale rotationally influenced geostrophic instabilities to meter-scale Kelvin-Helmholtz billows like that shown in Fig. 1. Here, we limit our consideration to small-scale, three-dimensional turbulence that acts to stretch and strain a fluid in an apparently random manner. Although an exact definition of turbulence is difficult to constrain, properties of turbulent flows are easily identified. Turbulent flows have high Reynolds number, defined as the ratio of inertial to viscous terms in the Navier-Stokes equation $Re = UL/\upsilon$, where U and L are characteristic velocity and length scales of the flow, respectively, and υ is the kinematic molecular viscosity. Turbulent flows are characterized by irregular, nonlinear motions that contain a range of space and time scales the largest of which is typically defined by the forcing and the smallest by molecular viscosity or diffusivity. Turbulent flows contain vortex filaments of a range of scales that interact with each other, straining the fluid and leading to a cascade of energy from large to small scales. These are often termed "turbulent eddies," and, as highlighted above, enhance diffusivity and dissipation relative to laminar flows. The diffusivity associated with the effects of turbulence is often referred to as an "eddy diffusivity" for obvious reasons.

1.1.1 Reynolds Decomposition of Stationary, Homogeneous, and Isotropic Flows

A common starting point is to consider isotropic (invariant with respect to orientation/rotation) turbulent motions that are also invariant to translation in time (stationary) and/or space (homogeneous). These assumptions allow for a clear separation between background and turbulent fluctuations, providing an analytic path forward that begins with Reynolds averaging of the momentum, kinetic energy, and/

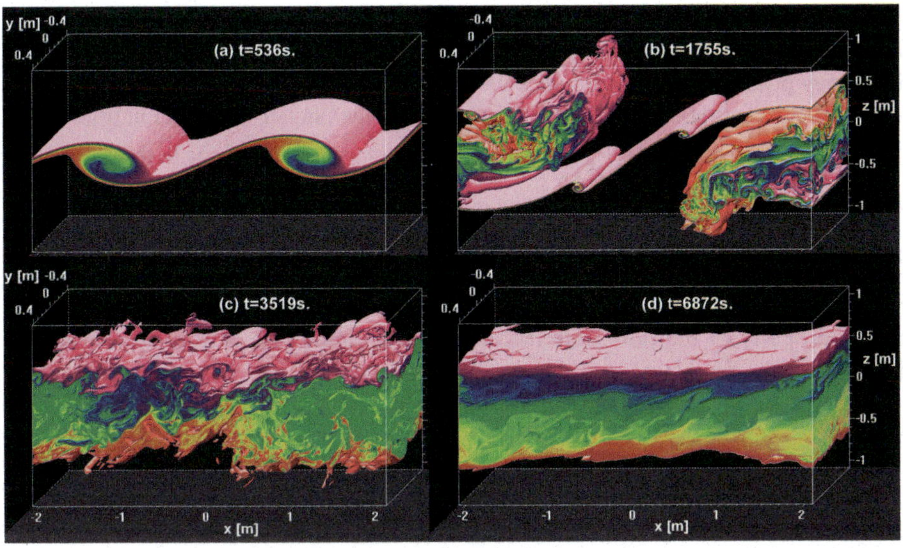

Fig. 1 Images show the time evolution of the density field in a direct numerical simulation of a Kelvin-Helmholtz billow. The color scale ranges from $-0.6\Delta\rho$ (*red*) to $0.6\Delta\rho$ (*purple*); outside this range, values are transparent. (Reproduced from Smyth and Thorpe [64])

or scalar conservation equations (e.g., see Chap. 2 in [Tennekes and Lumley, 69]). Reynolds decomposition relies on the ability for a state variable (θ) to be described by a combination of a mean quantity ($\overline{\theta}$) and perturbations (θ') about the mean, where $\theta = \overline{\theta} + \theta'$ and the mean is defined generally as an ensemble average represented by an overbar. For example, for stationary turbulence, the ensemble could be taken over time so that $\overline{\theta} = \lim_{T\to\infty} \frac{1}{T}\int_{t_0}^{t_0+T} \theta\, dt$ and $\overline{\theta'} = \lim_{T\to\infty} \frac{1}{T}\int_{t_0}^{t_0+T} \theta' dt = 0$. If instead the flow is considered to be homogeneous, a mean state could be defined in terms of spatial averages. In practice, the averaging time (or spatial) scale varies depending on forcing and the characteristics of turbulent fluctuations. That is, time scales of turbulent eddies need to be small compared to T, or, alternatively, length scales of the turbulent eddies need to be small compared to a spatial averaging scale, L, for a homogeneous flow.

We rely on Reynolds averaging since it is not practical to consider an exact description of random turbulent motions. Rather than attempt to predict specific details of the turbulent flow, Reynolds decomposition of a conservation equation yields a "new" equation that describes how mean conditions evolve with dependence on turbulent motions. This dependency is essentially the closure problem that troubles turbulence studies. For example, Reynolds decomposition of the nonlinear term θ^2 becomes $\overline{\overline{\theta}^2 + \theta'\theta'}$. We have taken one unknown, θ, and created two new terms, $\overline{\theta}$ and $\overline{\theta'\theta'}$, and so now require an additional equation to be solved or approximated to characterize the fluctuations. Useful application of Reynolds decomposition includes averaging of the energy equations, which demonstrates the connection between mean and fluctuating fields: the energy equation for the mean flow and the energy equation for turbulent fluctuations share two common terms, the shear and buoyancy production, that exchange energy between the mean flow and turbulence. As another example, Reynolds decomposition of the Navier-Stokes equation yields the Reynolds stress terms, $\frac{\partial}{\partial x_j}\overline{u_i' u_j'}$ for velocity component u_i. It is the Reynolds stress terms that need be parameterized in coarse resolution ocean models. A common method is to rewrite the Reynolds stress terms as $\frac{\partial}{\partial x_j} A_\nu \overline{u}_j$ using a turbulent (or eddy) viscosity A_ν as an analogue to the molecular viscosity. The turbulent viscosity is then a property of the flow state in contrast to the molecular viscosity, which is a property of the fluid state.

1.1.2 Dimensional Analyses and the Length Scales of Turbulence

Dimensional analyses provide a simple yet powerful tool for interpreting turbulent motions. A useful illustration of the approach is to use dimensional analyses to arrive at the various length scales that are important in turbulent flows. For example, the scale at which molecular viscosity (ν, m^2s^{-1}) dissipates energy is determined from ν and the dissipation rate (ε, m^2s^{-3}). Combining these two parameters to yield a length scale gives the Kolmogorov microscale, $\eta = (\nu^3/\varepsilon)^{1/4}$, or alternatively the Kolmogorov wavenumber, $k_K = (\varepsilon/\nu^3)^{1/4}$. The Kolmogorov microscale defines the smallest scales contained within a turbulent flow. The scale at which a turbulent fluid acts to dissipate tracer variance (the Batchelor wavenumber) can also loosely be derived from scaling assumptions using ε, ν, and the diffusivity of the tracer κ as $k_B = (\varepsilon/\nu\kappa^2)^{1/4}$ [Batchelor, 2], which for heat diffusing in seawater, is approximately five times smaller than the Kolmogorov microscale. We can also use dimensional scaling to estimate the dissipation rate in terms of the kinetic energy (u^2) contained within the larger-scale (relative to η) turbulent eddies and the rate of transfer of energy (u/ℓ) as $\varepsilon = u^3/\ell$ the length scale of the large turbulent eddies, ℓ, is often referred to as the integral scale.

We often treat turbulence in the ocean as being stationary, homogeneous, and isotropic primarily to simplify analysis and make the problem tractable. However, these assumptions are often violated, since the ocean experiences time-varying forcing with inhomogeneous background states. For a geophysical fluid at the scales considered here, we can neglect the influence of rotation; however, considerations related to stratification must be taken into account. Stratification imposes a vertical scale that limits the assumptions of isotropy and homogeneity. The Ozmidov scale, $L_O = (\varepsilon N^{-3})^{1/2}$, determines the vertical scale at which stratification, N, of the background fluid starts to limit turbulent motions (i.e., Dillon [10]). A similar scale limit exists due to shear; the Corrsin scale is defined as $L_C = (\varepsilon S^{-3})^{1/2}$ where S is the vertical gradient in the lateral velocity of the mean flow. Proximity to boundaries imposes yet another scale limit to isotropic motions. The scales associated with stratification (L_O), shear (L_C), and boundaries impose an upper bound on the integral scale, ℓ, of turbulent motions.

1.2 Theoretical Spectra and Subranges

The ratio between the Kolmogorov microscale and the integral scale is $\eta/\ell = Re^{-3/4}$. For a turbulent flow with large *Re*, η is much smaller than ℓ. Furthermore, as *Re* increases the separation between the microscale and integral scale widens. The range of wavenumbers, *k*, that fall between the integral and Kolmogorov microscale is referred to as the inertial subrange (Fig. 2) when considering velocity fluctuations or the inertial-convective subrange when considering tracer variations. The dependence of the length of the inertial subrange on *Re* hampered early efforts to measure turbulence within laboratory studies, as obtaining sufficiently large Reynolds number flows with a clear separation between micro and integral scales was limited by the physical constraints of a lab setting. This limitation was first overcome with oceanic measurements from a tidal channel (see Sect. 1.3), which provided a natural laboratory characterized by very large *Re* ($\sim 10^8$).

Figure 2 (upper curve) shows wavenumber spectra for kinetic energy for a high Reynolds number flow. Nonlinear interactions introduce a cascade of energy, primarily directed to smaller length scales. The strain associated with large eddies acts to stretch smaller eddies. The vorticity associated with the eddies causes a "vortex stretching," whereby the energy of the smaller eddies increases at the expense of the larger eddies (i.e., a downward cascade of energy). As turbulent eddies decrease in size they eventually approach the Kolmogorov microscale where molecular viscosity effectively dampens motions, dissipating energy and causing the roll-off in spectral energy (Fig. 2, upper curve). Kolmogorov argued that for $Re \rightarrow \infty$ the transfer of energy within the inertial subrange should be independent of viscosity [30, 31], and dimensional reasoning then implies that $E = C_k \varepsilon^{2/3} k^{-5/3}$ for a constant C_k.

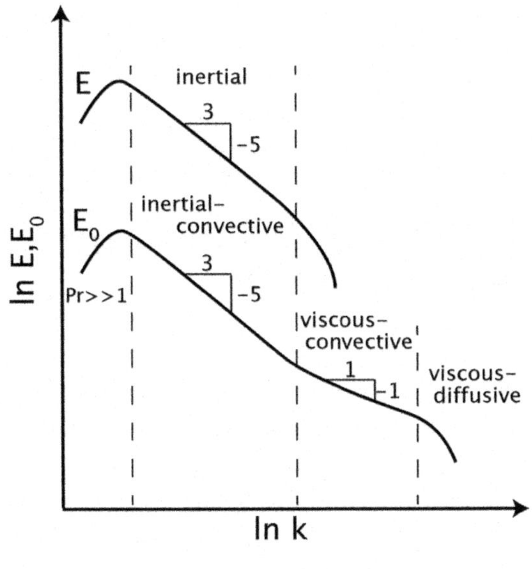

Fig. 2 Energy and temperature variance spectra in a fluid with large Prandtl number. (After Tennekes and Lumley [69] Fig. 8.11)

An analogous scale separation can exist due to combined effects of viscosity and diffusivity. The Prandtl (or Schmidt) number is the ratio of kinematic viscosity (υ) to the diffusivity (κ) of a tracer: $Pr = \nu/\kappa$. If Pr is much greater than one or equivalently $k_B \gg k_K$, i.e., the case for tracers in the ocean, there exists a range of scales over which molecular viscosity acts to dampen tracer fluctuations without a significant contribution from molecular diffusion. For example, the oceanic Prandtl number for temperature is roughly 7, indicating that the molecular viscosity of the ocean and diffusivity of temperature are sufficiently different to separate out a so-called viscous-convective and viscous-diffusive subrange (Fig. 2, lower curve). The viscous-convective subrange is bounded by the Kolmogorov and Batchelor scales ($k_K^{-1} > \lambda \gg k_B^{-1}$ for length scales λ). The viscous-convective subrange occurs at a range of scales smaller than the Kolmogorov scale, so that the evolution of tracer variance is governed by these Kolmogorov-scale eddies, which moves tracer variance to smaller scales. However, in this subrange, molecular diffusion is not yet sufficient to weaken tracer gradients over the appropriate time scales. It is within the viscous-diffusive subrange ($k_B^{-1} > \lambda$) that molecular diffusivity effectively dissipates tracer variance over time scales relevant to strain-rate fluctuations associated with turbulent eddies.

Tracer variance spectra (for large Prandtl numbers) differs from kinetic energy spectra in several ways, since the distribution of the variance of the tracer will depend both on the size distribution of the eddies themselves (equivalent to the energy cascade discussed above) as well as how the eddy strain fields deform the tracer gradients. Within the inertial-convective subrange both of these processes contribute to tracer variance and the spectra follow a form similar to the kinetic energy spectra: $E_\theta = C_\theta \chi \varepsilon^{-1/3} k^{-5/3}$, where χ is the dissipation rate of tracer variance. Within the viscous-convective subrange, the downscale cascade is slowed, and the spectral slope is reduced from $-5/3$ to -1 (Fig. 2, lower curve).

Note that the scale separation between the viscous-convective and viscous-diffusive subranges will depend on the diffusivity of the tracer. Accordingly, this separation is more pronounced for salinity compared to temperature, as the molecular diffusivity of salt is roughly two orders of magnitude smaller than κ_T. Observations of velocity, temperature, and salinity microstructure reflect the various magnitudes of molecular viscosity, temperature diffusivity, and salt diffusivity in which fluctuations in salinity have a smaller scale than fluctuations in temperature, and fluctuations in temperature have a smaller scale than fluctuations in velocity (Fig. 3). We rely on scaling theoretical spectra of ocean microstructure (either velocity or tracer) to quantify ocean turbulence.

1.3 *Early Developments in the Measurement of Ocean Microstructure and Turbulence*

In order to measure turbulent microstructure in the ocean, the sensing element must be small and be able to sample with a high sensitivity at high frequency. The mechanical bathythermograph [Spilhaus, 66] provided the first reliable records of

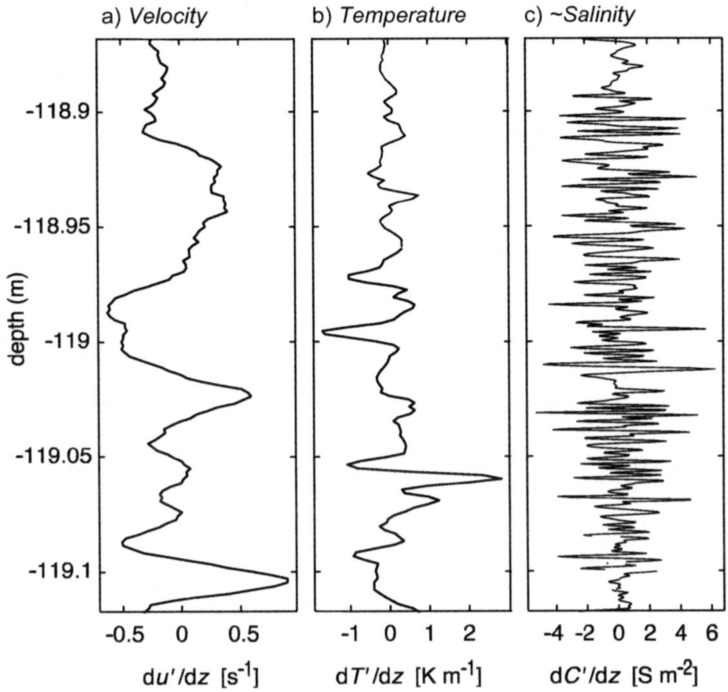

Fig. 3 A 25-cm vertical segment of (**a**) shear, (**b**) temperature gradient, and (**c**) conductivity gradient from offshore of Oregon, USA. (Reproduced from Nash and Moum [51], © Copyright 2002 American Meteorological Society (AMS)

continuous temperature profiles in the ocean, which up until its creation had been collected using discrete-depth estimates collected with reversing thermometers. Measurements from freshwater lakes revealed temperature microstructure roughly two decades before the development of the bathythermograph (see historical development by Gregg [20]). Despite these earlier efforts, similar sharp fluctuations in bathythermograph recordings from the ocean were first attributed to friction by the recording stylus rather than ocean microstructure [Munk and Garrett, 50]. However, within a decade of the bathythermograph's development, temperature measurements from thermocouples had sensitivities of a few thousandths of a degree Celsius [73, 74]. Shortly thereafter, Liebermann [38] mounted rapid-response platinum resistance thermistors to a submarine, yielding estimates of temperature variability at roughly 10-cm resolution. Grant, Stewart, and Moilliet [17] adapted atmospheric hot-film anemometers [6, 39] for use in the measurement of ocean turbulence when they examined velocity spectra (Fig. 4) from a horizontally towed platform within a tidal channel offshore of Vancouver Island, Canada. These landmark observations

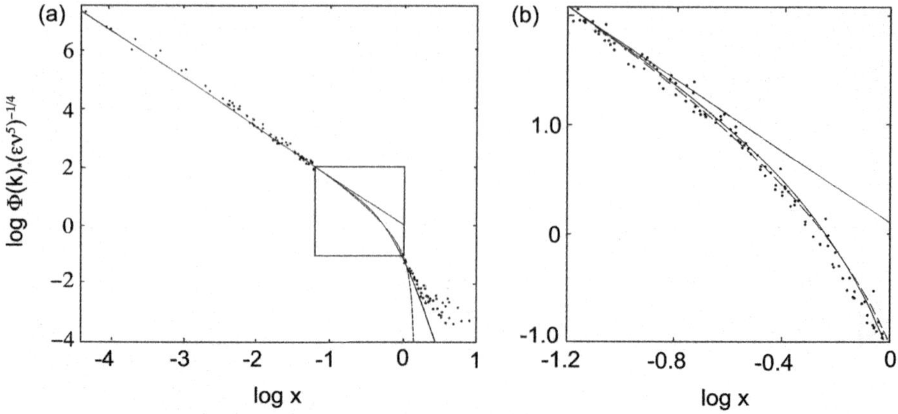

Fig. 4 Wavenumber spectra compared to theories of Kolmogorov (*straight solid line*), Heisenberg (*curved solid line*), and Kovasznay (*dashed line*). (Reproduction from Grant et al. [17] Figs. 12 and 13)

were the first to be made in a flow of sufficiently high Reynolds number to allow for resolution of the inertial subrange and confirmation of Kolmogorov's theories.

Microstructure signatures from salinity-temperature-depth recorders, the ancestor of the bathythermograph, were noted in several works from the late 1960s and early 1970s [7, 67, 68, 80–82]. For example, Woods [80] combined STD measurements with dye streaks to examine turbulent billows at the interface of sharp temperature jumps. Around the same time, Cox recognized that ship heave severely restricted the study of microstructure from lowered systems, and he developed the first free-falling microstructure profiler (the MSR or Microstructure Recorder) [Cox et al., 8], which was equipped with glass rod thermistors that originally recorded internally on paper charts [20, 54]. Osborn and Cox used these data to estimate diffusivity from temperature microstructure. In the early 1970s, Osborn [55] presented the first estimates of turbulent kinetic energy dissipation using one component of velocity microstructure calculated from airfoils mounted on a free-falling profiler. Development efforts expanded across many institutions and groups around this time [Lueck et al., 40], and several different systems then evolved independently of one another. Microstructure measurements from free-falling systems remain a "gold standard" in direct measurements of turbulence today. More recent efforts have developed new approaches in sampling; these include measurements of wavenumber spectra from acoustic Doppler currents [Veron and Melville, 75], measurements of temperature microstructure from high-frequency acoustics [26, 34], in situ particle imaging and tracking [Doron et al., 12], and dye release experiments [Ledwell et al., 36]. For in-depth reviews of measurement developments, the reader is referred to Gregg [20] and Lueck et al. [40].

2 Quantifying Turbulence with Ocean Measurements

Several different approaches are used to quantify turbulent mixing. We review a few of these approaches according to scale and consider methods that rely on (1) integral estimates of mixing, (2) finescale parameterizations, and (3) direct measurements of both turbulence and resultant microstructure independently. Note that the first and last of these approaches are direct measurements of mixing, while the second depends on semi-empirical relationships between finescale structure and mixing. Before reviewing these methods, we first present detailed measurements from the Oregon Shelf [Moum et al., 48] that demonstrate the richness of oceanic turbulent motions and some of the sampling challenges.

An example spectrogram of near-bottom, cross-shore velocity shows time-varying spectral peaks associated with a range of phenomena at high frequencies (Fig. 5, middle panel). The most energetic signals are due to two swell components—one that remains constant at about 12 s and the other that varies in period from about 22 to 15 s

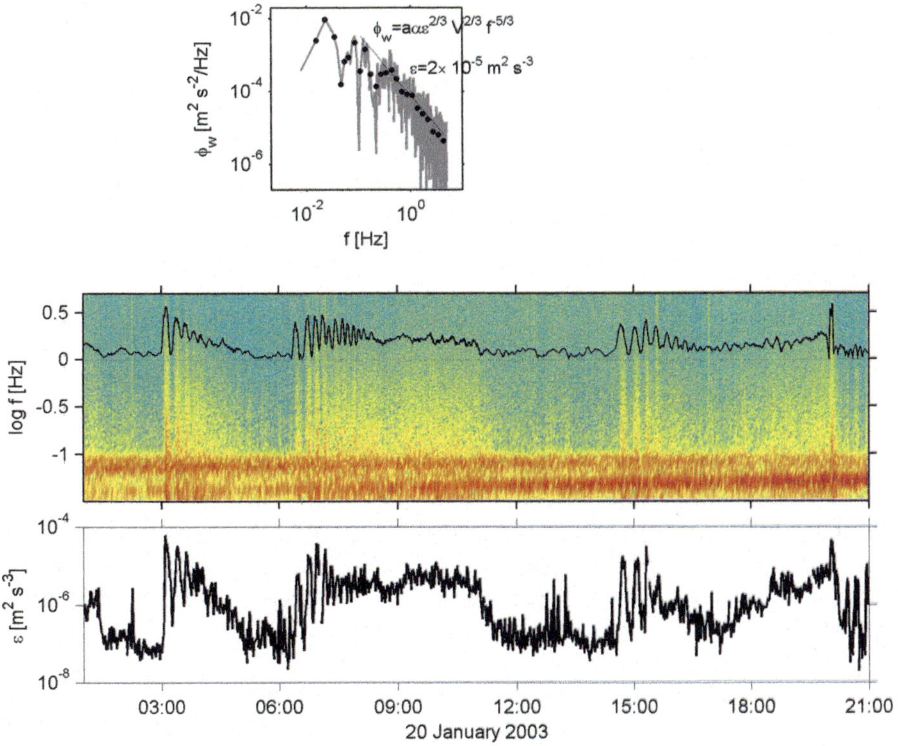

Fig. 5 Spectrogram of the cross-shore component of velocity recorded by an ADV at 1 m height with low-passed current speed over plotted in *black*. Spectra were computed over 2048 points (204.8 s). The upper panel shows one example spectra and the scaled inertial subrange estimate of dissipation. The bottom panel shows time series of dissipation derived from spectra. (Reproduced from Moum et al. [48], © Copyright 2007 American Meteorological Society (AMS))

over the record shown. The longer period swell component also increased in energy over the record. At this water depth (~125 m), the signature of wind waves (2–5 s) is not evident; instead, the broadband signal at frequencies greater than the swell is associated with turbulence. Even over this short record, turbulence dissipation (Fig. 5, bottom panel) varies by multiple orders of magnitude. The intermittency and variability of the turbulence are clearly seen. Adequate sampling of intermittent, patchy turbulent events is one of the primary difficulties in determining reasonable average estimate of turbulence. Note also that this record suggests an approach to overcoming this daunting sampling problem in that the overall behavior of the turbulence is closely associated with larger-scale motions (in this case nonlinear internal waves), which are to an extent deterministic, i.e., the turbulence is a property of the flow. If we can understand how larger-scale processes create and modulate turbulence, then we can appropriately parameterize the turbulence.

This record also illustrates technical challenges in measuring turbulence in the ocean. For example, the rapid onset of turbulence as the velocity pulses pass the sensor indicates either rapid generation of turbulence within each pulse or advection of turbulence by the waves – an important distinction. The problem of interpreting a spectrum when inertial subrange turbulence is advected by a random wave field past a point was first addressed by Lumley and Terray [41] to interpret the relative contribution of wind wave orbital velocities to near-surface turbulence in Lake Ontario. This effort was refined by Trowbridge and Elgar [72] who first estimated turbulence from ADV measurements in the surf zone where short-period waves are energetic. Observational strategies must overcome both statistical problems related to sampling intermittent flows with highly skewed distributions, as well as technical challenges related to cleanly sampling over the necessary wavenumber or frequency range.

2.1 Integral Approaches

While turbulence represents the mechanism for enhanced mixing and modification of bulk properties of a fluid, direct turbulence measurements can only be made *in situ* at a point in space and time. And as illustrated in Fig. 5, at geophysical values of *Re*, the paths to turbulence are many and turbulence statistics are characterized by a high skewness and kurtosis. Taken together, these properties bode ill for inferring representative estimates of mixing from sparse turbulence measurements.

From an oceanographic perspective, we are concerned with the large-scale thermodynamic transformation of water masses caused by mixing or, alternatively, the irreversible change in potential energy created when down-gradient, Fickian-like diffusion moves the fluid's center of mass upward. For example, consider the initial and final states illustrated in the direct numerical simulations (DNS) of the Kelvin Helmholtz instability shown in Fig. 1. Characterization of mixing based on the evolution of the bulk state reduces uncertainties associated with transient, reversible processes like internal waves that may contaminate point measurements. From the complete knowledge of the flow field produced by a properly constrained DNS, we can assess the net result of the mixing as a time-space integral. The net change in potential energy can be used to estimate an average turbulent diffusivity needed to represent the complexities of the modeled turbulent flow.

It is impossible to reproduce such a calculation in natural geophysical flows, simply because the density of time-space observations in the model cannot be replicated in the field. However, in recent years, J. Ledwell (Woods Hole Oceanographic Institution) and colleagues pioneered efforts to circumvent this limitation through the use of neutrally buoyant dyes (SF_6) [Ledwell et al., 36]. The dye is first injected on an isopycnal and then tracked over time (as long as several years depending on the environment). A full inventory of the initial dye patch is attempted at each successive mapping. Inevitably, the dye spreads across isopycnals at rates greater than would be predicted from molecular diffusion alone, producing a roughly Gaussian shape about the original injection isopycnal. The time evolution of the second moment of this Gaussian distribution (roughly, the variance) then represents the average turbulent diffusivity.

Results from such experiments have yielded considerable insight into the role that mixing plays in the abyssal ocean. For example, dye released in the deep Brazil Basin in 1996 was surveyed in 1997, 1998, 2000, and 2009 [Rye et al., 62]. The initial survey in 1997, part of the original funded study, led to an estimate of turbulent diffusivity, K, not much different from that estimated from the 2009 survey. The consistency of these results over more than a decade suggests steady mixing rates in that region. While this method yields a true integral of the mixing, it cannot tell us about the processes that contribute to mixing, necessitating detailed process experiments in support of bulk efforts. Furthermore, this approach is best applied to regimes with relatively low diffusivity where a census of the dye can be completed, and therefore is challenging to apply in energetic regimes where turbulent diffusivity may not be the leading control on dye evolution and where rapidly evolving conditions limit the robustness of mapping of dye [Ledwell, 35].

2.2 *Finescale Parameterizations*

Finescale parameterizations are those which rely on the fact that most of the variability in the oceanic shear and strain are caused by internal waves. The rate of energy transfer through the internal wave spectrum which eventually leads toward dissipation has been parameterized [19, 24, 25, 60] such that turbulent dissipation rates can be inferred from variances on the order of 10–100 m scale in vertical profiles of shear and strain. This transfer of energy is thought to be due to nonlinear interactions between individual internal waves. Finescale parameterizations allow for the calculation of the average dissipation rate expected over several wave periods, and therefore are helpful in assessing the spatial and temporal mean dissipation rate or diffusivity. These parameterizations assume a latitudinal dependence of such interactions (decaying toward the equator) [Gregg, 19] and a background internal wave spectral shape, the Garrett-Munk (GM) wave field [13–15]. It is this energy level that varies depending on the intensity of the nonlinear interactions and

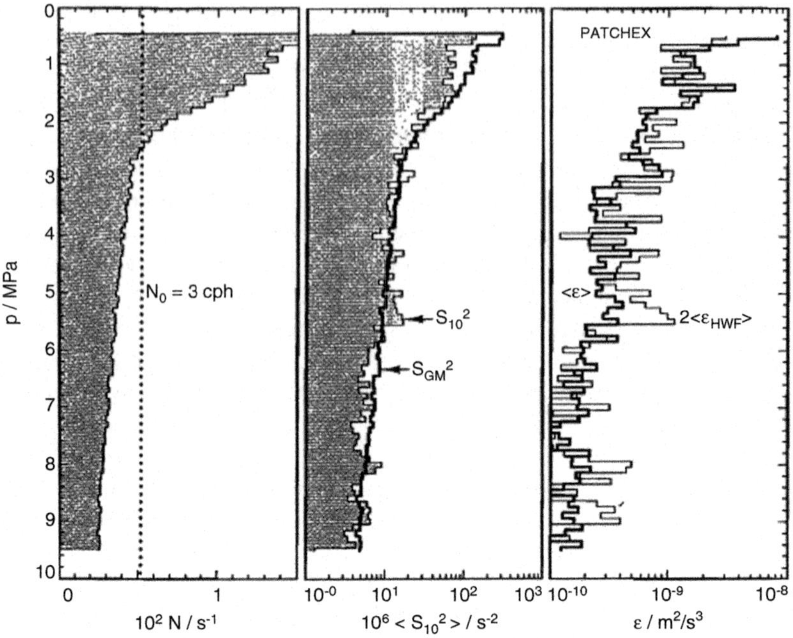

Fig. 6 Observations from approximately 700 microstructure profiles from the PATCHEX experiment showing average profiles of stratification (*left*), shear squared (*middle*), and dissipation rate (*right*). Dissipation rate is calculated directly from the microstructure profiler (*dark line*) and the Henyey et al. [25] parameterization (*light line*). (Reproduced from Gregg [19])

downscale energy cascade. For example, Gregg [19] used vertical profiles of stratification (Fig. 6, left panel), shear-squared (Fig. 6, middle panel), and dissipation rate (Fig. 6, right panel) from the PATCHEX experiment to evaluate the parameterization of Henyey et al. [25]. This parameterization shows good agreement between the parameterized dissipation rate and the observed dissipation rate.

Since the original implementation, further work has improved the finescale parameterization (see [Polzin et al., 60] for a recent review). In regions where the spectral energy levels are modified by regions of enhanced mixing, a ratio of horizontal kinetic to potential energy or shear to strain variance, R, is used to reduce this distortion [Polzin et al., 59]. Additionally, the accuracy of shear- and strain-based parameterizations is limited outside regions where the assumptions behind the parameterization apply [Polzin et al., 60]. These regions include continental shelves [MacKinnon and Gregg, 43], strong geostrophic flow regimes over rough topography [Waterman et al., 77], and regions with very large overturning internal waves [Klymak et al., 29]. Readers are referred to Polzin et al. [60] for the most recent advances in the finescale parameterization and MacKinnon et al. [42] for implementation of various finescale parameterization and theories in global ocean climate models.

Due to the wide variety of vertical profiles of shear and strain from various observational platforms, the finescale parameterization has allowed for the estimation of turbulent dissipation rate from a larger set of observations than microstruc-

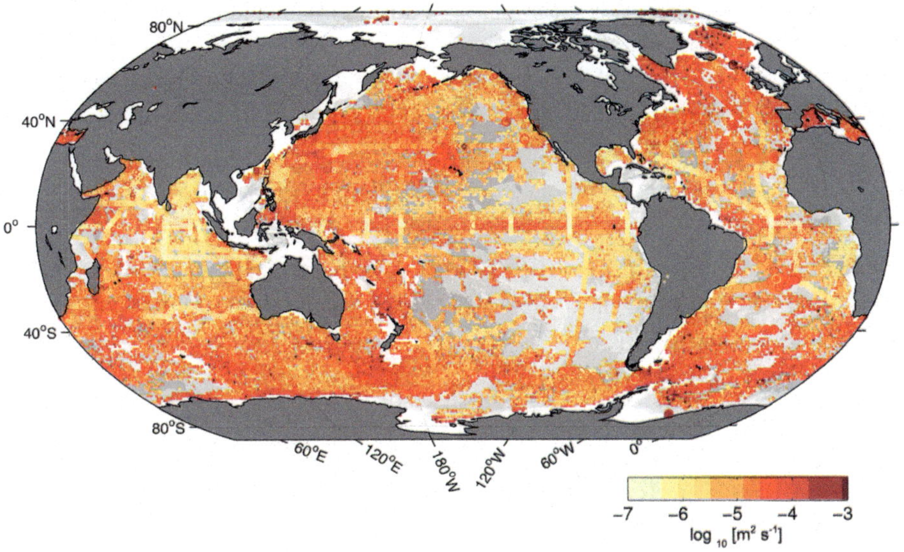

Fig. 7 Diapycnal diffusivity in the upper 1000 m inferred from a compilation of microstructure and finescale estimates. (Adapted from Waterhouse et al. [76])

ture profiles alone. When compared to microstructure observations, the finescale parameterization is often between a factor of 2–3 in agreement with microstructure estimates in open-ocean conditions [19, 59, 60, 79].

In recent years, our understanding of the geography of diapycnal mixing has grown due to various implementations of the finescale parameterization, including strain-based estimates from profiling Argo floats [78, 83] of which there are currently over 2000 as well as implementation of the finescale parameterization on CTD-LADCP profiles of shear and strain from GO-SHIP repeat hydrography surveys [27, 33, 58]. An updated compilation of all microstructure and finescale inferences from Waterhouse et al. [76] includes finescale inferences from strain-based estimates from Argo floats as well as shear-based estimates from ship-board shear (Fig. 7). This map highlights the variability in diapycnal diffusivity, K, such that diapycnal diffusivity is related to the dissipation rate by the Osborn relation $K = \gamma\varepsilon / N^2$, [Osborn, 56], where a dissipation flux coefficient γ is typically taken to be 0.2, although known to vary [Smyth et al., 65] and N^2 is the depth-averaged buoyancy frequency. Additionally, this map demonstrates regions where elevated diapycnal diffusivity has been associated with more complex topography (as shown by Decloedt and Luther [9] and Whalen et al. [78]) as well as the usefulness of the finescale technique.

2.2.1 Calculation from Vertical Shear

Dissipation rate can be calculated from the finescale parameterization following various derivations and implementations, including Gregg [19], Polzin et al. [59], Gregg et al. [23], Kunze et al. [33], and Whalen et al. [78] among many others.

Typically, vertical profiles of shear and/or strain are divided into overlapping vertical segments of several hundreds of meters (e.g., Kunze et al. [33] uses 320 m overlapping segments for CTD-LADCP data). Following Gregg et al. [23], dissipation rate can be written as

$$\varepsilon = \varepsilon_0 \left(\frac{N}{N_0} \right)^2 \hat{E}^2 L(R_w, \theta)$$

where ε_0 is a constant of 6.73×10^{-10} W kg^{-1}, N is the stratification of the segment, N_0 is a constant stratification of 5.24×10^{-3} rad s^{-1}, and $\hat{E}$ is the spectral energy level

$$\hat{E}^2 = \frac{\langle U_Z^2 \rangle^2}{\langle U_Z^2 \rangle_{GM}^2} \left(\frac{1 + 1/R_w}{4/3} \right).$$

In the above expression, $\langle U_Z^2 \rangle$ is shear integrated between a maximum (k_{max}) and minimum (k_{min}) wavenumber

$$\langle U_Z^2 \rangle = \int_{k_{min}}^{k_{max}} \phi_{U_Z}(k_Z) dk_Z$$

based on a buoyancy-normalized shear spectrum (ϕ_{U_Z}). The maximum integration wavenumber is calculated by requiring a set value of observed shear variance, e.g. $\langle U_z^2 \rangle = 0.66$ [Gregg et al., 23]. The corresponding Garrett-Munk shear spectrum from a similar stratification is calculated over the same wavenumber range. Finally, the variable $L(R_w, \theta)$ describes the theoretical dependence on downscale energy transfer rate on both average wave field content (through R_w) and latitude (where f is the Coriolis frequency) [MacKinnon et al., 44] such that

$$L(R_w, \theta) = \sqrt{\frac{2}{R_w - 1}} \left[\frac{f \cosh^{-1}\left(\frac{N}{f} \right)}{f_{30} \cosh^{-1}\left(\frac{N_0}{f_{30}} \right)} \right]$$

and R_w is the shear-to-strain ratio, the ratio of horizontal kinetic to potential energy, defined as

$$R_w = \frac{\langle U_Z^2 \rangle}{\zeta_Z^2}$$

which, in the absence of strain (or shear) data, can be set as a constant ranging between 3 and 7 [33, 60].

2.2.2 Calculation from Strain

Strain is calculated over segments of several hundreds of meters (e.g., Whalen et al. [78] uses 200 m overlapping segments for Argo strain data) using

$$\zeta_Z = \frac{N^2 - N_{\text{fit}}^2}{\overline{N^2}}$$

where N^2 is the stratification from the segment, $\overline{N^2}$ is depth-averaged stratification of the segment, and N_{fit}^2 is the quadratic fit of the buoyancy frequency for each segment. The strain variance, $\langle \zeta_Z^2 \rangle$, is then calculated as

$$\langle \zeta_Z^2 \rangle = \int_{k_{\min}}^{k_{\max}} \phi_{\zeta_Z}(k_Z)\, dk_Z$$

where ϕ_{ζ_Z} is the strain spectra over a particular segment. Similarly as for shear, the strain spectra is integrated between a maximum and minimum cutoff wavenumber. In this case, integration is to a maximum wavenumber such that $\langle \zeta_Z^2 \rangle \le 0.2$ to avoid underestimating the variance due to spectral saturation.

A strain-based estimate of dissipation can then be written as

$$\varepsilon = \varepsilon_0 \left(\frac{N}{N_0} \right)^2 \hat{E}_\zeta^2 L(R_w, \theta)$$

where the constants ε_0 and N_0 are the same as given above, while the spectral energy level for the strain spectrum is

$$\hat{E}_\zeta^2 = \frac{\langle \zeta_Z^2 \rangle^2}{\langle \zeta_Z^2 \rangle_{\text{GM}}^2}$$

where $\langle \zeta_Z^2 \rangle_{\text{GM}}^2$ is the Garrett-Munk strain variance over the same wavenumber range. In this case, the variable $L(R_w, \theta)$ is

$$L(R_w, \theta) = \frac{1}{6\sqrt{2}} \frac{R_w(R_w + 1)}{\sqrt{R_w - 1}} \left[\frac{f \cosh^{-1}\left(\frac{N}{f}\right)}{f_{30} \cosh^{-1}\left(\frac{N_0}{f_{30}}\right)} \right]$$

2.3 Direct Microstructure Measurements

The most reliable means of quantifying the turbulent transports of momentum or the mixing of scalars are by measuring the rates of processes that happen at the molecular level. This is because these rates describe the irreversible processes that are the ultimate consequences of large-scale turbulent motions that cascade energy or thermal variance to the smallest scales, where turbulent kinetic energy is dissipated and where scalar variance created by turbulent stirring at large scales is diffused away by molecular processes. This dissipation rate of turbulent kinetic energy (ε) and the dissipation rate of thermal variance (χ) represents the two fundamental quantities that determine these rates.

We note that it is also possible to infer these quantities from turbulent motions of the largest eddy scales (i.e., the integral scale) [Tennekes and Lumley, 69], or from the density structure that arises when large eddies act in the presence of stratification, creating statically unstable fluid, permitting "Thorpe-scale" estimates of turbulent overturning to be made [10, 70].

However, it was noted at an early stage [see [Gregg, 18] for a review] that there is much more variability in the spectral shapes of geophysical turbulence at the largest scales, owing to the influences of internal waves, stratification, and other imposed length scales and energy sources. These effects can act to either increase or decrease variance in the wavenumber band of turbulent motions, as compared to theory or isolated lab studies. In contrast, the smallest scales tend to be more isotropic and less influenced by details and/or geometric constraints of the forcing, and as a result, "universal" forms of turbulent spectra tend to be more robust at the smallest scales where the dissipation actively occurs. For example, Oakey [53] showed that velocity gradient spectra closely followed the empirical universal form of Nasmyth [52] in the dissipative subrange. Similarly, Dillon and Caldwell [11] found good agreement with the theoretical spectrum of Batchelor [2] for temperature gradient in the viscous diffusive turbulent subrange.

In summary, the smallest scales represent the scales where (1) the dissipation actively occurs, (2) the influences of internal waves and stratification are the weakest, and (3) assumptions of isotropy and homogeneity are more likely to be satisfied. Moreover, measurement of velocity and temperature gradient at the smallest scales provides statistically more independent realizations of a process, because more zero-crossings of fluctuations are measured. As a result, the most reliable metric for characterizing turbulence is by directly measuring ε and/or χ.

2.3.1 Calculating the Turbulent Kinetic Energy Dissipation Rate

The dissipation rate of turbulent kinetic energy (ε) is defined through the Reynolds-averaged turbulent kinetic energy equation, and in principle depends on 12 products of velocity gradients. Assuming isotropic turbulence, these can be reduced to as few terms as desired. In practice, shear probes installed on the nose of a free-falling turbulence profiler such as the Advanced Microstructure Profiler (AMP) [Gregg,

19] or Chameleon [Moum, 47] measure the along-path spatial gradient of transverse velocity at very high spatial resolution (O(1 cm)). Often two shear probes are mounted in orthogonal directions, which permit the measurement of $\partial u/\partial z$ and $\partial v/\partial z$. With the assumption of isotropy, the dissipation rate becomes

$$\varepsilon \approx \frac{15}{4}\nu\left[\left\langle\frac{\partial u}{\partial z}^2\right\rangle + \left\langle\frac{\partial v}{\partial z}^2\right\rangle\right]$$

if two components of velocity shear are measured, or

$$\varepsilon \approx \frac{15}{2}\nu\left\langle\frac{\partial u}{\partial z}^2\right\rangle$$

if only one is. Here, <> represents a suitable average over some region of quasi-homogeneous turbulence, and assumes all shear variance has been measured, which is seldom the case because the Kolmogorov scale is often much smaller than the physical size of a shear probe. Assumptions must be made to account for shear variance not resolved by the measurement.

In practice, measurements of velocity shear are often obtained continuously along a vertical line, allowing wavenumber spectra of transverse shear to be computed and compared to the universal forms of Nasmyth or Panchev [52, 57]. Typically the observed spectra match the theoretical forms over the resolved range of wavenumbers (below ~10–50 cycles/meter), but may deviate at the higher wavenumbers, where the sensor's size and sensitivity/noise affect the measured spectrum. Also, a profiler may vibrate at a particular narrowband frequency that produces a physical peak in the shear spectrum. By comparing shear probe and accelerometer spectra [Moum, 47], contaminated regions can be identified, and in cases, can be removed by subtracting the coherent part of the accelerometer-induced motions from the shear measurements [Levine and Lueck, 37]. The universality of shear spectra (in which both the spectral amplitude and its wavenumber extent scale with ε, e.g., Fig. 4) allows the dissipation rate to be computed even when only a fraction of the shear variance is resolved by the shear probes. When dissipation rates are small, shear spectra can be integrated out to near the Kolmogorov wavenumber (or the wavenumber where the shear spectrum intersects the sensor's noise floor; Gregg [21]); at higher dissipation rates, the spectrum must be extrapolated using the universal form because the scales of dissipation are much smaller than that of the probe and cannot be measured. Moum et al. [47] show that standard methods, when applied to multiple instrument platforms and techniques, produce consistent results.

2.3.2 Calculating the Dissipation of Thermal (Scalar) Variance

The dissipation rate of temperature variance is another fundamental turbulence quantity, which has the benefit that it can be used to directly estimate mixing through the Cox-Osborn relationship without the need for assuming a mixing efficiency

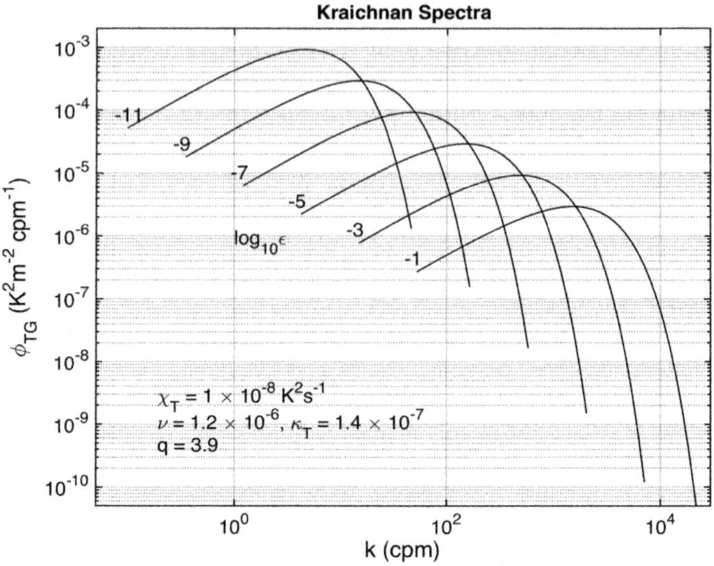

Fig. 8 Temperature gradient spectra for varying turbulent kinetic energy dissipation rates (ε) and a set dissipation rate of thermal variance (χ). (After from Gregg [21])

[Osborn, 54]. In addition, it depends on the variance of only three quantities, which, assuming isotropy, can be reduced to one.

$$\chi \approx 6\kappa_T \left\langle \left(\frac{\partial T}{\partial z} \right)^2 \right\rangle$$

Moreover, while the shear probe measurements are easily contaminated by platform vibrations, microscale temperature gradient is not, permitting it to be estimated from more platforms, such as moorings, AUVs and gliders, towed bodies, and traditional CTD rosettes [16, 28, 45, 61]. The major complication, however, is illustrated in Fig. 8, in that the amplitude and wavenumber extent of the temperature gradient spectrum depend on ε and not just χ alone (also see Sect. 1.2). As a result, if the amplitude of the temperature gradient spectrum is only measured at low wavenumbers, it is not possible to simply fit a universal form and extrapolate it to compute the total variance of $\partial T/\partial z$ without knowing a-priori ε, because the wavenumber extent is controlled by the Batchelor wavenumber, k_B, which depends on ε, and is even smaller in scale than the Kolmogorov wavenumber. Because even the fastest temperature probes are relatively slow [responding at 20–30 Hz; [Gregg and Meagher, 22]], temperature gradient spectra are seldom resolved, unless profiles are acquired at very slow profiling speeds. Alford and Pinkel [1] and Moum and Nash [45] show ways of estimating the ε from the temperature gradient spectrum alone, but this approach involves assumptions.

Two "universal" forms for the temperature gradient spectrum have been used, one proposed by Kraichnan [32] and one by Batchelor [2]. Both have been shown to represent scalar variance adequately, although there has been recent evidence to suggest that Kraichnan's form might be more representative [Bogucki et al., 3]. In any event, both forms are relatively complicated to apply due to their multiparameter dependencies, yet this approach offers significant strengths related to lower power sampling and relative insensitivity to platform vibrations leading to long-term (>1 year) sampling capabilities [Moum et al., 49].

3 Summary

Elucidating geophysical turbulence is challenging because the fluid's nonlinearity leads to interactions between motions at a range of scales, both in space and time. The resultant flows are not deterministic and so must be addressed using a stochastic approach. In high Reynolds number flows, like those that occur in the ocean, a range of processes and instabilities lead to turbulence. The evolving turbulence is often highly intermittent in time and space, leading to complications related to appropriate flow sampling and density of observations in both time and space.

We are often interested in characterizing quantities such as the mean turbulent heat flux, the total energy dissipation, or the average turbulent stress acting on a volume of fluid. But because of turbulent intermittency, the statistics of these quantities are approximately log-normally distributed in space and time, meaning that one must sample the relatively rare but extreme events to quantify the mean [Moum and Rippeth, 46]. Understanding the relevant processes at play and their frequency of occurrence is key to quantifying the effects of a process. It is paramount that sampling strategies be developed to produce convergence in turbulence estimates given the space/time scales of intermittency within the dynamical regime of interest. For example, even if the most direct/accurate turbulence measurements can be made, these may not be of any use in defining the mean if an insufficient number of independent samples are made. A handful of turbulence profiles are seldom useful even for the steadiest of flows. Because of this, and the expense and challenges of traditional turbulence profiling, it is often beneficial to include less direct means of characterizing turbulence if these are more available and have a better hope of capturing the extreme events.

Although vertical profiles of dissipation rates from microstructure systems are considered a community standard, these observations are necessarily limited due to constraints and costs related to operating profilers at sea. Other approaches (e.g., dye release studies, glider measurements, time series on bottom landers or moorings) help to close sampling gaps, but ultimately a processes-based understanding that relates turbulence to larger-scale processes through the governing dynamics remains essential. For example, profiles collected near the ocean-atmosphere boundary (in both the ocean and atmosphere) using a variety of platforms can be related to the surface buoyancy flux over the mixed layer in unstable conditions (i.e., cooling of the ocean from

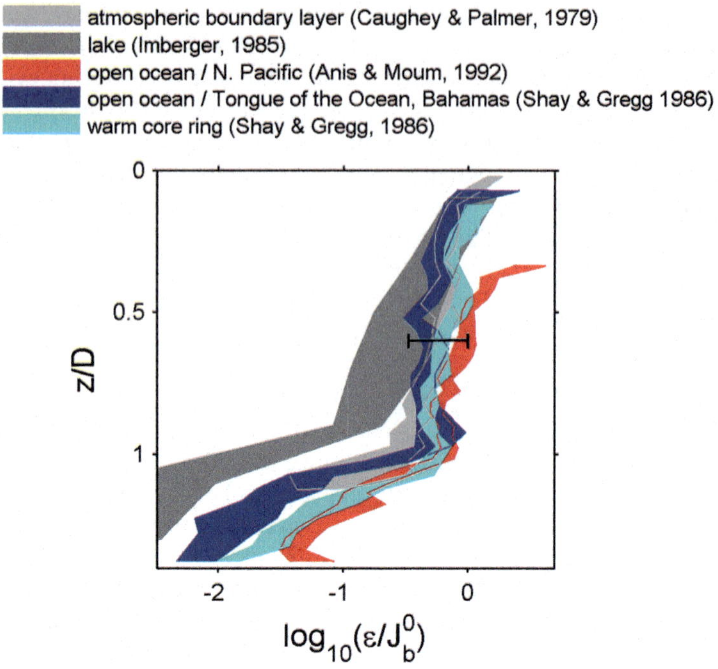

Fig. 9 Profiles of the turbulent kinetic energy dissipation rate normalized by the surface buoyancy flux, J_b^0, and measured using a variety of techniques. The vertical depth (height) is normalized by the mixed layer thickness in the ocean (atmosphere). (Reproduced from Moum and Rippeth [46])

above or warming of the atmosphere from below, Fig. 9). In such cases, the buoyancy production provides the main source for turbulence and the dissipation rate (as well as buoyancy flux) is linear. Application of finescale parameterizations to the global Argo dataset demonstrates another strength of this type of approach in that for the first time we are able to assess variations in mixing at a global scale. Despite its appeal, this particular approach is also limited by sensitivity to implementation details, violation of assumptions due to environmental state (e.g., near boundaries), and contributions from other processes leading to energy transfer in the finescale wavenumber band [Polzin et al., 60]. Because of the strengths and weaknesses associated with different methods, we continue to rely on multiple approaches, and our understanding of turbulence in the ocean continues to develop through intuition gained from a combination of laboratory studies, fieldwork, theory, and numerical models.

References

1. Alford MH, Pinkel R (2000) Observations of overturning in the thermocline: the context of ocean mixing. J Phys Oceanogr 30(5):805–832
2. Batchelor GK (1959) Small-scale variation of convected quantities like temperature in turbulent fluid part 1. General discussion and the case of small conductivity. J Fluid Mech 5(1):113–133

3. Bogucki DJ, Luo H, Domaradzki JA (2012) Experimental evidence of the Kraichnan scalar spectrum at high reynolds numbers. J Phys Oceanogr 42(10):1717–1728
4. Caldwell DR (1983) Small-scale physics of the ocean. Rev Geophys 21(5):1192. https://doi.org/10.1029/RG021i005p01192
5. Caldwell DR, Moum JN (1995) Turbulence and mixing in the ocean. Rev Geophys 33(S2):1385–1394. https://doi.org/10.1029/95RG00123
6. Chabai AJ, Emrich RJ (1955) Measurement of wall temperature and heat flow in the shock tube. J Appl Phys 26(6):779. https://doi.org/10.1063/1.1722091
7. Cooper J, Stommel H (1968) Regularly spaced steps in the main thermocline near Bermuda. J Geophys Res 73:5849–5854
8. Cox C, Nagata Y, Osborn T (1969) Oceanic fine structure and internal waves. Bull Japan Soc Fish Ocean 3:67–71
9. Decloedt T, Luther DS (2012) Spatially heterogeneous diapycnal mixing in the abyssal ocean: a comparison of two parameterizations to observations. J Geophys Res Ocean 117(11). https://doi.org/10.1029/2012JC008304
10. Dillon T (1982) Vertical overturns: a comparison of Thorpe and Ozmidov length scales. J Geophys Res Ocean 87:9601–9613
11. Dillon T, Caldwell D (1980) The Batchelor spectrum and dissipation in the upper ocean. J Geophys Res 85:1910
12. Doron P, Bertuccioli L, Katz J, Osborn TR (2001) Turbulence characteristics and dissipation estimates in the coastal ocean bottom boundary layer from PIV data. J Phys Oceanogr 31(8):2108–2134
13. Garrett C, Munk W (1971) Internal wave spectra in the presence of fine-structure. J Phys Oceanogr 1:196–202
14. Garrett C, Munk W (1972) Oceanic mixing by breaking internal waves. In: Deep sea research and oceanographic abstracts, vol 19. Elsevier, Oxford, pp 823–832
15. Garrett C, Munk W (1975) Spacetime scales of internal waves: a progress report. J Geophys Res 80(3):291–297
16. Goodman L, Levine ER, Lueck RG (2006) On measuring the terms of the turbulent kinetic energy budget from an AUV. J Atmos Ocean Technol 23(7):977–990
17. Grant HL, Stewart RW, Moilliet A (1962) Turbulence spectra from a tidal channel. J Fluid Mech 12(2):241. https://doi.org/10.1017/S002211206200018X
18. Gregg MC (1987) Diapycnal mixing in the thermocline: a review. J Geophys Res 92(C5):5249. https://doi.org/10.1029/JC092iC05p05249
19. Gregg MC (1989) Scaling turbulent dissipation in the thermocline. J Geophys Res 94(C7):9686–9698. https://doi.org/10.1029/JC094iC07p09686
20. Gregg MC (1991) The study of mixing in the ocean: a brief history. Oceanography 4(1):39–45
21. Gregg MC (1999) Uncertainties and limitations in measuring ϵ and χT. J Atmos Ocean Technol 16(11):1483–1490. https://doi.org/10.1175/1520-0426(1999)016<1483:UALIMA>2.0.CO;2
22. Gregg MC, Meagher TB (1980) The dynamic response of glass rod thermistors. J. Geophys. Res. Ocean. 85(C5):2779–2786
23. Gregg MC, Sanford TB, Winkel DP (2003) Reduced mixing from the breaking of internal waves in equatorial waters. Nature 422(6931):513–515
24. Henyey FS, Pomphrey N (1983) Eikonal description of internal wave interactions: a non-diffusive picture of "induced diffusion". Dyn Atmos Ocean 7(4):189–219
25. Henyey FS, Wright J, Flatté SM (1986) Energy and action flow through the internal wave field: an eikonal approach. J Geophys Res Ocean 91(C7):8487–8495
26. Holbrook WS, Fer I, Schmitt RW, Lizarralde D, Klymak JM, Helfrich LC, Kubichek R (2013) Estimating oceanic turbulence dissipation from seismic images. J Atmos Ocean Technol 30(8):1767–1788. https://doi.org/10.1175/JTECH-D-12-00140.1
27. Huussen TN, NaveiraGarabato AC, Bryden HL, McDonagh EL (2012) Is the deep Indian Ocean MOC sustained by breaking internal waves? J. Geophys. Res. Ocean 117(C8):C08024
28. Klymak JM, Moum JN (2007) Oceanic isopycnal slope spectra. Part II: turbulence. J Phys Oceanogr 37(5):1232–1245. https://doi.org/10.1175/JPO3074.1

29. Klymak JM, Pinkel R, Rainville L (2008) Direct breaking of the internal tide near topography: Kaena ridge, Hawaii. J Phys Oceanogr 38(2):380–399
30. Kolmogorov AN (1941) The local structure of turbulence in incompressible viscous fluid for very large Reynolds numbers. In: Dokl Akad Nauk SSSR, vol 30 JSTOR, pp 301–305. http://www.jstor.org/stable/51980?origin=JSTOR-pdf&seq=1#page_scan_tab_contents
31. Kolmogorov AN (1991) Dissipation of energy in the locally isotropic turbulence. Proc R Soc A Math Phys Eng Sci 434(1890):15–17. http://www.jstor.org/stable/51981?seq=1#page_scan_tab_contents
32. Kraichnan RH (1959) The structure of isotropic turbulence at very high Reynolds numbers. J Fluid Mech 5(4):497–543
33. Kunze E, Firing E, Hummon JM, Chereskin TK, Thurnherr AM (2006) Global abyssal mixing inferred from lowered ADCP shear and CTD strain profiles. J Phys Oceanogr 36(8):1553–1576
34. Lavery AC, Chu D, Moum JN (2009) Measurements of acoustic scattering from zooplankton and oceanic microstructure using a broadband echosounder. ICES J Mar Sci J du Cons 67(2):fsp242. https://doi.org/10.1093/icesjms/fsp242
35. Ledwell JR (2004) Mixing in a coastal environment: 1. A view from dye dispersion. J Geophys Res 109(C10):C10013. https://doi.org/10.1029/2003JC002194
36. Ledwell JR, Watson AJ, Law CS (1993) Evidence for slow mixing across the pycnocline from an open-ocean tracer-release experiment. Nature 364(6439):701–703. https://doi.org/10.1038/364701a0
37. Levine ER, Lueck RG (1999) Turbulence measurements from an autonomous underwater vehicle. J Atmos Ocean Technol 16(2):1533–1544. https://doi.org/10.1175/1520-0426(1999)016<1533:TMFAAU>2.0.CO;2
38. Liebermann L (1951) The effect of temperature inhomogeneities in the ocean on the propagation of sound. J Acoust Soc Am 23(5):563. https://doi.org/10.1121/1.1906805
39. Ling S-C (1955) Measurement of flow characteristics by the hot-film technique. University of Iowa, Ann Arbor
40. Lueck RG, Wolk F, Yamazaki H (2002) Oceanic velocity microstructure measurements in the 20th century. J Oceanogr 58(1):153–174. https://doi.org/10.1023/A:1015837020019
41. Lumley JL, Terray EA (1983) Kinematics of turbulence convected by a random wave field. J Phys Oceanogr 13(11):2000–2007
42. MacKinnon JA, Alford MH, Ansong JK, Arbic BK, Barna A, Briegleb BP, Bryan FO, Buijsman MC, Chassignet EP, Danabasoglu G, Diggs S, Griffies SM, Hallberg RW, Jayne SR, Jochum M, Klymak JM, Kunze E, Large WG, Legg S, Mater B, Melet AV, Merchant LM, Musgrave R, Nash JD, Norton NJ, Pickering A, Pinkel R, Polzin K, Simmons HL, St. Laurent LC, Sun OM, Trossman DS, Waterhouse AF, Whalen CB, Zhao Z (2017) Climate process team on internal-wave Driven Ocean mixing. Bull Amer Meteor Soc. https://doi.org/10.1175/BAMS-D-16-0030.1
43. MacKinnon JA, Gregg MC (2003) Mixing on the late-summer New England Shelf-solibores, shear, and stratification. J Phys Oceanogr 33(7):1476–1492
44. MacKinnon JA, Alford MH, Pinkel R, Klymak J, Zhao Z (2013) The latitudinal dependence of shear and mixing in the Pacific transiting the critical latitude for PSI. J Phys Oceanogr 43(1):3–16
45. Moum JN, Nash JD (2009) Mixing measurements on an equatorial ocean mooring. J Atmos Ocean Technol 26(2):317–336. https://doi.org/10.1175/2008JTECHO617.1
46. Moum JN, Rippeth TP (2009) Do observations adequately resolve the natural variability of oceanic turbulence? J Mar Syst 77(4):409–417. https://doi.org/10.1016/j.jmarsys.2008.10.013
47. Moum JN, Gregg MC, Lien RC, Carr ME (1995) Comparison of turbulence kinetic energy dissipation rate estimates from two ocean microstructure profilers. J Atmos Ocean Technol 12(2):346–366. https://doi.org/10.1175/1520-0426(1995)012<0346:COTKED>2.0.CO;2
48. Moum JN, Klymak JM, Nash JD, Perlin A, Smyth WD (2007) Energy transport by nonlinear internal waves. J Phys Oceanogr 37(7):1968–1988. https://doi.org/10.1175/JPO3094.1
49. Moum JN, Perlin A, Nash JD, McPhaden MJ (2013) Seasonal sea surface cooling in the equatorial Pacific cold tongue controlled by ocean mixing. Nature 500(7460):64–67. https://doi.org/10.1038/nature12363

50. Munk WH, Garrett CJR (1973) Internal wave breaking and microstructure (the chicken and the egg). Boundary-Layer Meteorol 4(1):37–45. https://doi.org/10.1007/BF02265224
51. Nash JD, Moum JN (2002) Microstructure estimates of turbulent salinity flux and the dissipation spectrum of salinity. J Phys Oceanogr 32:2312–2333. https://doi.org/10.1175/1520-0485(2002)032<2312:MEOTSF>2.0.CO;2
52. Nasmyth PW (1970) Oceanic turbulence. University of British Columbia, Vancouver, CA
53. Oakey N (1982) Determination of the rate of dissipation of turbulent energy from simultaneous temperature and velocity shear microstructure measurements. J Phys Oceanogr 12:256–271
54. Osborn T, Cox C (1972) Oceanic fine structure, Geophys. Astrophys Fluid Dyn 3:321–345
55. Osborn TR (1974) Vertical profiling of velocity microstructure. J Phys Oceanogr 4(1):109–115. https://doi.org/10.1175/1520-0485(1974)004<0109:VPOVM>2.0.CO;2
56. Osborn TRT (1980) Estimates of the local rate of vertical diffusion from dissipation measurements. J Phys Oceanogr 10:83–89
57. Panchev S, Kesich D (1969) Energy spectrum of isotropic turbulence at large wavenumbers. CR Acad Bulg Sci 22:627–630
58. Polzin KL, Firing E (1997) Estimates of diapycnal mixing using LADCP and CTD data from I8S. Int WOCE Newsl 29:39–42
59. Polzin KL, Toole JMJ, Schmitt RW (1995) Finescale parameterizations of turbulent dissipation. J Phys Oceanogr 25(3):306–328
60. Polzin KL, Naveira Garabato AC, Huussen TN, Sloyan BM, Waterman S (2014) Finescale parameterizations of turbulent dissipation. J Geophys Res Ocean 119(2):1383–1419. https://doi.org/10.1002/2013JC008979
61. Rainville L, Winsor P (2008) Mixing across the Arctic Ocean: microstructure observations during the Beringia 2005 expedition. Geophys Res Lett 35:L08606. https://doi.org/10.1029/2008GL033532
62. Rye CD, Messias M-JJ, Ledwell JR, Watson AJ, Brousseau A, King BA (2012) Diapycnal diffusivities from a tracer release experiment in the deep sea, integrated over 13 years. Geophys Res Lett 39(4):n/a–n/a. https://doi.org/10.1029/2011GL050294
63. Smyth WD, Moum JN (2001) Three-dimensional (3d) turbulence A2- Steele. In: John H (ed) BT- encyclopedia of ocean sciences. Academic Press, Oxford, pp 2947–2955
64. Smyth WD, Thorpe SA (2012) Glider measurements of overturning in a Kelvin- Helmholtz billow train. J Mar Res 70(1):119–140
65. Smyth WD, Moum JN, Caldwell DR (2001) The efficiency of mixing in turbulent patches: inferences from direct simulations and microstructure observations. J Phys Oceanogr 31(8):1969–1992. https://doi.org/10.1175/1520-0485(2001)031<1969:TEOMIT>2.0.CO;2
66. Spilhaus AF (1938) A bathythermograph. J Mar Res:95–100
67. Stommel H, Fedorov KN (1967) Small scale structure in temperature and salinity near Timor and Mindanao. Tellus 19(2):306–325. https://doi.org/10.1111/j.2153-3490.1967.tb01484.x
68. Tait R, Howe M (1968) Some observations of thermo-haline stratification in the deep ocean. Deep Sea Res Oceanogr Abstr 15:275–280
69. Tennekes H, Lumley JL (1972) A first course in turbulence. The MIT Press, Cambridge, MA
70. Thorpe SA (1977) Turbulence and mixing in a Scottish loch. Philos Trans R Soc London Ser A, Math Phys Sci 286(1334):125–181
71. Thorpe SA (2007) An introduction to ocean turbulence. Cambridge University Press, Cambridge
72. Trowbridge J, Elgar S (2001) Turbulence measurements in the surf zone. J Phys Oceanogr 31(8):2403–2417. https://doi.org/10.1175/1520-0485(2001)031<2403:TMITSZ>2.0.CO;2
73. Urick RJ, Searfoss CW (1948) The microthermal structure of the ocean near Key West, Florida, part I, discussion. Nav Res Lab Rep S-3444
74. Urick RJ, Searfoss CW (1949) The microthermal structure of the ocean near Key West, Florida, part II, analysis. Nav Res Lab Rep S-3444
75. Veron F, Melville WK (1999) Pulse-to-pulse coherent Doppler measurements of waves and turbulence. J Atmos Ocean Technol 16(11):1580–1597

76. Waterhouse AF et al (2014) Global patterns of Diapycnal mixing from measurements of the turbulent dissipation rate. J Phys Oceanogr 44(7):1854–1872. https://doi.org/10.1175/JPO-D-13-0104.1
77. Waterman S, Polzin KL, Naveira Garabato AC, Sheen KL, Forryan A (2014) Suppression of internal wave breaking in the antarctic circumpolar current near topography. J Phys Oceanogr 44(5):1466–1492
78. Whalen CB, Talley LD, MacKinnon JA (2012) Spatial and temporal variability of global ocean mixing inferred from Argo profiles. Geophys Res Lett 39(17). https://doi.org/10.1029/2012GL053196
79. Winkel DP, Gregg MC, Sanford TB (2002) Patterns of shear and turbulence across the Florida current. J Phys Oceanogr 32(11):3269–3285. https://doi.org/10.1175/1520-0485(2002)032<3269:POSATA>2.0.CO;2
80. Woods J (1968) Wave-induced shear instability in the summer thermocline. J Fluid Mech 32:791
81. Woods J (1969) On Richardson's number as a criterion for laminar turbulent laminar transition in the ocean and atmosphere. Radio Sci 4:1289–1298
82. Woods J, Wiley R (1972) Billow turbulence and ocean microstructure. Deep Sea Res Oceanogr Abstr 19:87
83. Wu L, Jing Z, Riser S, Visbeck M (2011) Seasonal and spatial variations of Southern Ocean diapycnal mixing from Argo profiling floats. Nat Geosci 4(6):363–366

Underwater Gliders

Craig M. Lee and Daniel L. Rudnick

Abstract This chapter focuses on underwater gliders, placing them in the context of the recent surge in autonomous observing technologies, reviewing the underlying design philosophy and providing a brief history of their development. Gliders resolve scales of kilometers and hours, with the seasonal to annual endurance required to characterize climate variability and capture episodic events – a region of the spatial-temporal sampling spectrum that had previously been challenging to address. Examples of gliders applied to sustained studies of large-scale variability in boundary regions, to physical and biological/biogeochemical process studies, and to studies of polar regions illustrate strategies for efficient use that capitalize on the platform's strengths. Although gliders are a mature platform with demonstrated scientific output, improvements to reliability, ease of use, and range would have large impacts on platform efficiency, enabling broader adoption and application to a wider range of scientific and operational tasks.

1 Introduction

Oceanographic phenomena unfold over a broad range of temporal and spatial scales – seconds to decades and millimeters to ocean basins – posing severe challenges for investigations that rely on in situ observations. Advances in understanding often rest on resolving the interactions between processes that span diverse scales. Quantification of the ocean's role in climate, and prediction of its response to climate change, requires sustained, persistent measurements over the broadest spatial and temporal scales. Operational forecast modeling imposes the added requirement of real-time data delivery. Observational approaches have struggled to meet these competing demands while working within budgetary and logistical constraints.

C.M. Lee (✉)
Applied Physics Laboratory, University of Washington, Seattle, WA, USA
e-mail: craig@apl.washington.edu

D.L. Rudnick
Scripps Institution of Oceanography, University of California, San Diego, La Jolla, CA, USA

R. Venkatesan et al. (eds.), *Observing the Oceans in Real Time*, Springer Oceanography, DOI 10.1007/978-3-319-66493-4_7

Low-power electronics and advances in battery technologies and sensors, coupled with the wide availability of satellite geolocation (e.g., Global Positioning System) and communications (e.g., Iridium) have enabled major advances in robotic observing of the world's oceans. The Argo float array [1] provides the foremost example of this revolution. Departing from the established approach of sampling along designated lines, the Argo array relies on broad deployments and subsequent drift to provide distributed, quasi-uniform sampling suitable for quantifying the large-scale distribution of heat and salt within the basin interiors. Argo floats are designed to be low-cost, long-endurance, and simple to deploy (allowing flexible logistics), requirements that are central to achieving and sustaining the operational scale required of the array. The array currently consists of nearly 4000 floats profiling the upper 2000 m of the water column with 10-day temporal and nominal 300-km spatial resolution. When at the surface, floats use GPS to geolocate and transmit data to shore via Argos and Iridium satellite systems. Argo has operated since 1999, successfully fulfilling its primary objective of quantifying basin-scale variability in circulation, heat and salt content at seasonal to decadal timescales, while also serving a wide range of additional research and operational needs. The persistence and scalability provided by low-cost profiling floats allow oceanographers to access a combination of spatial (regional to global) and temporal (monthly to decadal) scales that had previously been challenging or impractical to sample. This is important because these scales are critical to understanding the ocean's role in the climate system.

Complementary approaches are required to resolve features at shorter temporal and spatial scales, including boundary currents, frontal zones, and the suite of processes that govern vertical exchange. Stommel [2] presented a compelling vision of an autonomous sampling network composed of buoyancy-driven underwater gliders – essentially profiling floats that adjust vehicle attitude to use wings to project vertical motion into the horizontal, enabling them to navigate from place to place. Gliders were envisioned to be cost-effective to procure and operate, thus enabling a sustained network of a thousand vehicles. They would communicate via satellite to a centralized mission control facility, retuning profiles of temperature, salinity, oxygen, nutrients, and other biogeochemical properties several times per day. Gliders carrying designated reference sensors would deliberately visit and cross-calibrate other gliders within the network, explicitly addressing the need for intercalibration within large sensor networks. The plan called for half the fleet to sustain observations of the ocean interior by conducting monthly occupations of the 48 hydrographic sections established by the World Hydrographic Program (WHP) Experiment – an observational requirement that has since been fulfilled in a different fashion by Argo. Instead, two decades of experience with gliders suggests that the platform's strengths lie in the second half of Stommel's vision, which focused on exploiting mobility and extended endurance to repeatedly sample critical sites that exhibit strong lateral gradients, such as boundary currents, recirculation regions, and overflows, and to conduct focused process studies.

2 Design and Development History

2.1 Challenges and Design Philosophy

Design of the current generation of underwater gliders strove to optimize vehicle properties, including size, power, speed, and payload, around a suite of potential missions. Envisioned uses included serving as self-deploying virtual moorings (station-keeping), sustained occupations of sections, and as survey platforms in limited-duration process studies. The latter two applications drove the design toward persistent sampling of the oceanic mesoscale, resolving scales of kilometers and hours, with the seasonal to annual endurance required to characterize climate variability and capture episodic events. This particular region of the spatial-temporal sampling spectrum had been challenging to address, as existing tools provided either persistence with excellent temporal, but coarse spatial, resolution (mooring arrays) or excellent spatial resolution over limited time spans (ship-based surveys). Recognizing that scalability would factor strongly into the platform's utility for sustained, climate-scale observing, gliders were envisioned as inexpensive and simple to operate, enabling deployment in quantity. Low capital and operating cost also enable the use of fleets of gliders to expand the range of accessible time and space scales. Glider designs also addressed the challenges encountered when planning observations in remote, difficult-to-access locations, and harsh operating conditions.

These mission requirements led to design decisions common to existing operational gliders. Design choices trade off size (battery and payload capacity), cost, range, and speed. Size and range drive capital and operating costs. Robust vehicles that are small enough to be carried by hand, without assistance from cranes, davits or other powered lifting gear, allow a broad range of logistics, including operations from small vessels such as fishing boats and rigid hull inflatables, with minimal personnel requirements. This obviates reliance on large, dedicated research vessels, reducing cost and dramatically increasing flexibility. Extended range, best quantified as total distance traversed through the water, can translate to long missions that reduce overall operating costs by reducing both the number of vehicles and lengthening the service interval (reducing the total number of deployments and recoveries) required to sustain a given sampling scheme.

Hydrodynamic drag mediates the relationship between speed through the water, range, and energy consumption. Gliders pay a stiff penalty to move at high speeds, as energy loss to drag is proportional to speed through the water cubed. Because propulsion represents the largest fraction of the glider energy budget, increased speed comes at significant cost to range. Although this could, to some degree, be mitigated by increasing battery capacity, this impacts vehicle size and cost.

Facing these constraints, three initial glider developments (Slocum, Spray, and Seaglider; Fig. 1) arrived at similar overarching designs. To meet handling requirements and constrain cost, these gliders are roughly 2 m long, with a mass of 50 kg. They propel themselves by changing buoyancy, deflating (inflating) an oil-filled

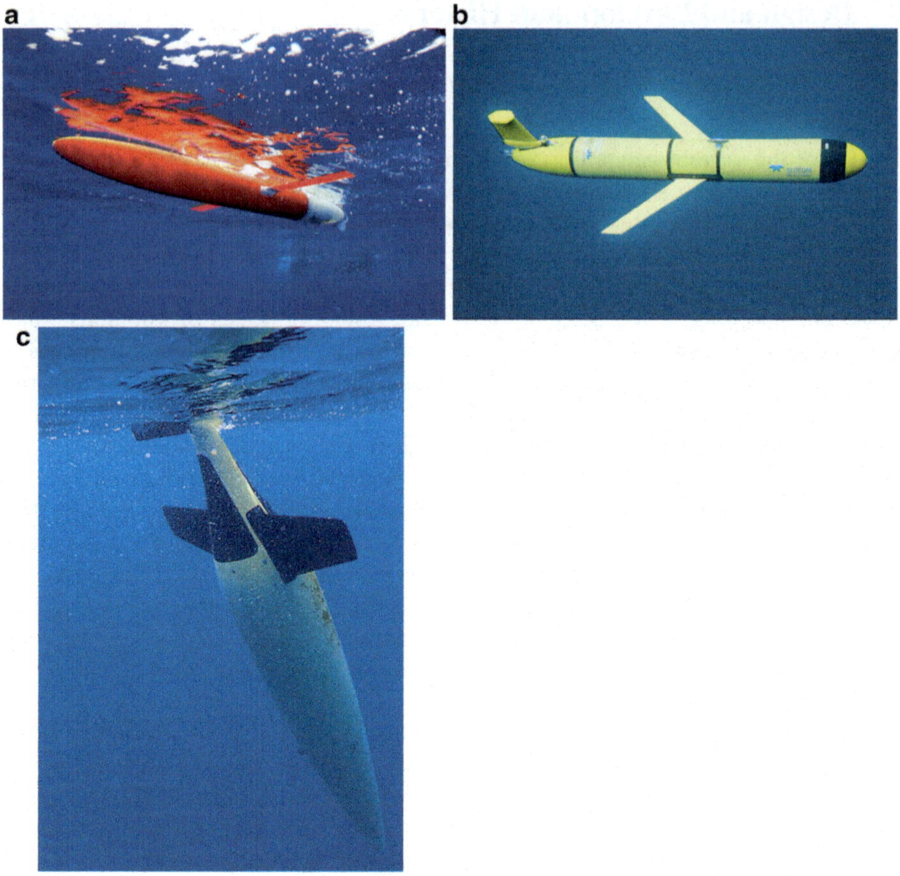

Fig. 1 The three initial underwater glider designs: (**a**) Spray, (**b**) Slocum, (**c**) Seaglider

external bladder to sink (rise) through the water column, with hull and wings providing lift to translate vertical motion into the horizontal. Gliders steer by controlling attitude through a combination of buoyancy change, shifting battery mass fore-and-aft and side-to-side, and, in the case of Slocum, employing an active rudder.

Gliders geolocate using the Global Positioning System (GPS) and communicate to shore via Iridium satellite modem during the surface interval between dives. Two-way communications allow gliders to upload data and pilots to issue commands and, if necessary, upload new operating code. This allows users to exploit platform mobility by adapting sampling in response to observed environmental variability, and ensures data return regardless of the platform's fate, mitigating risk and reducing the need for platform recovery.

Glider size places constraints on operating depth, payload, and speed. Gliders typically profile to 1000 m (though they can perform shallower profiles) at a vertical speed of roughly 0.1 ms^{-1}, requiring about 6 h to complete a dive. Deeper depths require larger, heavier hulls – for example, a new 6000 m Deepglider weighs roughly 80 kg. Gliders operate at a typical horizontal speed through the water of 0.25 ms^{-1}, sufficient to stem 1000-m depth average currents in many, though not all, envisioned operating environments. Although faster speeds can be attained by applying additional buoyancy, gliders intended for long missions must operate at slow speeds to minimize drag. Mission durations range from weeks to nearly a year, dependent on a variety of factors including stratification, glider speed, dive frequency, and payload. Gliders can execute 500–1000 dives and traverse thousands of kilometers over the course of a single mission.

2.2 *Development History*

The trajectory of underwater glider development illustrates the magnitude and character of the effort required to bring a relatively complex oceanographic instrument into broad use. Glider development stretches over a 25+ year span (Fig. 2), beginning with Webb Research Corporation's early tests of a buoyancy-driven vehicle in Wakulla Springs, Florida in 1991 [3], and stretching to the present, where the technology is employed widely for scientific [4] and operational uses. Significant investment began in 1995, when the US Office of Naval Research (ONR) funded development of three vehicles, Slocum [3], Spray [5] and Seaglider [6], undertaken by three teams working in managed competition. The three teams collectively recognize 44 individuals for their contributions, pointing to the scope of skills and level of effort required to bring the platform to maturity. Although there are novel aspects to each of the resulting designs, fundamental similarities between the three vehicles speak to the strength of the physical and technological constraints.

Early multi-day ocean missions took place in 1998 and 1999, progressing to larger, multi-glider deployments in Monterey Bay in 2000 and 2003, as part of the ONR's Autonomous Ocean Sampling Network program. Commercial delivery of gliders began in 2002 with Webb Research Corporation's Slocum gliders, followed by the Office of the Secretary of Defense's Technology Transition Initiative (2005–2008) that brought gliders into operational use within the US Navy. During this period, maturing technology and growing experience extended glider missions from multi-week to multi-month durations, with process studies in the difficult operating environments of the wintertime Labrador Sea [7] and Gulf of Alaska [8] and sustained operations off the US west [9, 10, 11] and east [12] coasts, and in the western boundary currents of the Solomon Sea [13] and the Kuroshio [14, 15]. Early missions were conducted in regions that could be readily accessed in the event of platform failures, rapidly followed by operations in more remote, but scientifically important, areas that placed gliders beyond reach of rescue. Relatively low platform cost and real-time data return mitigated risk and enabled pioneering use of gliders in these challenging environments.

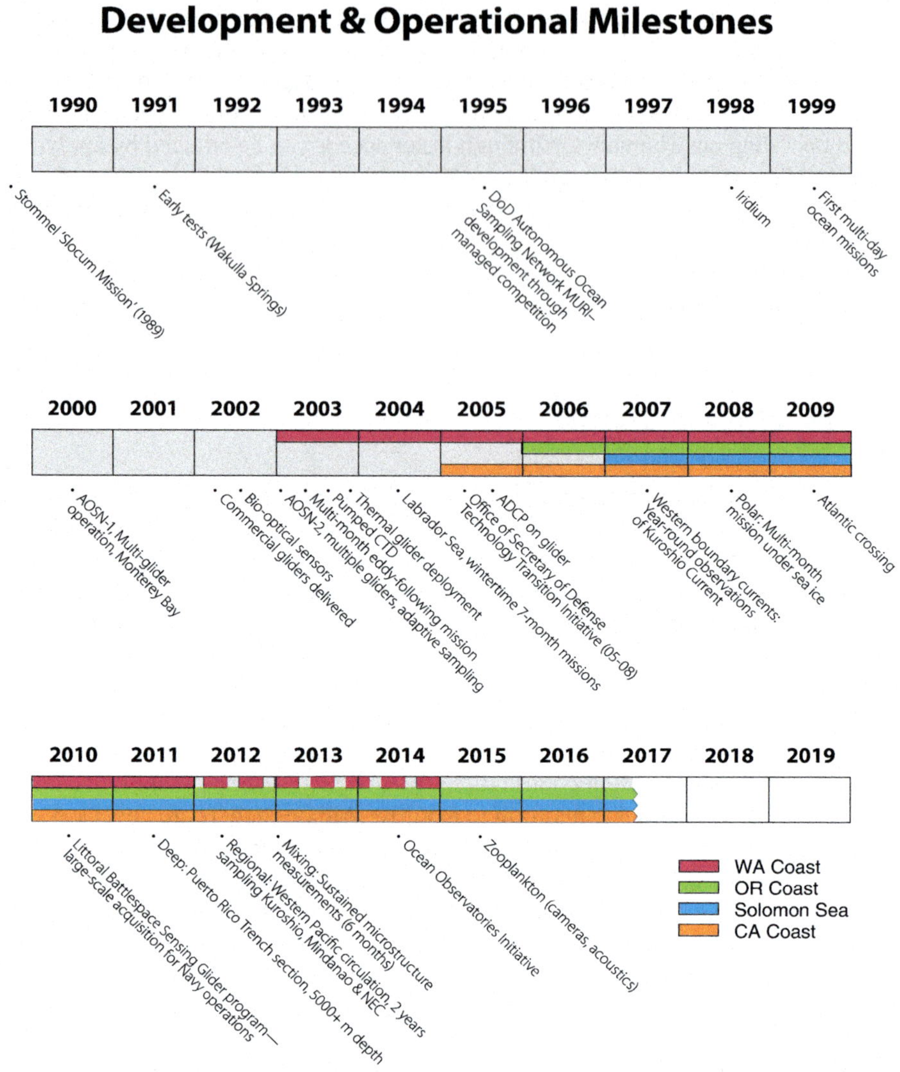

Fig. 2 Glider development timeline, with selected milestones. Colored bars mark the span of continuous (*solid*) or seasonal (*dashed*) sustained occupation of selected regions

The progression of sensor integration has paralleled the maturation of the platform. Volume and power place significant limits on glider payloads. Gliders began with a basic sensors suite for pressure, temperature, and conductivity, for estimating salinity. Early additions included dissolved oxygen, chlorophyll fluorescence, and optical backscatter [16], expanding glider-based sampling to include a limited set of biological and biogeochemical parameters. Ocean velocity averaged over the profile depth was first estimated by differencing horizontal displacement over a dive estimated from a hydrodynamic model from actual displacement derived from GPS positions [6, 17].

This depth-average current could then be used in conjunction with density computed along a section to estimate profiles of absolute geostrophic currents perpendicular to the track. Subsequent integration of acoustic Doppler current profilers [9, 18, 19] provides the ability to measure profiles of absolute velocity. More recently, the range of available sensors has expanded as developers respond to platform constraints. Available parameters now include nitrate, pH, acoustic backscatter, passive acoustics for sensing marine mammals and rain rate, light transmission, upwelling and downwelling irradiance, and microstructure temperature and shear.

3 Applications and Strategies

As envisioned by Stommel [2], gliders have been used both for sustained observations of large-scale variability and for focused investigations of processes that unfold at much shorter temporal and spatial scales. The history of glider operations in the Pacific (Fig. 3) illustrates the scope of large-scale glider-based sampling and reveals patterns that reflect the platform's strengths and constraints. A striking feature of the 15+ year record in the Pacific is the distribution of sections. In contrast

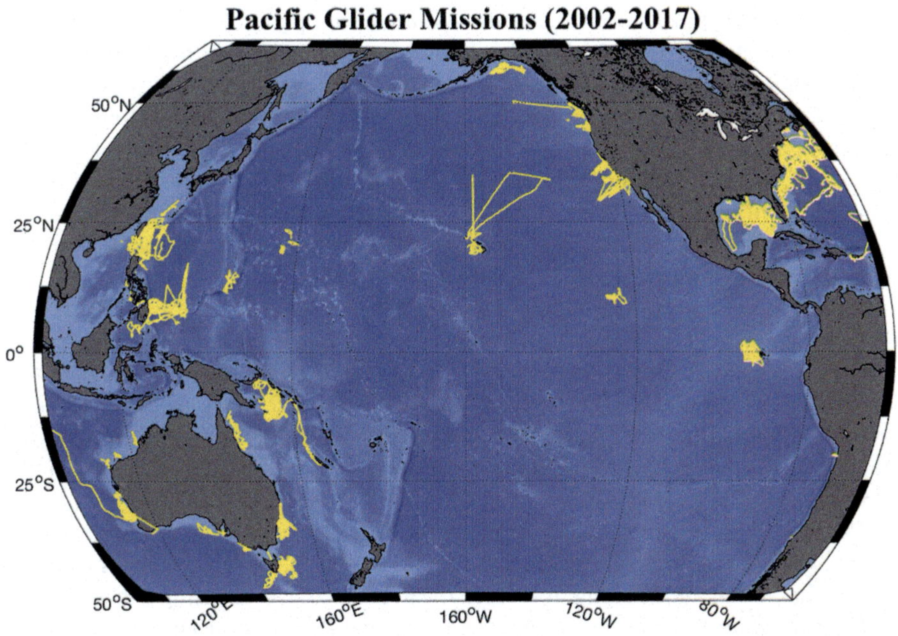

Fig. 3 Glider missions in the Pacific Ocean from the period 2002 to 2017, undertaken by Oregon State University, Rutgers University, Scripps Institution of Oceanography, University of Hawaii, University of Washington, University of Western Australia, and the Woods Hole Oceanographic Institution

to Stommel's picture of trans-basin WHP sections, most glider missions, including all of the sustained, multi-year efforts, have focused on the boundaries. With finely spaced profiles at relatively slow speed, gliders are well suited to measure horizontal scales of 10 km and larger, and timescales of weeks and longer. Glider operations maximize efficiency by operating locally from small, flexible vessels, often chartered fishing or tourist boats. Gliders are thus suited for characterizing boundary currents, where their persistence and resolving power are well matched to the important physical, biological, and biogeochemical scales, and local logistics are readily available. There have been a number of sustained boundary current observations using gliders across the world's oceans, including off California, along the western boundary of the North Pacific, and in the Gulf Stream, to name a few. In this sense, glider sampling has grown to complement Argo, which provides distributed observations within the ocean interiors.

3.1 Boundary Currents

The longest running sustained program of glider observations is likely the California Underwater Glider Network (CUGN). CUGN started in 2006, and has occupied three sections in the California Cooperative Oceanic Fisheries Investigations (CalCOFI; [20]) grid for over a decade. As a mature operational program, CUGN data are publicly distributed within hours of collection, and their results have been distilled in a gridded climatology [10]. The CUGN has proven useful for quantifying the regional responses to climate variability [18, 21, 22, 23], in constraining and validating models [21, 24, 25], and for exploring the interface between biology and physics [9, 26, 27]. The colocation with the CalCOFI ship survey has allowed a comparison of ship and glider sampling, which demonstrated the glider's advantages of sustained presence and sheer number of profiles (gliders provide over 50 times the profiles from CalCOFI ships on a given line [10]).

As a complement to that comparison, we address the sampling by Argo in the CUGN region. All glider and Argo profiles from a 3-year period (2010–2012) are shown (Fig. 4) in an area defined by a regional assimilating model [18]. There are 28,893 CUGN profiles and 2482 Argo profiles, hence 12 times as many profiles from gliders as from floats. More important is the distribution of the profiles, with the CUGN sampling connecting the coast to the open ocean where Argo coverage prevails.

The use of gliders for making repeated sections, as across a boundary current, can be evaluated through objective analysis [28]. As an example, consider the observation of a section 500 km long, for which the goal is to have sampling dense enough in distance along the section and in time to resolve the relevant length and timescales (Fig. 5). A glider traveling at 0.25 ms^{-1} can complete a section in 23 days, with a profile to 1000 m every 6 h in time and 6 km in distance. A sampling plan using moorings might consist of six moorings spaced at 100 km. To make this example concrete, assume that the covariance of the oceanographic signal is Gaussian with a

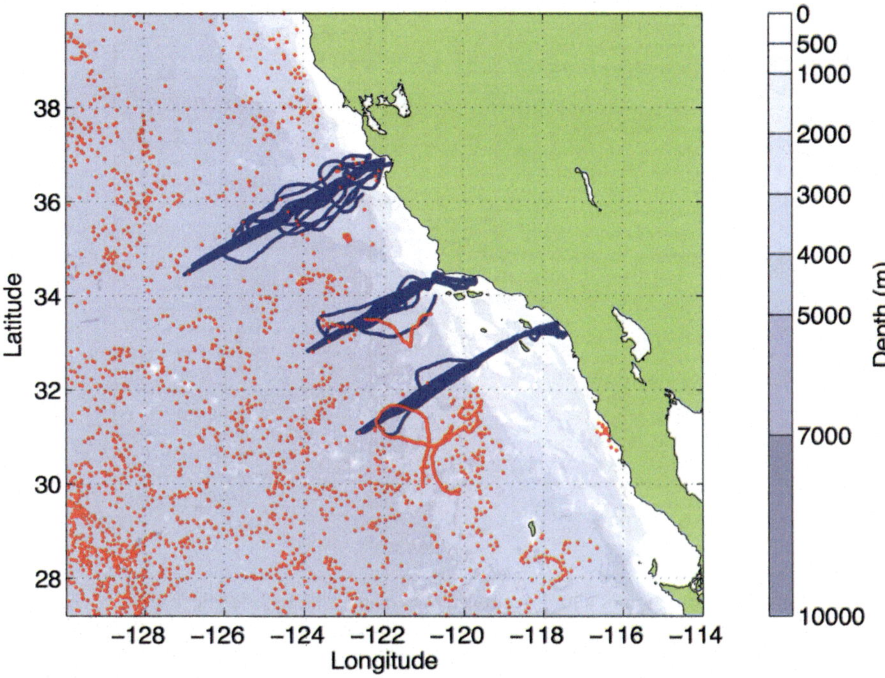

Fig. 4 Profiles by gliders in the CUGN (*blue dots*) and by floats in Argo (*red dots*) during 2010–2012. There are 28,893 profiles from CUGN and 2482 from Argo. Note the complementary distribution of profiles with Argo in the open ocean and the CUGN connecting the coast to the open ocean

length scale of 30 km and a timescale of 30 days, scales that are relevant to sustained observation of a boundary current. Further assume that there is a noise of 0.1 times the signal variance. An objective analysis yields the error map, as in Fig. 5, where the 0.3 contour of error variance is shown. The error in the map from glider measurements is less than 0.3 everywhere but in scallops at the edges of the line. The error in the map from mooring measurements is less than 0.3 in strips centered on the moorings. The map is sensitive to the length scale and timescale, but in this example, the one glider yields a more accurate map than six moorings.

Gliders can be used to accurately quantify boundary current transports. Sections of absolute geostrophic velocity can be derived by referencing geostrophic shear, calculated from cross-stream density gradients using thermal wind, to depth-average velocity estimated from glider displacement over a dive [6, 17]. Comparisons between velocity at points spanning the Kuroshio Current measured by moored, upward-looking 75 kHz acoustic Doppler current profilers and absolute geostrophic velocity calculated from glider-based sections show good agreement, both when averaged over multiple sections (Fig. 6, after [15]) and for contemporaneous vertical profiles derived by averaging over the time when gliders were with 2 km of individual moorings (not shown). Direct and glider-derived geostrophic velocities

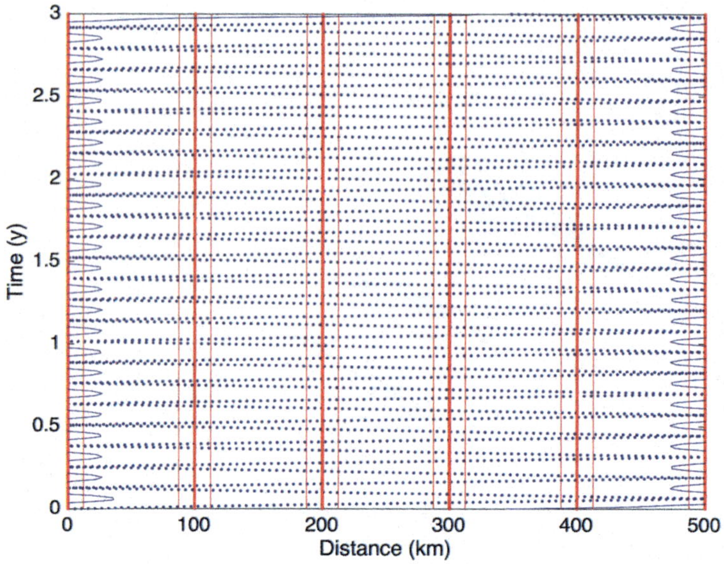

Fig. 5 Sampling as a function of a long section distance and time by one glider and six moorings on a line 500 km long. Glider profiles (*blue dots*) are separated by 6 km and 6 h, while the glider travels at 0.25 m s^{-1}. Moorings (*red dots*) are separated by 100 km, and can be taken to be sampling with arbitrarily fine resolution (so the red dots appear as a thick red line). An objective analysis using a Gaussian covariance with 30-day timescale, 30-km length scale, and a 0.1 noise-to-signal variance ratio, is used to estimate error maps. The 0.3 error-to-signal variance ratio from an objective map is shown for one glider (*thin blue line*) and six moorings (*thin red line*)

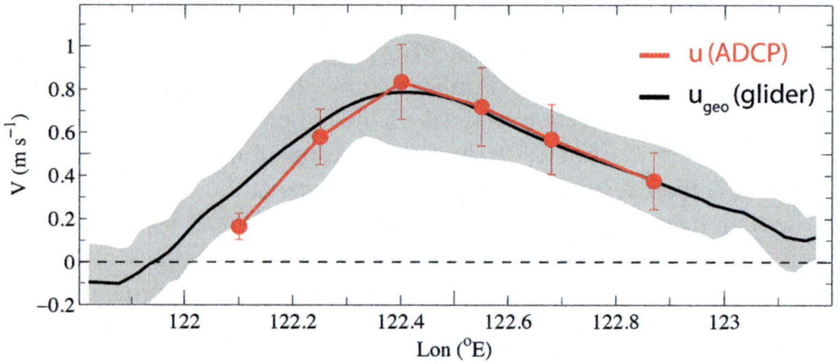

Fig. 6 Comparison between absolute geostrophic velocity computed from glider-based measurements (*black line* with 1 standard deviation interval plotted in *gray*) and velocity observed directly from moored acoustic Doppler current profilers (*red dots*, with 1 standard deviation interval). Data are averaged over the time span of 15 glider sections across the mooring array, and over the upper 200 m of the water column (After [15])

differ most on the inshore (western) flank of the Kuroshio, where models suggest the flow includes a significant ageostrophic component. The close correspondence between directly measured and glider-based geostrophic currents lends confidence when calculating geostrophic transports from sections sampled by gliders.

3.2 *Process Studies*

Gliders have been successfully applied to a range of process studies, often as components of integrated observing systems composed of several platforms that provide complementary capabilities. For example, gliders can quantify spatial structure surrounding fixed moorings, which provide high-resolution time series and, when needed, measurements in the atmospheric boundary layer. The US National Aeronautics and Space Administration (NASA) Salinity Processes in the Upper Ocean-Regional Study (SPURS) used this strategy to resolve the processes that govern upper ocean salinity in the region of the sea surface salinity maximum in the subtropical North Atlantic [29]. A trio of surface moorings collected meteorological measurements and highly resolved time series of the vertical structure of salinity, temperature, and velocity, while three underwater gliders and three autonomous surface vehicles surveyed a roughly 100 km by 100 km region around the central mooring array (Fig. 7). Farrar et al. [30] employed the resulting combination of high-resolution spatial and temporal sampling to assess mixed-layer salt and heat budgets over a 6-month period, computing surface flux, turbulent flux, entrainment, horizontal advection, and time rate of change of heat and salt within a mixed layer control volume (Fig. 8; [30]). Persistent spatial sampling provided by gliders allowed the quantification of advective fluxes over the entire 6-month span, demonstrating that horizontal advection controlled salinity variability at monthly timescales. Similar combinations of fixed platforms and gliders have yielded novel results

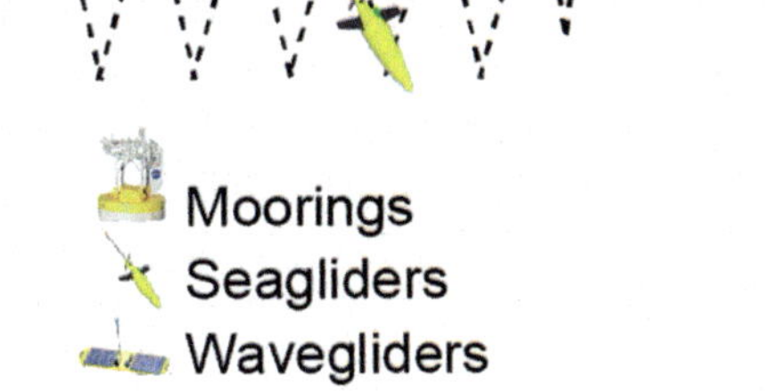

Fig. 7 SPURS sampling strategy, with underwater gliders and autonomous surface vehicles sampling spatial variability around heavily instrumented surface moorings. Sampling footprint was designed to roughly match the footprint of an Aquarius satellite pixel

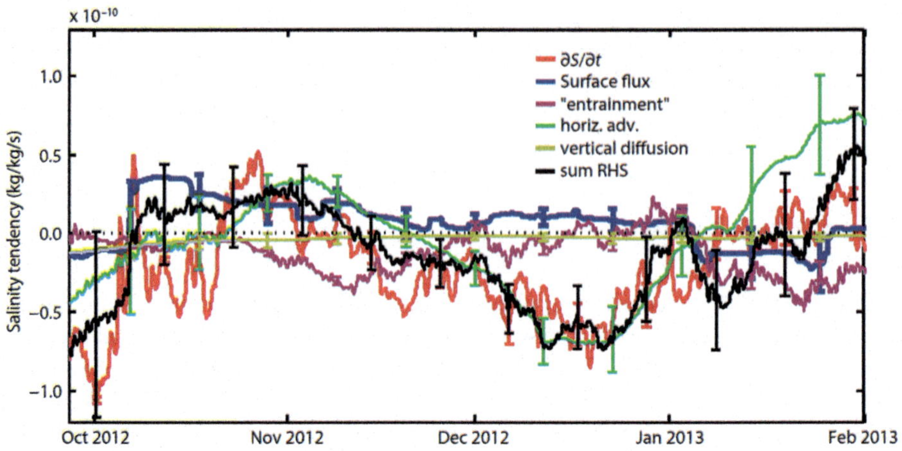

Fig. 8 Terms of the salinity balance integrated over the mixed layer. Periods where the *black* and *red* error bars overlap indicate closure to within error bounds (From [30])

by pairing the gliders' ability to make sustained measurements of spatial variability to the fine temporal resolution and large range of sensors provided by mooring arrays [15].

3.3 *Biology and Biogeochemistry*

Development of small, low-power sensors for nitrate, oxygen, and bio-optical properties has enabled gliders and floats for studies of biological and biogeochemical processes. Gliders sample horizontal and vertical scales that are sufficient to resolve important processes such as phytoplankton blooms and carbon export from the euphotic zone, and provide the persistent sampling required to capture episodic events. For example, a 2008 experiment employed four gliders sampling around a Lagrangian float, along with ship-based sampling, to characterize the spring phytoplankton bloom in the subpolar North Atlantic. Results include identification of the processes that drove bloom initiation [31], quantification of the associated net community production [32, 33], and characterization of subsequent export due to aggregation and sinking [34] and eddy-driven subduction [35]. Small numbers of detailed ship-based biological and chemical measurements were used to develop proxies for leveraging the far more numerous measurements collected by sensors carried on gliders and floats. Examples include fluorescence to chlorophyll concentration, optical backscatter and light transmission to particulate organic carbon [36] and the ratio of fluorescence to optical backscatter as a measure of phytoplankton community composition [37]. By using these proxies, biological and biogeochemical parameters could be estimated over the broad range of temporal and spatial scales sampled by the system of float and gliders. Quantitative results relied on the development of these proxies and on extensive efforts to ensure calibration and cross-calibration across the ensemble of sensors.

3.4 Polar Regions

The mobility and persistence of gliders can be used to overcome the challenges presented by difficult environments. The Arctic represents an extreme case, where geographical remoteness, harsh operating conditions, and ice cover, which curtails access to satellite services (for data/control telemetry and geolocation), make observational efforts risky, difficult, and costly. Science objectives often demand measurements in locations that are difficult or, during some parts of the year, impossible to access using conventional approaches. Many important processes occur near the ice-ocean interface, where instruments risk destructive interaction with deep keels formed by rafting ice and encounters with icebergs. Gliders adapted for working in ice-covered environments can sample from open water, through the marginal ice zone (which poses difficulties for instruments that rely on the ice for a platform) and deep into the pack ice. When operating beneath the ice, gliders navigate by multilateration from an array of O (1 kHz) acoustic beacons, which can either be bottom-moored or suspended from, and thus drifting with, the ice. Although acoustic data transfer is possible, ranges are short (kilometers) and transmission requires significant energy. Instead, gliders combine an altimeter and near-surface temperature measurements to distinguish between overhead ice and open water, thus determining when to surface to communicate via satellite modem.

Ice-capable gliders have been used to occupy year-round sections across Davis Strait, at the southern end of Baffin Bay [38], and as part of a drifting, integrated observing system employed by the US Office of Naval Research Marginal Ice Zone (MIZ) [39]. The MIZ experiment focused on understanding how the processes that govern sea ice melt vary as a function of open water fraction, distance from the ice edge, and open water extent to the south (which modulates surface wave activity). The program thus required sustained, multimonth sampling through an entire melt season, with the ability to follow the ice edge through its northward retreat while resolving variability under partial and full ice cover at scales of a few kilometers. Gliders, navigating in real time from an array of broadband 900 Hz acoustic sources drifting with the ice, occupied repeated sections from open water, through the MIZ and into the pack (Fig. 9a), providing finely-spaced profiles in the critical transition region to complement time series collected by ice-based instruments within the pack. A section taken across the MIZ in late September (Fig. 9b, c), when the ice is advancing southward, shortly after the start of freeze-up, reveals small-scale T-S variability near the ice edge, indicative of stirring, with the freshest surface waters to the north, likely the result of late summer melt capped by the advancing ice. Here, the mobility and endurance of gliders allow them to follow the MIZ, maintaining focus on this dynamic region through the multimonth span of the experiment.

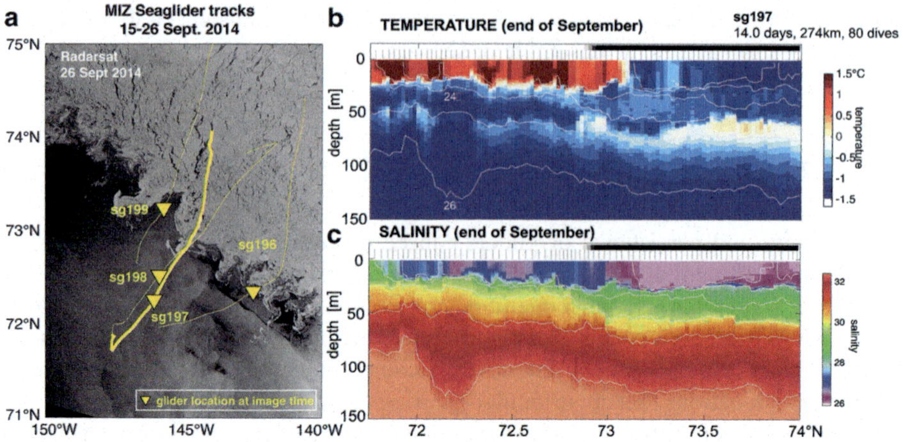

Fig. 9 (**a**) Glider sections through the marginal ice zone, with *yellow lines* marking the tracks of four gliders, and the *thick line* indicating the plotted section. *Yellow triangles* mark the position of gliders at the time of the 26 September Radarsat image plotted in the background, with all gliders moving southward. *Light gray* in the upper half of the image indicates ice, *while darker* colors denote open water. Late September sections of (**b**) temperature and (**c**) salinity for the upper portion of the glider profiles. The bar across the top of each panel indicates ice concentration, with *black* denoting full ice cover and *white* indicating open water. Thin *white lines* mark isopycnals

4 Lessons and Future Directions

Gliders provide access to spatial and temporal scales that have previously been difficult to address. Specifically, they permit sampling at scales of kilometers and hours, while maintaining a persistent presence over time spans of seasons to decades. This makes them well suited to tasks that include sampling across strong lateral gradients, such as boundary currents and eddies, capturing small-scale, episodic processes such as phytoplankton blooms and carbon export events, and quantifying climate variability. When used efficiently, gliders deliver profiles at per-profile costs that are roughly equivalent to floats, but finely spaced along track lines due to the nature of the platform. This complements the distributed sampling provided by profiling Argo floats, illustrated by the use of gliders to connect the coastal zone with the open basin in the CUGN (Fig. 4). Similar synergies exist between gliders and other platforms optimized toward different regions of the spatial-temporal sampling spectrum. Gliders are but one tool in the oceanographer's toolbox, and can be valuable components of integrated, multiplatform observing approaches.

Operational modalities, reliability, ease of use, and capital cost determine efficiency and scalability of gliders as an observing tool. The ability to sustain continuous operations in an efficient, cost-effective manner relies on the use of local logistics and small, flexible vessels for deployments and recoveries, thus favoring use of gliders in the boundary regions and near islands. Improvements to reliability and ease of use could ultimately allow gliders to adopt the diverse deployment strategies used by the Argo float and surface drifter communities, where volunteers with little training successfully deploy instruments from a wide range of vessels of opportunity. This would provide a path for deploying instruments over broad swaths of the global ocean. Improved reliability also reduces costs by facilitating longer missions. Efforts to reduce capital costs may ultimately be constrained by payload, as sensors can represent a significant fraction of vehicle cost.

Although gliders have demonstrated success in diverse science applications, efforts to establish national and international observing programs around gliders are still in their infancy. The key challenge is to identify and articulate compelling, coherent science objectives that require networks of gliders to achieve. Establishment of a boundary current observing system, complementing Argo's distributed observations of the basin interiors, offers a potential objective that has clear scientific and societal value. The fact that the existing, sustained glider operations focus on boundary regions speaks to the need for these observations.

Looking forward, development efforts directed at improving reliability, ease of use, and range/endurance would have the largest impacts on platform efficiency. A continued push for new sensor development aimed to meet the severe size and power constraints imposed by gliders would make them useful for an increasingly broad range of science. Finally, efforts should be invested to broaden access to the technology, both by improving ease of use and by making hardware and expertise more readily available.

Acknowledgments Initial glider development was supported by the US Office of Naval Research (ONR), with subsequent support from the US National Science Foundation, the US National Oceanographic and Atmospheric Administration (NOAA), and the US Office of Naval Research. The technology development and science reported here represent that work of a large number of people from institutions that include Oregon State University, University of Hawaii, University of Western Australia, Rutgers University, Woods Hole Oceanographic Institution, Scripps Institution of Oceanography, and the University of Washington. The authors thank Luc Rainville and Barry Ma for their contributions to this chapter. This chapter was prepared with support from the ONR under grants N00014-13-1-0478 (CML), N00014-13-1-0478 (CML), N00014-13-1-0483 (DLR), and NOAA Ocean Observing and Monitoring Division under grant NA15OAR4320071 (DLR).

References

1. Roemmich D, Johnson GC, Riser S, Davis R, Gilson J et al (2009) The Argo program observing the global ocean with profiling floats. Oceanography 22:34–43
2. Stommel H (1989) The Slocum mission. Oceanography 2(1):22–25
3. Webb DC, Simonetti PJ, Jones CP (2001) SLOCUM: an underwater glider propelled by environmental energy. IEEE J Ocean Eng 26:447–452
4. Rudnick DL (2016) Ocean research enabled by underwater gliders. Annu Rev Mar Sci 8:9.1–9.23. https://doi.org/10.1146/annurev-marine-122414-033913
5. Sherman J, Davis RE, Owens WB, Valdes J (2001) The autonomous underwater glider "Spray.". IEEE J Ocean Eng 26:437–446
6. Eriksen CC, Osse TJ, Light RD, Wen T, Lehman TW et al (2001) Seaglider: a long-range autonomous underwater vehicle for oceanographic research. IEEE J Ocean Eng 26:424–436

7. Eriksen CC, Rhines PB (2008) Convective to gyre-scale dynamics: Seaglider campaigns in the Labrador Sea 2003–2005. In: Dickson RR, Meincke J, Rhines PB (eds) Arctic-Subarctic ocean fluxes: defining the role of the northern seas in climate. Springer, Dordrecht, Neth, pp 613–628
8. Martin JP, Lee CM, Eriksen CC, Ladd C, Kachel NB (2009) Glider observations of kinematics in a Gulf of Alaska eddy. J Geophys Res 114:C12021. https://doi.org/10.1029/2008JC005231
9. Davis RE, Ohman MD, Rudnick DL, Sherman JT, Hodges B (2008) Glider surveillance of physics and biology in the southern California Current system. Limnol Oceanogr 53:2151–2168
10. Rudnick DL, Zaba KD, Todd RE, Davis RE (2017) A climatology of the California Current System from a network of underwater gliders. Prog Oceanogr 154:64–106
11. Pelland NA, Eriksen CC, Lee CM (2013) Subthermocline eddies over the Washington continental slope as observed by Seagliders, 2003–09. J Phys Oceanogr 43:2025–2053
12. Castelao R, Glenn S, Schofield O, Chant R, Wilkin J, Kohut J (2008) Seasonal evolution of hydrographic fields in the central Middle Atlantic Bight from glider observations. Geophys Res Lett 35(L03617). https://doi.org/10.1029/2007GL032335
13. Davis RE, Kessler WS, Sherman JT (2012) Gliders measure western boundary current transport from the South Pacific to the Equator. J Phys Oceanogr 42:2001–2013
14. Rudnick DL, Jan S, Centurioni L, Lee C, Lien R-C et al (2011) Seasonal and mesoscale variability of the Kuroshio near its origin. Oceanography 24(4):52–63
15. Lien R-C, Ma B, Cheng Y-H, Ho CR, Qiu B, Lee CM, Chang M-H (2014) Modulation of Kuroshio transport by Mesoscale eddies at the Luzon Strait entrance. J Geophys Res Oceans 119. https://doi.org/10.1002/2013JC009548
16. Perry MJ, Sackmann BS, Eriksen CC, Lee CM (2008) Seaglider observations of blooms and subsurface chlorophyll maxima off the Washington coast. Limnol Oceanogr 53:2169–2179
17. Rudnick DL, Cole ST (2011) On sampling the ocean using underwater gliders. J Geophys Res 116:C08010
18. Todd RE, Rudnick DL, Davis RE, Ohman MD (2011a) Underwater gliders reveal rapid arrival of El Niño effects off California's coast. Geophys Res Lett 38:L03609
19. Todd RE, Rudnick DL, Sherman JT, Owens WB, George L (2017) Absolute velocity estimates from autonomous underwater gliders equipped with doppler current profilers. J Atmos Ocean Technol 34:309–333. https://doi.org/10.1175/JTECH-D-16-0156.1
20. McClatchie S (2014) Regional fisheries oceanography of the California Current System. Springer, New York, 235 pp
21. Todd RE, Rudnick DL, Mazloff MR, Davis RE, Cornuelle BD (2011b) Poleward flows in the southern California Current System: glider observations and numerical simulation. J Geophys Res 116:C02026
22. Zaba KD, Rudnick DL (2016) The 2014–2015 warming anomaly in the Southern California Current System observed by underwater gliders. Geophys Res Lett 43:1241–1248
23. Jacox MG, Hazen EL, Zaba KD, Rudnick DL, Edwards CA et al (2016) Impacts of the 2015–2016 El Niño on the California Current System: early assessment and comparison to past events. Geophys Res Lett 43:7072–7080
24. Todd RE, Rudnick DL, Mazloff MR, Cornuelle BD, Davis RE (2012) Thermohaline structure in the California Current System: observations and modeling of spice variance. J Geophys Res 117:C02008
25. Kurapov AB, Pelland NA, Rudnick DL (2017) Seasonal and interannual variability in along-slope oceanic properties off the US West Coast: inferences from a high-resolution regional model. J Geophys Res Oceans 122. doi: 10.1002/2017JC012721.
26. McClatchie S, Cowen R, Nieto K, Greer A, Luo JY et al (2012) Resolution of fine biological structure including small narcomedusae across a front in the Southern California Bight. J Geophys Res 117:C04020
27. Ohman MD, Rudnick DL, Chekalyuk A, Davis RE, Feely RA et al (2013) Autonomous ocean measurements in the California Current Ecosystem. Oceanography 26:18–25
28. Bretherton FP, Davis RE, Fandry CB (1976) A technique for objective analysis and design of oceanographic experiments applied to MODE-73. Deep-Sea Res 23:559–582

29. Lindstrom E, Bryan F, Schmitt R (2015) SPURS: salinity processes in the upper-ocean regional study—the North Atlantic experiment. Oceanography 28(1):14–19. https://dx.doi.org/10.5670/oceanog.2015.01
30. Farrar JT, Rainville L, Plueddemann AJ, Kessler WS, Lee C, Hodges BA, Schmitt RW, Edson JB, Riser SC, Eriksen CC, Fratantoni DM (2015) Salinity and temperature balances at the SPURS central mooring during fall and winter. Oceanography 28(1):56–65. https://dx.doi.org/10.5670/oceanog.2015.06
31. Mahadevan A, D'Asaro E, Lee CM, Perry MJ (2012) Eddy-driven stratification initiates the North Atlantic spring bloom. Science 337:54–58. https://doi.org/10.1126/science.1218740
32. Alkire MB, D'Asaro E, Lee CM, Perry MJ, Gray A, Cetinic I, Briggs N, Kallin E, Kaiser J, Gonzalez-Posada A (2012) Estimates of net community production and export using high-resolution, Lagrangian measurements of O2, NO3−, and POC through the evolution of a spring diatom bloom in the North Atlantic. Deep Sea Res Part 1 Oceanogr Res Pap. https://doi.org/10.1016/j.dsr.2012.01.012
33. Alkire MB, Lee CM, D'Asaro EA, Perry MJ, Briggs N, Cetinić I, Gray A (2014) Net community production and export from Seaglider measurements in the North Atlantic after the spring bloom. J Geophys Res 119. https://doi.org/10.1002/2014JC010105
34. Briggs N, Perry MJ, Cetinić I, Lee C, Gray AM, Rehm E (2011) High-resolution observations of aggregate flux during a sub-polar North Atlantic spring bloom. Deep Sea Res Part 1 Oceanogr Res Pap. https://doi.org/10.1016/j.dsr.2011.07.007
35. Omand MM, D'Asaro EA, Lee CM, Perry MJ, Briggs N, Cetinic I, Mahadevan A (2015) Eddy-driven subduction exports particulate organic carbon from the spring bloom. Science. https://doi.org/10.1126/science.1260062
36. Cetinić I, Perry MJ, Briggs NT, Kallin E, D'Asaro EA, Lee CM (2012) Particulate organic carbon and inherent optical properties during 2008 North Atlantic Bloom Experiment. J Geophys Res 117:C06028. https://doi.org/10.1029/2011JC007771
37. Cetinić I, Perry MJ, D'Asaro E, Briggs N, Poulton N, Sieracki ME, Lee CM (2015) A simple optical index shows spatial and temporal heterogeneity in phytoplankton community composition during the 2008 North Atlantic Bloom experiment. Biogeosciences. https://doi.org/10.5194/bg-12-1-2015
38. Curry B, Lee CM, Petrie B, Moritz R, Kwok R (2013) Multi-year volume, liquid freshwater, and sea ice transports through Davis Strait, 2004-2010. J Phys Oceanogr. https://doi.org/10.1175/JPO-D-13-0177.1
39. Lee, C.M., J. Thomson, and the Marginal Ice Zone and Arctic Sea State Teams. 2017. An autonomous approach to observing the seasonal ice zone in the western Arctic. Oceanography 30(2):56–68, https://doi.org/10.5670/oceanog.2017.222

Advances in In-Situ Ocean Measurements

David Murphy and Carol Janzen

Abstract This chapter focuses on recent advances in in-situ ocean measurements. Recent interest in the ocean's response to and impact on climate change has encouraged the development of improved sensor technologies for measuring oceanic parameters such as conductivity, temperature, dissolved oxygen and pH. It introduces various sensors used for measuring oceanic parameters, the underlying principles of these sensors and their respective calibration parameters. It also discusses what is still needed in the development of sensors to achieve the oceanographic need.

1 Introduction

Recent interest in the ocean's response to and impact on climate change has encouraged the development of improved sensor technologies for measuring ocean pressure, conductivity, temperature, dissolved oxygen and pH. The state-of-the-art design criteria for these sensors must consider not only the capability of the underlying sensing technology, but also its practical implementation in the field. These practicalities include the ability to perform adequate static calibrations in a lab as well as acceptable response characteristics and adequate sensor stability (calibration drift) during long deployments.

As with all technologies, continued improvement is driven by customer needs. For more mature sensing technologies that measure the core oceanographic parameters of pressure and temperature, recent improvements in static calibration, response characteristics as well as electronic and calibration stability promise to deliver higher accuracy and longer-term stability. New sensing technologies offer improvements in both our ability to make measurements and in sensor implementations.

Platform technology developments in recent decades have also enhanced our ability to observe the oceans on large scales and to monitor ocean conditions with an enhanced temporal and spatial resolution. The transition to platforms capable of

D. Murphy (✉)
Sea-Bird Scientific, Bellevue, WA, USA
e-mail: dmurphy@seabird.com

C. Janzen
The Alaska Ocean Observing System, Anchorage, AK, USA

R. Venkatesan et al. (eds.), *Observing the Oceans in Real Time*, Springer Oceanography, DOI 10.1007/978-3-319-66493-4_8

making autonomous ocean measurements has moved the oceanographic community from relying on calibrated instrument systems onboard ships to unattended installations used in long-term deployments.

Platform technologies currently deployed as part of most modern ocean observing programs include the following:

- Shipboard underway continuous sampling devices run for days to months
- Gliders, AUVs and drifters deployed for weeks to months (surface and subsurface)
- Moorings deployed for up to 2 years
- Argo profiling floats, which have proposed life spans of 5+ years and include recent Deep Argo floats that can measure to full ocean depth

Developing sensors that can maintain static calibration on these platforms poses a big challenge to sensor manufacturers. The question this brings to bear is how good is good enough for measurements made as part of these deployments? Long-term trend and climate change monitoring for both the coastal and deep ocean will require part per million accuracies over many years. This is well illustrated by salinity changes in Gulf of Alaska near Seward which experiences the largest freshwater input in the North Pacific. Based on a 35 year observation record, the practical salinity has freshened by 0.14, which equates to a practical salinity slope of −0.004/year [16]. Oceanic oxygen trends reported thus far are on the order of few umol kg^{-1} $year^{-1}$, which though significant to ocean chemistry, is less than a 1% change in dissolved oxygen per year [8].

What this translates to for sensor developers is that data quality requirements have gone from three decimal place initial static accuracy (e.g. 0.002 in salinity) for a shipboard research cruise (typically with daily water samples to augment annual calibrations) to attempting to achieve three decimal place stability over 5 years without any direct validation or calibration. This requirement is at least 100 times more demanding in stability than typical ship-based sensors. Enhanced electronic stability and durability are additional requirements. To achieve accuracy in all measurements, it must be possible to correct data from each sensor on moving platforms for dynamic errors. Dynamic errors occur whenever continuous measurements are made while traversing spatial gradients, such as a thermocline.

This chapter offers a brief review of the state-of-the-art technologies used for measuring these core parameters and discusses what is still needed to achieve the oceanographic needs.

2 Conductivity Sensors

Oceanographic conductivity sensors do not measure conductivity directly, but rather measure the conductance of a volume of seawater from which conductivity is calculated (Eq. 1):

$$R = \frac{\rho L}{A} \quad (1)$$

where

R = resistance = 1/conductance
ρ = resistivity = 1/conductivity
L = length of sampled water volume
A = cross-sectional area of sampled water volume

Existing conductivity sensors use electrodes or transformers to make an electrical connection with seawater. The electrode method is simpler, but requires either four electrodes or electrodes with stable and low inherent resistance. The transformer (inductive) method uses one transformer to couple a known voltage signal to the seawater and a second to detect the flow of current in its core. The inductive method is more complex, because the transformers have unknown characteristics arising from the copper and core components.

The challenges to making an accurate conductivity measurement are as follows:

- A stable electrical connection to the seawater, which is difficult but not limiting
- An accurate resistance measurement for the seawater in the cell, which is not difficult
- A stable cell geometry over the oceanographic temperature and pressure range, which is a limiting factor

The 'cell' is the volume of seawater that contributes to the measurement. Therefore, maintaining stable cell geometry is the most limiting factor in seawater conductivity measurements. This is especially true for sensors having external electrical fields, such as the most inductive cell technologies.

Conductivity cells can be constructed to have either internal or external electrical fields. Figure 1 shows the cross section of the internal conductivity cell, such as those manufactured by Sea-Bird Electronics, where the sample volume and the electrical field are contained within the glass tube and the two outer electrodes.

The two outer electrodes are electrically connected; so, no voltage difference can exist between them and no external electrical field can develop outside this volume or the cell. The sensor is constructed as a two-terminal device in which the electrode resistance is in series with the cell (seawater) resistance. The electrode resistance is low and the cell resistance is high; therefore, errors caused by changes in electrode properties are minimal, and the electrical field between the

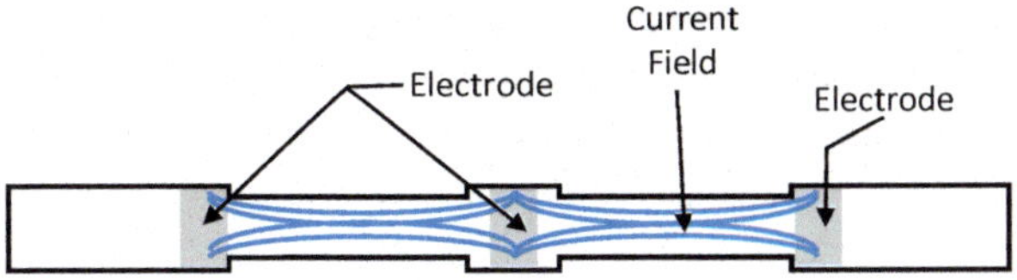

Fig. 1 Cross section of an internal electrode conductive conductivity cell

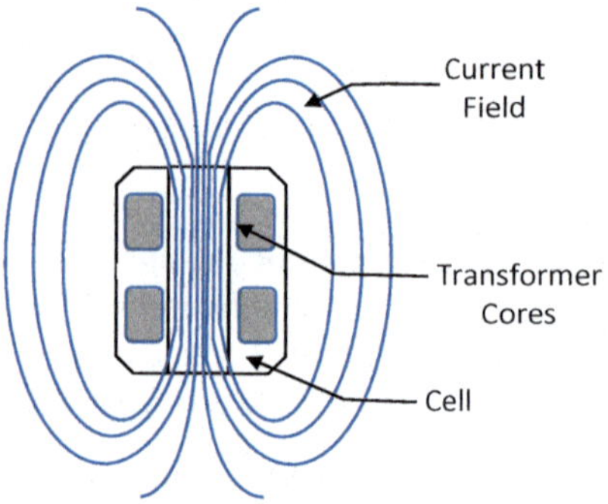

Fig. 2 Cross section of an inductive type conductivity cell

electrodes is completely contained within the tube, making the sensor immune to errors caused by the proximity to external objects.

Figure 2 shows a cross section of a typical inductive conductivity cell.

The inductive sensor operates by passing a known current through the drive transformer which induces an electric current in the seawater that passes through the central hole in the cell. The induced current in the seawater in turn induces a current in the signal transformer which is measured and is proportional to the conductance of the seawater. The magnitude of both induced currents depends solely on the electrical field density of the field paths through the hole in the cell. Typically, 20% of the seawater conductance is measured external to the central hole of the inductive cell. This significant dependence on the external portion of the electrical field to the overall accuracy of the measurement can result in significant errors in conductivity caused by nearby objects that distort the external field. To illustrate the impact of this, if the goal of salinity accuracy is 0.0035, a dimensional stability of 1 part in 10,000 would be required, which is challenging to achieve. This dimensional stability must be maintained despite distortion from temperature and pressure fluctuations and coatings from sea surface oils, mineral deposits, marine growth and the application of antifoulants.

2.1 Conductivity Sensor Metrology and Calibration

Ocean salinity (via conductivity) metrology relies on International Association for the Physical Sciences of the Ocean (IAPSO) standard seawater manufactured by Ocean Science International Limited. Bacon et al. [2] found the uncertainty in IAPSO standard seawater conductivity ratio to be 1×10^{-5} through batch P144 with no discernible batch-to-batch variability. Although other researchers have observed some batch-to-batch variability, the performance of the standard still exceeds the ±0.002 accuracy estimate for the practical salinity scale.

2.2 Conductivity Sensor Response Characteristics

Conductivity sensor response time is dependent on the flow speed of water passing through the cell. This can be reported as a direct relationship to the speed at which the sensor is moving through the water column, or 'fall speed'. For example, many inductive sensor manufacturers report the response time of their conductivity cells at a fixed fall speed of 1 m/s. However, it is difficult to control the speed of a profiling CTD or other moving platforms like a glider, making it problematic to maintain a fixed response time for conductivity sensors that depend on the speed of the platform carrying it.

Since inductive conductivity cells maintain part of their measurement field outside of the central core of the cell, attaching plumbing and pumps to inductive sensors to provide a constant flow of water through the sensor is not possible, as doing so will cause errors in the overall measured signal.

Internal electrode conductive sensors, on the other hand, have their entire electrical field contained inside the cell. Therefore, this type of sensor can have plumbing and pumps attached to them to provide flow control that constrains the response time of the sensor. Internal electrode sensors are typically pumped externally on moving platforms for this reason, and can take advantage of constant pump speeds to control the actual response time of the sensor while it is moving through the water column.

Pumping water through a sensor provides a number of other benefits that improve the dynamic response of the sensor. Dynamic errors are largely a problem on moving platforms, but can also occur on moorings where water conditions are changing on rapid timescales in and around the mooring (e.g. passage of internal wave packets). Dynamic errors are important to recognize because computed parameters (variables) require that different sensor measurements be made simultaneously on the same parcel of water. A mismatch in sensor response time or in the water parcel being sampled between two sensors can cause significant errors in computed quantities. For example, temperature (T), conductivity (C) and pressure (P) of a given water parcel are required to compute salinity (S); T, S, P and partial pressure of oxygen gas are needed to derive dissolved oxygen concentration; T, S, P and pH-voltage are necessary to compute an accurate pH value. The inherent mismatch in sensor response times or a mismatch in the actual water parcel being measured by each sensor is a significant cause of dynamic errors in oceanographic measurements. Salinity spiking is a common dynamic error and is caused by mismatches in T and C measurements (Fig. 3).

Conductivity cells experience another kind of dynamic error while traversing a thermal gradient due to a separate, time-dependent process caused by the sensor body material's ability to retain heat. The conductivity of seawater is very much temperature dependent, with up to 90% of the overall conductivity signal being temperature-driven. Only about 10% of a conductivity change is actually due to a change in salinity. Therefore, if the sensor does not equilibrate instantly to a new temperature encountered in the environment, the conductivity measurement will be skewed by the temperature memory of the sensor materials for as long as it

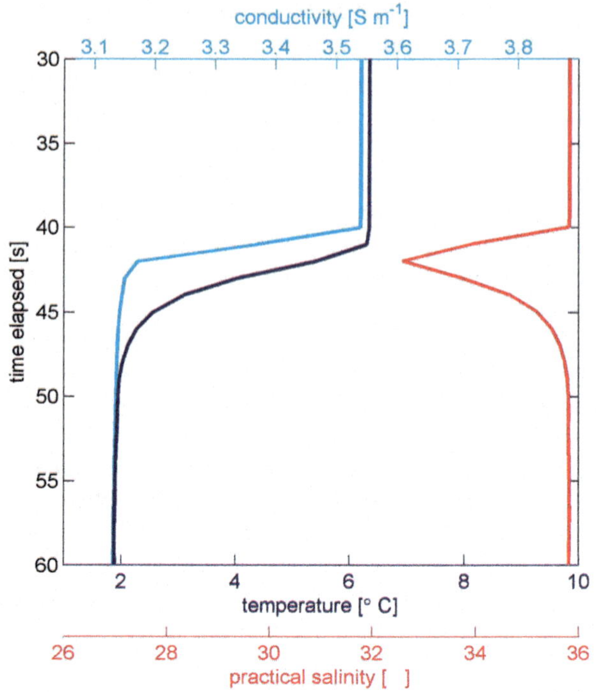

Fig. 3 Salinity spiking due to a mismatch in sensor response times between conductivity and temperature. In this example, conductivity leads temperature on a CTD, resulting in a negative salinity spike

takes for the sensor to equilibrate. This characteristic is referred to as the thermal mass error or conductivity thermal lag, and it causes errors in amplitude of the measurement and temporal response of the sensor. The thermal lag effect is exacerbated if flow through the conductivity cell is not constant, as the time lag of the thermal change experienced by the cell itself is also flow-dependent. Thermal lag errors in conductivity sensors are realized in conductivity but are more easily observed in computed salinity. The conductivity thermal lag error that results is therefore most often observed as a mismatch between upcast and downcast salinity profiles in a fixed location (Fig. 4).

2.3 *Conductivity Sensor Drift and Calibration Stability*

Conductivity cells are susceptible to fouling, resulting in unpredictable conductivity drift. Biofouling occurs in the form of organisms and oil films encountered on the water surface. Biofouling does not occur on every deployment. On moored platforms, biofouling can appear either as a linear change in the sensor accuracy with time due to steady biological growth on or near the sensors, or as an episodic change due to biofouling caused by biological settling or an increase in biological productivity.

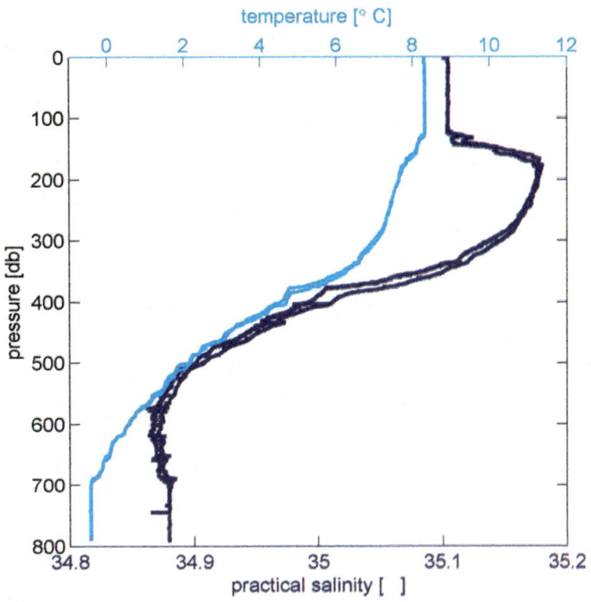

Fig. 4 Example of what a conductivity thermal lag error looks like in computed salinity. As the conductivity sensor traverses a thermal gradient from warm to cold water, it will read higher than true conductivity (and subsequently salinity) and lower than true conductivity when moving from cold to warm water (and hence lower in salinity). This creates a gap between the upcast and downcast

3 Temperature Sensors

High-accuracy temperature sensors can be made from platinum wire or semiconductor thermistors; both of these types change resistance in response to a change in temperature.

In the case of the platinum wire, a length of wire is coiled at the end of the probe and the resistance of the wire is measured using a bridge circuit and an analog-to-digital converter. This type of thermometer offers high stability and low drift but is subject to shifts in calibration due to a physical impact to the sensor. It also has a slower response time compared to thermistors and limitations in how small platinum thermometers may be constructed.

Thermistor-based thermometers are not susceptible to drift caused by physical impact and can be much smaller than platinum wire thermometers. The small size allows sensors to be fabricated that have very fast response times. Conversion of the resistance of the thermistor to digital form can be done either with an oscillator circuit and a frequency counter or with a bridge circuit and an analog-to-digital converter.

3.1 Temperature Sensor Metrology and Calibration

Two primary standards are used in oceanographic thermometry – a triple point of water cell and a gallium melt cell. The triple point of water cell is composed of pure water in three phases – vapor, liquid and solid – and with all three phases present is defined by the International Temperature Scale 1990 to be 0.01 °C. Gallium melt cells are used for the high end of the oceanographic temperature range, as pure gallium melts at a temperature of 29.7646 °C when operated with two phases, solid and liquid.

3.2 *Temperature Sensor Response Characteristics*

The temperature response time is largely determined by the physical size and construction of the sensor sheath surrounding the semiconductor. Figure 5 shows two sizes of thermistors used in oceanographic temperature sensors and Fig. 6 shows each sensor's step response. The sensor on the right side is made of a larger semiconductor bead and has a thicker sheath, thus giving it a slower response, with a 63% time constant on the order of 0.5 s. The smaller thermistor bead shown on the

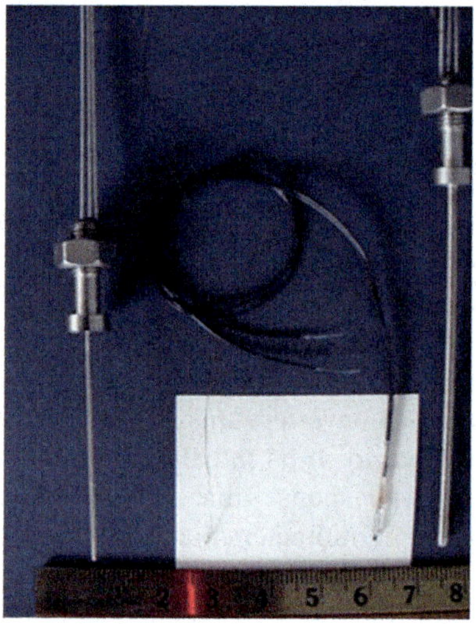

Fig. 5 Two thermistor-type temperature sensors: a larger semiconductor with a longer, thicker protective sheath (*right*) and a smaller semiconductor with a thinner shorter sheath (*left*)

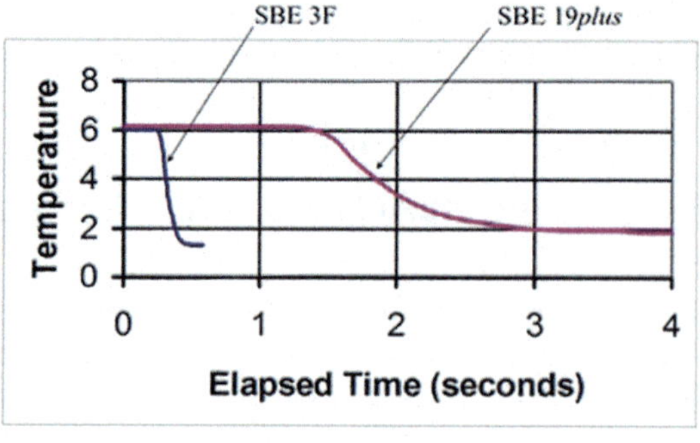

Fig. 6 The time response curves for the two temperature sensors shown in Fig. 5, (Sea-Bird Electronics SBE 3F and SBE 19plus) with *pink* representing the slower sensor shown on the right, and *blue* the faster sensor shown on the left

white background on the left side in Fig. 5 produces a faster response curve with a 63% time constant of approximately 0.070 s (70 ms).

3.3 Temperature Sensor Drift and Calibration Stability

Temperature sensors typically drift in offset with time. In a normal sensor, this drift will be unidirectional with age, and will be uniform across the measurement range, though this will depend on the manufacturer and the electronics in the sensor. When exposed to high heat (over about 50 °C), temperature sensors can experience a calibration drift in offset that is separate from age-related inherent drift in a sensor.

4 Pressure Sensors

Pressure measurements are used to compute in-situ depth in the ocean. The types of pressure sensors most commonly used in oceanographic instruments are strain gauge sensors (e.g. Druck, Kistler brands) and quartz oscillator devices (e.g. Paroscientific Digiquartz ®).

Quartz-based sensors are larger, require more power and are more expensive than comparable strain gauge sensors, and are chosen when accuracy requirements exceed what can be achieved with a strain gauge. They are routinely deployed on research vessel profiling systems, tide gauges, bottom pressure recorders and tsunami detectors.

The Paroscientific Digiquartz ® pressure sensors are considered the most accurate pressure sensors available on oceanographic instruments. Figure 7 shows the sensor's exterior and a cutaway view of the basic internal elements of the sensor.

The force of pressure from the environment acts on the interior of the bourdon tube either extending the tube or allowing it to relax, and this in turn applies force to the crystal tuning fork causing its frequency of oscillation to change. The balance weights shown in Fig. 7 are placed to locate the centre of inertia at the centre of the bourdon tube to minimize errors caused by changing the orientation of the sensor. The sensor has a secondary response to temperature, therefore the interior temperature of the pressure sensor must be measured.

Strain gauge sensors offer very good accuracy, compact size, low power requirements and better resistance to physical shock. They are commonly used in smaller shipboard profiling systems, moored time series sensors, moored profilers, Argo floats, gliders and autonomous underwater vehicles. As the name implies, strain gauge pressure sensors make use of a bridge circuit that incorporates a pressure or strain sensitive element as part of the bridge. These are typically not as accurate as a quartz sensor but have the advantages mentioned above. Figure 8 shows the exterior of a strain gauge pressure sensor and a cutaway view of the interior.

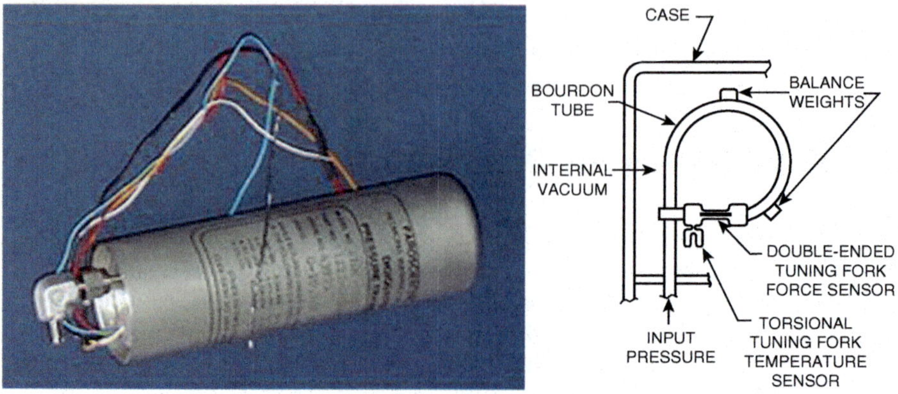

Fig. 7 External view and cutaway of Paroscientific Digiquartz ® pressure sensor

Fig. 8 External view and cutaway of strain gauge sensor

The pressure-sensitive part of the strain gauge bridge is made of piezoresistive material. A physical deflection of this material causes the bridge voltage to change proportionally. The pressure bridge has a secondary response to temperature; so, to achieve the most accurate pressure measurement, the interior temperature of the pressure sensor must be measured. These pressure devices are typically installed inside the instrument housing, and the changing pressure is transferred through a port on the exterior of the instrument.

4.1 Pressure Sensor Metrology and Calibration

In theory, the most rigorous standard to use for calibrating ocean pressure sensors is a National Institute of Standards certified dead weight tester. However, this equipment cannot be scaled up to meet the manufacturing requirements for producing the quantity of ocean sensors needed to satisfy community demand. Paroscientific Digiquartz ® pressure sensors are typically used as secondary standards for calibrating strain gauge sensors.

Calibrating pressure sensors in the lab and the field is complicated by their combined response to pressure and temperature. Pressure sensor accuracy is usually expressed as a percentage of full-scale range (FSR), for example, 0.1% FSR (Sea-Bird Electronics). To illustrate what this means, for Argo CTDs that operate to 2000 decibars, the specified error is ±2 decibars. For Deep Argo CTDs that operate to full ocean depth of 7000 decibars, the specified error is ±7 decibars. Figure 9 is a comparison of a shipboard SBE 9 CTD Digiquartz sensor with three Deep Argo CTDs (SBE 61) equipped with 7000 decibar strain gauge pressure sensors. These pressure sensors were calibrated over the full range of oceanic temperature and pressure conditions. The observed difference between the SBE 9 and three Argo pressure observations down deep is within ±4.5 decibars over 12 ocean casts to 6000 m, and at 0.05% of full-scale range.

4.2 *Pressure Sensor Response Characteristics*

Large step changes in pressure are not typically seen in the ocean environment when making measurements; therefore, in most cases, the pressure sensor time constant is not an issue. However, all pressure sensors are temperature-dependent; so, when moving through a water column with thermal gradients, dynamic errors in pressure are possible. On moving platforms, it is imperative to mitigate this dynamic response through instrument design and installation and by lessening temperature effects through calibration and validation checks (such as barometer checks in the lab and field).

Characterization of the sensor's static response to pressure and internal temperature is determined during factory calibrations. Strain gauge type sensors exhibit a quadratic response for both temperature and pressure, although this may be difficult to discern for the graph of sensor output versus pressure (Fig. 10).

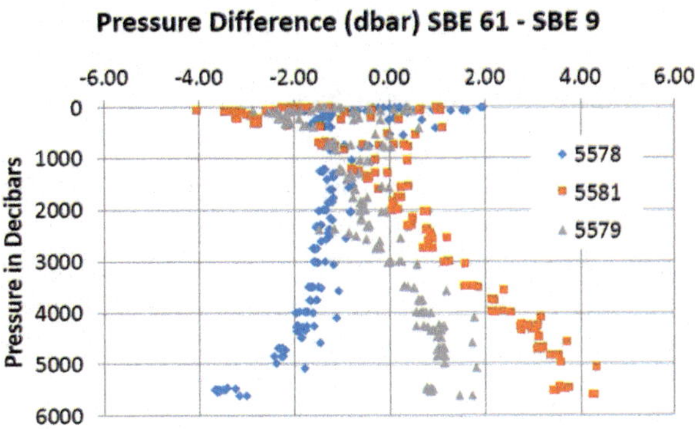

Fig. 9 Pressure sensor comparisons expressed as the difference between a calibrated shipboard CTD (*SBE 9*) equipped with a Paroscientific Digiquartz pressure sensor and three Deep Argo CTDs (*SBE 61*) equipped with strain gauge pressure sensors rated to 7000 dbar

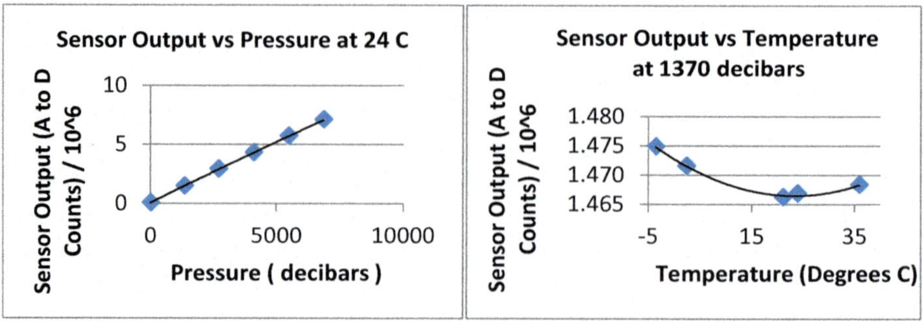

Fig. 10 Pressure sensor output versus pressure changes at 24 °C (*left*) and pressure sensor output versus temperature changes at 1370 decibars (*right*)

In oceanographic applications, pressure sensors are always mounted inside a waterproof housing. The pressure sensor's internal temperature depends on environmental conditions both inside and outside the housing and usually lags changes in in-situ ambient temperature. The strain gauge itself is subject to stress caused by uneven temperature across the gauge, and this effect shows up as a data artefact in conditions of rapidly changing temperature. This effect is also present in quartz sensors and is illustrated in Fig. 11, which shows the nearly equal and opposite response as the sensor is moved from warm to cold and from cold to warm conditions.

4.3 Pressure Sensor Drift and Calibration Stability

Well characterized and calibrated pressure sensors typically drift only in offset, meaning that the observed drift is constant across the measurement range. Drift is usually due to mechanical and material changes with time and use, and most of the sensor drift will occur during the early years of the sensor's life. Figure 12 shows typical Digiquartz sensor drift over a time period of 240 days amounting to a drift of less than a 1 decibar/year.

5 Dissolved Oxygen Sensors

Sensing technology for measuring dissolved oxygen concentrations in the ocean has improved dramatically in the past decade, and is advancing our understanding of climate change impacts on ocean circulation and ecology. There are two methods commonly used for sensing dissolved oxygen in aquatic environments and seawater. These are the polarographic method (electrochemical Clark cell) and the luminescent optical method.

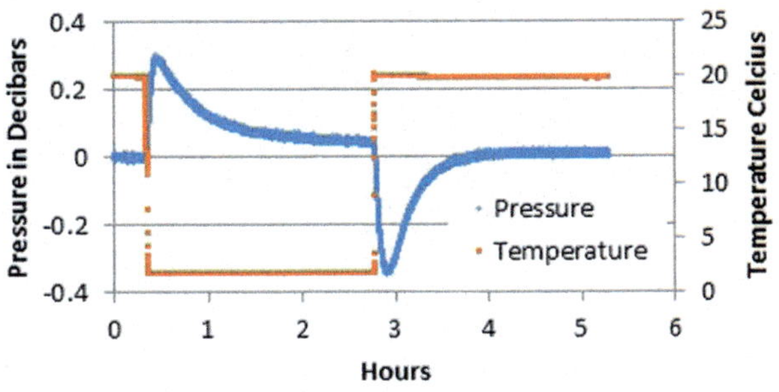

Fig. 11 Response of a quartz pressure sensor to a sudden temperature change. In this experiment, the quartz sensor is allowed to equilibrate in a water bath at room temperature, then is plunged into cold water and allowed to equilibrate for an equal amount of time. The sensor is then returned to the room temperature bath

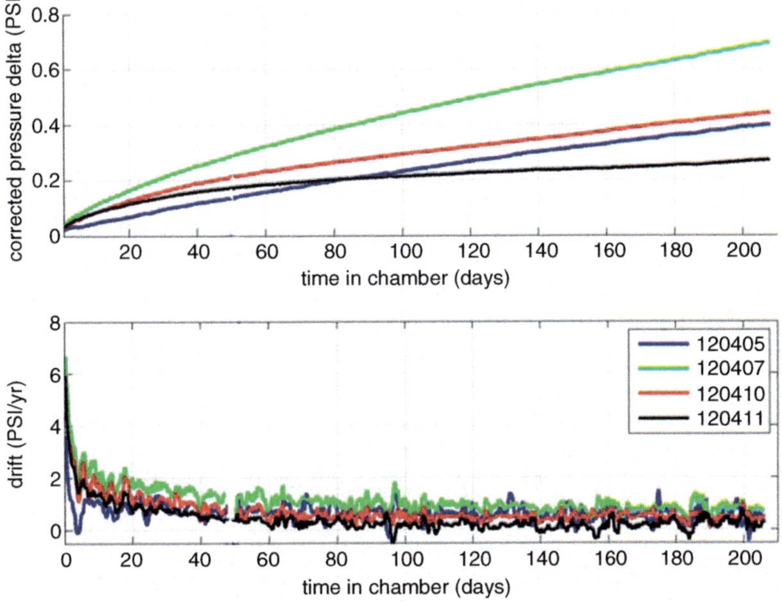

Fig. 12 Digiquartz pressure sensor offset drift for four sensors measured at 1 atmosphere. This plot shows a typical drift rate of less than 1 dbar per year

The Clark oxygen sensor has been in use for many years [5]. These sensors determine dissolved oxygen concentration by measuring the number of oxygen molecules per second (flux) that diffuse through a gas-permeable membrane from the ocean environment to the working electrode. At the working electrode (cathode), oxygen gas molecules are converted to hydroxyl ions (OH-) in a series of chemical reaction steps that produce four electrons. The sensor measures oxygen molecules

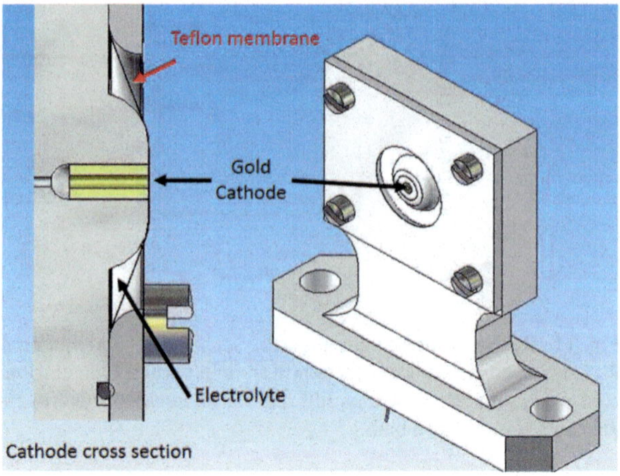

Fig. 13 Cutaway drawing of polarographic (Clark cell) oxygen sensor made by Sea-Bird Electronics (the SBE 43)

as electrons per second (amperes) delivered by the reaction. The measured flux of oxygen molecules is proportional to the partial pressure of oxygen in the ambient water, and this relationship is used to convert the sensor output to a meaningful oxygen concentration.

The cathode and electrolyte are covered by a film that has high gas permeability. Figure 13 shows a diagram of a Clark-type electrochemical sensor made by Sea-Bird Electronics (SBE 43), which uses a Teflon film for its membrane. Configurations among the various manufactured Clark electrode sensors vary in size, shape and materials, but the basic operating principles are very similar.

Optical oxygen sensors use a luminescent material that responds to excitation with blue light by emitting red light. These sensors have become more widely used in oceanographic work in the past decade as the technology has improved and sensor characterizations are better developed.

There are several types of luminescent materials that may be used in optical sensors. A common feature of each is that the specific luminescent material, or luminophore, is an organometallic molecule that absorbs energy in blue wavelengths and emits energy in red wavelengths. This emission is quenched by oxygen molecules that absorb the energy from the luminophore rather than emitting red photons. Figure 14 shows a cutaway diagram of Sea-Bird Electronic's optical oxygen sensor, the SBE 63.

While the intensity of the emitted red light can be measured and is proportional to the partial pressure of oxygen present, this is not the preferred measurement technique. The intensity measurements depend on the amount and condition of the luminophore on the optical window, which can change with time. A more robust measurement technique uses a sine wave modulated excitation signal and measures the phase difference between the excitation sine wave and the received emission sine wave (Fig. 15). This method produces measurements less effected by fouling and photobleaching.

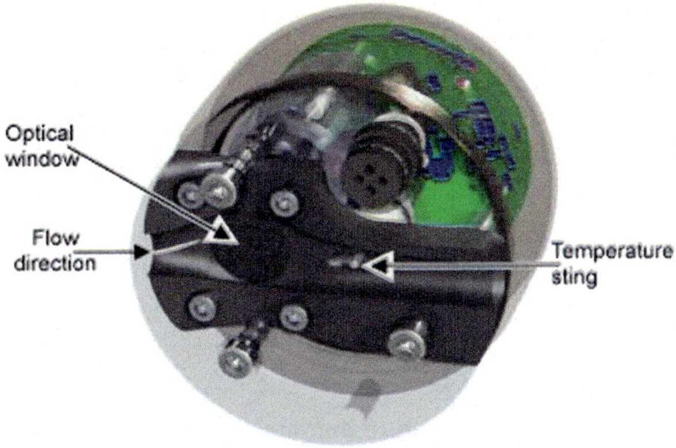

Fig. 14 Internal diagram of an optical oxygen sensor manufactured by Sea-Bird Electronics, the SBE 63

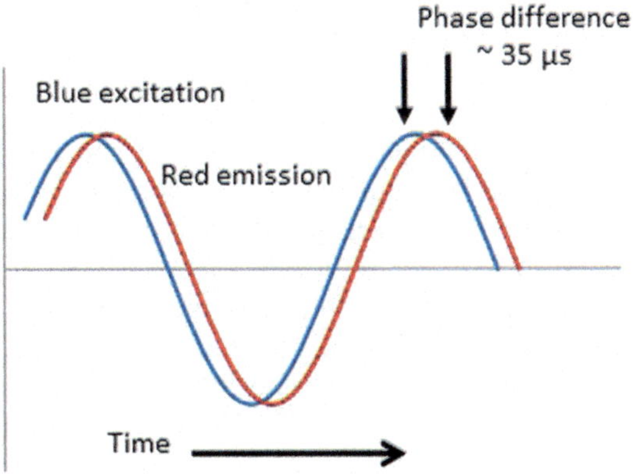

Fig. 15 Diagram of the phase-based measurement of oxygen concentration using an optical oxygen sensor

5.1 Dissolved Oxygen Sensor Metrology and Calibration

The standard to which all dissolved oxygen sensors are calibrated against in both the laboratory and the field is the Winkler titration of a water sample [12]. In factory calibrations, stable secondary standard oxygen sensors are designed for use in calibration baths which allow numerous calibration points to be made on a production scale. These secondary standards are calibrated frequently (weekly) with Winkler titrated water samples and cross-referenced with other secondary standards to ensure a specified level of accuracy for the given technology.

5.2 Dissolved Oxygen Response Characteristics

Both the electrochemical and the optical oxygen sensors have complex responses to oxygen partial pressure with secondary dependencies on temperature, pressure and salinity. In electrochemical sensors, the permeability of the membrane to oxygen is

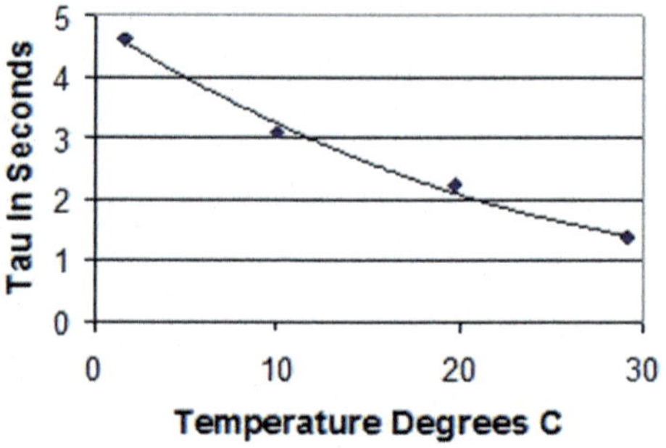

Fig. 16 Electrochemical sensor time constant versus temperature for the SBE 43

a function of temperature and ambient pressure, and these dependencies are taken into account in the sensor's calibration equation [7].

The electrochemical oxygen sensor response is flow-dependent and requires a fresh sample flowing over the membrane to make accurate and meaningful measurements. This requires that the sensor either be moving continuously through the water it samples, or that the sensor is pumped. Further, the time response of the sensor is also dependent on pressure and temperature [7]. To illustrate, Fig. 16 shows typical dependency of the electrode sensor's time constant on temperature.

Optical oxygen sensors are typically slower in response time than electrochemical sensors. Other response characteristics are similar, with one exception: the response time of the optical oxygen sensors is less flow-dependent and does not require a continual flow of water past the sensor to measure properly. That said, there is an equilibration time for a change in oxygen and temperature for optical sensors. Flushing optical sensors has been found to improve their response [3].

Plumbing and flushing of sensors were discussed in the previous section, but the same benefits hold here because flushing optimizes flow control, response characteristics of oxygen sensors, sensor response matching and water parcel alignment for other independent measurements required for making accurate oxygen calculations (T, S, P). The solubility of oxygen in the water is a function of temperature, salinity and pressure, and these dependencies are considered in the final algorithm used to compute oxygen concentration from both electrochemical and optical sensor output. The sum of these dependencies requires colocated measurements on the same water parcel of water temperature, salinity and pressure, with the oxygen sensor output. This can only be accurately achieved by plumbing sensors together and flushing with a constant pumped flow.

5.3 Dissolved Oxygen Sensor Drift and Calibration Stability

Electrochemical sensors typically drift over time towards lower sensitivity for two primary reasons: fouling of the gas-permeable membrane, which is the predominant cause, and changes in the electrolyte chemistry due to the accumulation of silver ions (Ag^{+}) dissolved into the solution as oxygen is consumed. The electrochemical

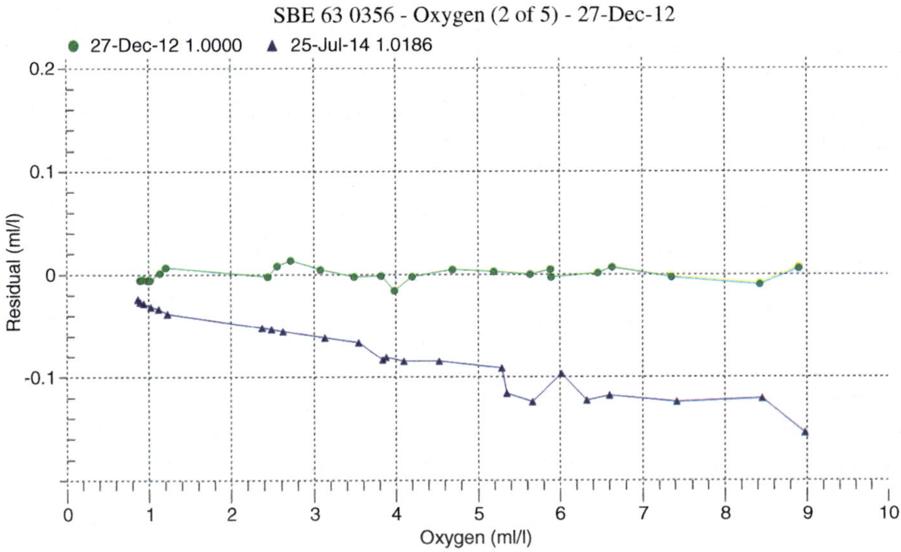

Fig. 17 A calibration drift plot for an SBE 63 optical oxygen sensor that was deployed on a marine mooring for 18 months

drift is accelerated (though still slow) at high oxygen concentrations and falls to zero when no oxygen is being consumed. Accordingly, sensor storage and deployment strategies that produce zero or near-zero oxygen environments when the sensor is not deployed can substantially reduce electrochemical drift.

Optical sensors have the advantage for long-term deployments due to their calibration stability both on the shelf and in the field (Fig. 17). Furthermore, the luminophore is less sensitive to fouling. The main drift mechanism in these sensors is photobleaching.

Although most optical sensors have slower response times (7–10 s versus 1–3 s for a modern electrochemical sensor), the facts that they incur less error caused by inherent and ubiquitous biofouling and experience no electrochemical drift outweigh this shortcoming for long-term deployments on moorings and slow-moving platforms.

6 pH Sensors

Measuring ocean pH in aquatic environments and in particular, the ocean, has benefited from the recent development of Ion-Specific Field Effect Transistors (ISFET), microfluidic and optical sensing techniques. In addition, research into seawater buffers for calibration has brought improved accuracy.

The technologies for the practical measurement of pH using commercially available instruments are still somewhat limited. The traditional electrode, optical and ISFET sensors are appropriate to all oceanographic platforms, while microfluidic sensors are being used primarily for collecting time series on moorings. At this time, a profiling microfluidic sensor is under development.

Measuring in-situ ocean pH with traditional bulb electrodes has the longest history of the current pH technologies on the market. These sensors are made of a thin bulb of a specific type of glass that has contact with the environment on the outside and a conductive solution surrounding an internal reference electrode made of silver chloride or calomel. The reference electrode has electrical contact with the external environment through a porous glass frit. The potential difference measured between these two electrodes exhibits a linear dependence on pH, and this relationship is used to calculate the pH value. Advantages with bulb electrodes are that they are well known, easy to calibrate and relatively inexpensive. The leading issue with this technology is its accuracy, which is not good enough for most oceanographic applications. Another major disadvantage is that although the response of the electrode is linear with respect to pH, the offset is not stable. The bulbs are also subject to fouling and are fragile, making them vulnerable in the harsh oceanic environment. Figure 18 shows a bulb-type pH electrode with the reference junction.

Recent research has led to the development of sensors using pH-sensitive ISFETs. These are specially constructed field effect transistors that have a hydrogen ion (H^+) sensitive gate that is exposed to the environment. The ISFET technology is based on the Honeywell Durafet™ pH sensor. Development efforts are focused on a pressure-tolerant mounting of the ISFET and the addition of a silver/silver chloride electrode. Because the reference electrode is sensitive to chloride ions, the salinity of the measured water must also be known. Figure 19 shows an ISFET and a reference electrode that are mounted for use on an Argo float.

The sensor has three components exposed to the environment, the ISFET, the reference electrode and a bias electrode. In operation, the ISFET and the external bias electrode are continuously powered in a feedback circuit. The electrical potential at the ISFET gate is the sum of the contribution from H^+ ions and the voltage developed by the electrical field of the bias electrode.

The measured pH signal is the voltage developed by the silver chloride reference electrode as the sum of the contribution from the chloride concentration in the environment and the electrical field voltage of the bias electrode. The chloride concentration is calculated using the salinity, and this allows the concentration of H^+ to be calculated [9].

Fig. 18 Bulb-type pH sensor

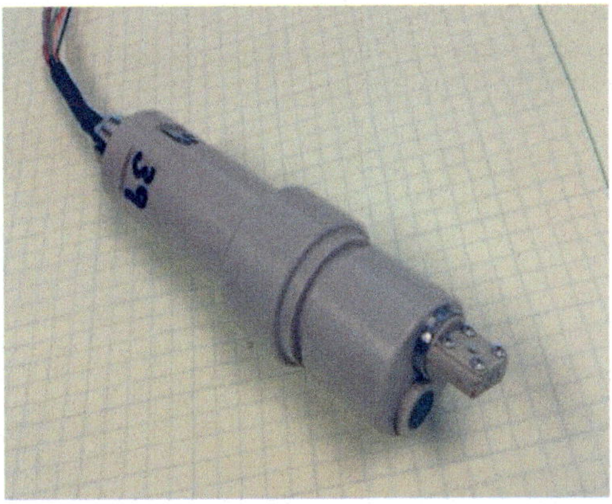

Fig. 19 ISFET pH sensor and reference mounted for deployment on an Argo float. The reference is the round *black part* that is beside the upright ISFET holder

Advantages of ISFET pH sensors are stability and sensitivity compared to bulb electrode pH sensors. The main disadvantage is that the reference electrode must equilibrate with local seawater, a process that requires up to a week after deployment.

The measurement of pH with microfluidics uses an indicator dye and spectrophotometric method for measurement of pH [13, 17]. A seawater sample is drawn into the instrument system and injected with meta-cresol purple indicator. The absorbance of the seawater-indicator solution is measured at two wavelengths that represent the protonated and unprotonated forms of the indicator. The ratio of these absorbances, along with temperature, salinity and pressure dependent properties of the indicator, is used to calculate pH. This technology is not well suited for freshwater or brackish water, and works best in a salinity range of 25–40. Microfluidics offers a highly accurate and stable measurement of in-situ pH requiring little or no field calibration. However, there are reagents required to make the measurement, and this limits the field time and imposes some handling constraints on the sensor. Figure 20 shows a microfluidic pH sensor.

Optical luminescence pH sensors, though not currently commercially available, are analogous to the optical oxygen sensing technology and show promise for use in oceanic and coastal applications. To date, optical pH has not performed well in field trials, and has experienced problems with salinity correction, photobleaching of the luminophore used for pH sensing and pressure hysteresis.

Several papers are available that provide additional details on these technologies, applications and successes, and quality requirements. A few of the more recent references are provided in references [1, 4, 11, 14, 15].

6.1 pH Sensor Metrology and Calibration

Current laboratory pH standards include freshwater buffer solutions made to a range of pH values and single pH seawater buffer solutions. Freshwater buffers are commonly available from scientific supply houses while seawater buffers are prepared in a few national laboratories and are offered as Certified Reference Material (CRM) [6].

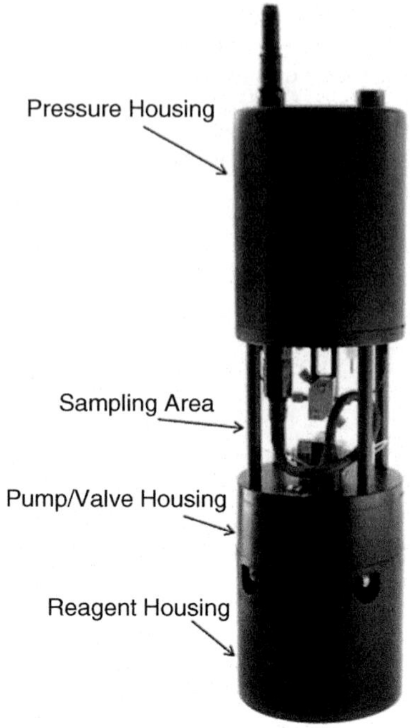

Fig. 20 Microfluidic pH sensor manufactured by Sunburst Sensors

Bulb electrode sensors are typically calibrated to an accuracy of ± 0.1 using either freshwater pH buffers in a range suitable for the environment in which the sensor will be used, or a CRM. For oceanographic applications, seawater buffer is required to achieve the best accuracy when using these sensors.

Calibrations of ISFET sensors require that the sensor be powered for 2–3 days before calibration to ensure that the ISFET is stable. Once stable, the ISFET is calibrated over the oceanographic temperature and pressure range in 0.01 N HCl solution. This is because the thermodynamic properties of HCl are well known compared to seawater. Once the temperature and pressure characteristics of the particular ISFET are measured, the sensor is equilibrated in seawater similar to which it will sample, and an offset is measured against spectrophotometric determination of the pH using CRM as a primary standard. This seasoning process can take up to a week and is thought to be caused by the substitution of bromine in the seawater for some portion of the chloride in the silver/silver chloride reference electrode. Accuracy specifications are ± 0.03 pH units, and precision is 0.004 pH units.

Microfluidic methods for both moored and flow-through applications are capable of measuring pH to an accuracy of ± 0.003 pH units with a precision less than 0.001 pH units. The instrument is capable of carrying CRM standards on board and can perform in-situ calibrations.

Optical luminescence technologies are capable of measuring pH to ± 0.05 pH units. At the time of this writing, there are development efforts underway to characterize the sensor response over a range of oceanographic temperature, pressure and salinity. However, as mentioned previously there is no oceanographic sensor currently available.

6.2 pH Sensor Response Characteristics

pH electrodes typically have slow response times (20–40 s), making fast-profiling measurements with these sensors in high gradient regions impossible. ISFET pH sensors are more stable and sensitive compared to bulb electrode types, and have been reported to have a response time of 0.6 s [18]. Microfluidics-based sensors must pump a seawater sample and mix it with an indicator solution and fill a flow cell before acquiring the spectrophotometric measurement. This process requires 3 min between samples. Optical luminescence based pH sensors are under development, and currently there is no available literature on sensor response characteristics.

6.3 pH Sensor Drift and Calibration Stability

Bulb electrodes unfortunately have always presented problems in stability and durability, and tend to shift in offset with time both on the shelf and in the field. Some scientists have successfully used these sensors to enhance spatial sampling resolution, but this requires frequent field calibrations against water samples and reference checks with the appropriate seawater buffers to maintain a meaningful time series measurement. Much of the historical pH data collected using these bulb sensors were calibrated against commercially available freshwater buffer standards, which are not reliably accurate for sensor applications in seawater. Bulb sensors are generally not recommended for moored applications, as the internal electrolyte rapidly becomes degraded. Field endurance is typically less than 90 days for modified bulb applications, and a few days to weeks for simple exposure deployments.

ISFET pH sensors exhibit much better stability and sensitivity compared to bulb types. They are calibrated over oceanographic temperature and pressure ranges and are accurate for changing salinity environments. ISFET sensors require equilibration time on the order of two to three days in the medium in which they will measure. Once they equilibrate to the environment in which they are deployed, they must be field-calibrated for the best accuracy. These sensors have been shown to return high-quality data for up to three months on moorings [10] and have demonstrated stability on Argo floats for over a year.

As mentioned above, microfluidic pH instruments can perform in-situ calibration using CRM as part of their sampling protocol. Field demonstrations have shown a mean error from CRM of -0.0009 ± 0.003 over a five month period.

Optical luminescence sensors currently demonstrate significant drift of about 0.05 pH units per day at a pH of 7. More work is necessary to make this technology suitable for long-term deployments in the ocean.

References

1. Assmann S, Fram C, Koertzinger A (2011) Spectrophotometric high-precision seawater pH determinations for use in underway measuring systems. Ocean Sci 7:597–607. https://doi.org/10.5194/os-7-597-2011. https://www.ocean-sci.net/7/597/2011/
2. Bacon S, Culkin F, Higgs N, Ridout P (2007) IAPSO standard seawater: definition of the uncertainty in the calibration procedure, and stability of recent batches. J Atmos Ocean Technol 24(10):1785–1799. https://doi.org/10.1175/JTECH.20811
3. Bittig HC, Fiedler B, Scholz R, Krahmann G, Körtzinger A (2014) Time response of oxygen optodes on profiling platforms and its dependence on flow speed and temperature. Limnol Oceanogr Methods 12(2014):617–636
4. Bresnahan J, Philip J, Martz TR, Takeshita Y, Johnson KS, LaShomb M (2014) Best practices for autonomous measurement of seawater pH with the Honeywell Durafet. Methods Oceanography 9:44–60
5. Clark LC (1959) Electrochemical device for chemical analysis, US Patent 2,913,386, issued 17 Nov 1959
6. Dickson AG (2010) Standards for ocean measurements. Oceanography 23:34–47
7. Edwards B, Janzen C, Murphy D, Larson N (2010) Calibration, response and hysteresis in Deep-Sea dissolved oxygen measurements. J Atmos Ocean Technol 27(5):920
8. Gruber N, Doney SC, Emerson SR, Gilbert D, Kobayashi T, Körtzinger A, Johnson GC, Johnson KS, Riser SC, Ulloa O (2007) The ARGO-oxygen program – a white paper to promotethe addition of oxygen sensors to the international Argo float program. Supporting documents for the 8th meeting of the International Argo Steering Team, pages 40–99
9. Johnson KS, Jannasch HW, Coletti LJ, Elrod VA, Martz TR, Takeshita Y, Carlson RJ, Connery JJ (2016) Deep-Sea DuraFET: a pressure tolerant pH sensor designed for global sensor networks. Anal Chem. https://doi.org/10.1021/acs.analchem.5b04653
10. Kapsenberg L, Hofmann GE (2016) Ocean pH time-series and drivers of variability along the northern Channel Islands, California, USA. Limnol Oceanogr 61(3):953–968. https://doi.org/10.1002/lno.10264
11. Kapsenberg L, Kelley AL, Shaw EC, Martz TR, Hofmann GE (2015) Near-shore Antarctic pH variability has implications for the design of ocean acidification experiments. Sci Rep 5:9638. (9 p) http://www.nature.com/scientificreports. https://doi.org/10.1038/srep09638
12. Langdon C (2010) Determination of dissolved oxygen in seawater by Winkler titration using the amperometric technique, IOCCP Report No. 14, ICPO Publication Series No. 134, Version 1
13. Martz TR, Carr JJ, French CR, DeGrandpre MD (2003) A submersible autonomous sensor for spectrophotometric pH measurements of natural waters. Anal Chem 75:1844–1850
14. Martz TR, Daly KL, Byrne RH, Stillman JH, Turk D (2015) Technology for ocean acidification research: needs and availability. Oceanography 28(20):40–47. https://doi.org/10.5670/oceanog.2015.30
15. Newton JA, Feely RA, Jewett EB, Williamson P, Mathis J (2015) Global Ocean Acidification Observing Network: Requirements and Governance Plan, 2nd edn, October 2015. http://www.goa-on.org/docs/GOA-ON_2nd_edition_final.pdf
16. Royer TC, Grosch CE (2006) Ocean warming and freshening in the northern gulf of Alaska. Geophys Res Lett 33:L16605. https://doi.org/10.1029/GL.026767
17. Seidel MP, DeGrandpre MD, Dickson AG (2008) A sensor for in-situ indicator-based measurements of seawater pH. Mar Chem 109:18–28
18. Woias P, Meixner L, Amandi D, Schönberger M (1995) Modelling the short-time response of ISFET sensors. Sensors Actuators B Chem 24(1–3):211–217. https://doi.org/10.1016/0925-4005(95)85045-7

Part IV
Remote Sensing

Ocean Remote Sensing: Concept to Realization for Physical Oceanographic Studies

Tapan Misra, Rashmi Sharma, Raj Kumar, and Pradip K. Pal

Abstract In this chapter, we briefly describe various space-borne sensors which have become the backbone of oceanographic research and applications. Operating in the electromagnetic region (mainly optical to microwave), these sensors provide measurements of various physical oceanographic parameters such as sea surface temperature, height, salinity, wave, winds, sea ice extent, thickness, and concentration on a global scale. This chapter also describes remote sensing techniques, measurement principles, retrieval of geophysical parameters, and their applications.

1 Introduction

Remote sensing by space-borne sensors has become an extremely important component of ocean observing system. Major programs like Global Ocean Data Assimilation (GODAE) and GODAE OceanView have established the role of satellites in observing oceans for research and operational needs. The vast expanse of ocean presents a formidable task to be studied at many scales using in-situ measurements by ships, buoys, and floats. Space-borne sensors provide repetitive measurements with synoptic view at a glance. Satellite Oceanography encompasses oceanographic research and technological development resulting from systems in Earth's orbit. Remote sensing technology makes use of electromagnetic (*em*) radiations of certain wavelength (ranging from visible to microwave) to distinguish different objects. Satellite observations are based on measurements of energy either emitted from earth and atmosphere (*passive sensing*) or returned as backscatter from earth–atmosphere system when a satellite-based pulse source illuminates (*active sensing*) the target. The absorption by atmospheric gases and reflection/emission from the earth's surface is the backbone of these remote sensing methods. Oceanographic parameters observed and measured by space-borne sensors are sea surface winds, sea surface temperature (SST), ocean surface waves, sea surface

T. Misra • R. Sharma (✉) • R. Kumar • P.K. Pal
Space Applications Centre, Ahmedabad, India
e-mail: rashmi@sac.isro.gov.in

R. Venkatesan et al. (eds.), *Observing the Oceans in Real Time*, Springer Oceanography, DOI 10.1007/978-3-319-66493-4_9

salinity, sea surface height, and ocean color. Solar radiation reflected/scattered from the ocean surface mainly relates to ocean color measurements, and thermal infrared (*IR*) emitted from the surface provides information about SST. While emitted microwave radiation is related to both temperature and roughness of the sea, backscattered energy from the surface provides measurements of roughness, slope, and height of the sea surface. Oceanographic parameters are estimated by suitable retrieval algorithms utilizing the underlying physics of the process of observations. Due to this, it needs detailed calibration and validation, prior to the estimation of the parameters. Processes of interest in the ocean span a horizontal length scale ranging from 1 mm to 1000 km, and a timescale ranging from seconds to years. Observations of such wide ranges must necessarily employ a variety of experimental strategies in terms of wavelength selection and space–time sampling.

Satellite Oceanography began with the launch of first artificial satellite Sputnik-1 by the USSR in 1957. Since then, in the last 60 years, spectacular advances have been made in this field. First civilian oceanographic satellite, SEASAT, was launched by NASA in the year 1978. The satellite carried radar altimeter, scatterometer, visible and infrared radiometer, microwave radiometer, and synthetic aperture radar to monitor oceans. Although this mission lasted only for 105 days, it provided immensely valuable data to understand oceans and their role in climate. India's tryst with meteorological and oceanographic satellites started with the launch of its first experimental remote sensing satellite, Bhaskara-1, in the year 1979. IRS–P3 (launched in March 1996) with the sensor MOS onboard and IRS–P4 (or Oceansat-1, launched in May 1999) with multifrequency scanning microwave radiometer (MSMR) and ocean colour monitor (OCM) payloads gave significant fillip to these activities in terms of real-time utilization of satellite-based geophysical information and enhanced user interactions. There are a large number of ocean optical and microwave instruments on the anvil in international arena – assuring uninterrupted supply of data for ocean studies. ISRO's own missions, viz. Oceansat-2, RISAT, INSAT–3D, Megha Tropiques, and SARAL/AltiKa have contributed significantly towards the understanding of oceans.

The book by Robinson [57] gives basic concepts of ocean remote sensing in greater detail. Hence, in this chapter, we provide a brief account of satellite technologies, sensors, and applications, with suitable examples wherever possible from satellites launched in the recent past.

2 Remote Sensing of Sea Surface Temperature

Sea surface temperature (SST) is one of the first oceanographic parameters to be measured from the space and is widely used by the ocean and climate researchers. SST can be measured from both Infrared (IR) and passive microwave radiometers, each with its own advantages and drawbacks. The SST varies on diurnal, seasonal, interannual, and on climate scale. Diurnal variability in SST has been observed up to 6°C [23]. The first global composite of SST from the satellite measurements was prepared in 1970s [42]. Since then numerous satellites have been launched for the measurement of SST by several space agencies.

2.1 *Measurement Principle: Thermal IR and Microwave Regime*

Radiometers which can be imaging or non-imaging are passive sensors that operate in the visible, infrared, and microwave regions of *electromagnetic spectrum*. These radiometers detect naturally emitted or reflected radiation from the earth's surface. Thermal emission and absorption from atmospheric constituents mainly contribute to the *em* energy in the thermal *IR* and microwave regions, whereas in the visible and near *IR* range it is the reflection/scattering of the incident solar radiation which is prominent. That is why the satellite measurements in the spectral bands within the visible region are sensitive to the reflectance/absorption properties of water constituents over oceanic regions, whereas in the infrared/microwave region, it is sensitive to the emission/absorption from the ocean surface as well as the atmospheric constituents. Reflectance of seawater is sensitive to the surface roughness, bathymetry, and presence of tracers such as salinity, chlorophyll, turbidity, etc.

The basic principle behind the passive radiometry is Planck's law which describes a relationship between thermal emission and the physical temperature of an ideal blackbody (with emissivity as unity):

$$L_\lambda = \frac{2hc^2}{\lambda^5\left(e^{hc/\lambda kT} - 1\right)}$$

where h is Planck's constant and k is Boltzmann's constant.

The relationship between the wavelength at which a blackbody emits the maximum radiation, λ_{max}, and the physical temperature of the blackbody, *T*, is given by Wein's displacement law (i.e., λ_{max} T = constant). At larger wavelengths, that is in the microwave region (1–40 GHz), Planck's law becomes the Raleigh-Jean approximation which states that the emitted radiation is directly proportional to the temperature of the emitting surface. The above relation is much simpler for microwave than the one for *IR* radiometry where the full Planck function must be used. For this reason, emitted radiation is sometimes simply referred to as the brightness temperature.

Since the aim of the radiometer is to measure the SST, a suitable spectral band is chosen such that the atmospheric attenuation is minimum and there is sufficiently large amount of energy received at the satellite sensor. These spectral bands in the electromagnetic spectrum are known as the atmospheric windows. There are two important atmospheric windows in the infrared spectrum, 3.8 μm midwave infrared (MWIR) window and 10–12 μm longwave or thermal *IR* (LWIR or TIR) window that are used for the SST retrieval. The peak of the emitted radiation from the sea surface having SST around 300 K is in the wavelength range 10–12 μm which is a window region. This allows to obtaining high spatial resolution SST with highest accuracy. On the other hand, the MWIR window has the advantage in terms of maximum sensitivity of the observed radiances with respect to the changes in the surface temperature due to shorter wavelengths. In the microwave region, the window region exists below 18 GHz where there is significantly smaller attenuation due to atmosphere even in the presence of the cloud. However, due to the longer wavelengths the sensitivity of the microwave radiometer observations to the changes in the surface temperature is smaller than that in the infrared. In addition to this, a small amount of radiated energy in this region of the *em* spectrum causes a large noise equivalent ΔT (NEΔT) which necessitates a coarser spatial resolution or larger antenna to obtain a meaningful signal for the SST retrieval. The C-band (4–8 GHz) in the microwave spectrum is best suited for the SST retrieval due to its higher sensitivity and lower impact due to variable wind-induced surface roughness as well as other atmospheric attenuations. Keeping in mind the advantages they provide in the infrared and microwave parts of the *em* spectrum, a blended product is possible by suitably combining the best features of both the sensors.

2.2 *Retrieval of Geophysical Parameters*

We first start with the SST retrieval from radiances measured in the infrared region of the *em* spectrum. For SST retrieval, mainly the atmospheric windows in the MWIR (3.8–4 μm) and LWIR (10–12 μm) are used. However, due to the contamination of the emitted radiation by the reflected solar radiation in the MWIR band during daytime, this band is used to retrieve SST only during nighttime, hence the name given to it as the nighttime SST channel. During daytime, the LWIR window channels are used for SST retrieval. However, absorption in this band due to highly variable atmospheric water vapor makes SST retrieval erroneous. To correct for the water vapor absorption, the split window channels (i.e., 10.3–11.3 μm or T_{11} and 11.5–12.5 μm or T_{12}) observations are employed. Absorption in the second split window channel is higher than the first channel; therefore, the difference of brightness temperature observations in these two channels gives a quantitative estimate of the atmospheric water vapor that is required for correction in the SST computation. Due to the weak water vapor absorption in these split window channels, the weighting function for these channels lies very close to the surface. Therefore, the amount of water vapor estimated from their differences is equivalent to the total column water vapor as more than 90% of the water vapor lies in the lowest few kilometers of the atmosphere.

A simple form of the dual channel algorithm is given as follows:

$$\mathrm{SST} = \mathrm{A_1 T_{11}} + \mathrm{A_2 T_{12}} + \left[\mathrm{A_3}\left(\mathrm{T_{11}} - \mathrm{T_{12}}\right) + \mathrm{A_4}\right]\sec\theta + \mathrm{A_5}$$

where A_1, A_2, A_3, A_4, and A_5 are coefficients derived using regression analysis between actual SST and the collocated satellite observations. Since water vapor absorption is strongly dependent on the observation zenith angle (θ), the relation needs correction for the zenith angle variation. The MWIR channel at 3.8 μm is highly sensitive to the surface temperature variations; so, this channel can replace T_{11} during nighttime when contamination from reflected solar radiation is absent. During nighttime, this channel along with the split window channels provides the accurate SST retrieval.

Passive microwave radiometer measurements of the sea surface from space functions in essentially the same way as do the infrared radiometers. Normally operating at electromagnetic radiation between 1 and 200 GHz frequencies, these radiometers observe the thermal radiation emitted in the microwave part of the spectrum by the sea surface, atmosphere, and that reflected by the sea surface. At these comparatively longer wavelengths, there is no scattering by the atmosphere or aerosols, haze, dust, or small water particles in the clouds. This provides all weather-sensing capability from a microwave sensor, although liquid water in the form of precipitation does scatter the radiation and can render the atmosphere opaque at microwave frequencies. On the other hand, there are certain disadvantages of the microwave sensors. The emitted radiation is very weak at these wavelengths, which leads to poor signal-to-noise ratio (SNR). To improve the SNR, emission from a larger field of view must be viewed which leads to the coarser spatial resolution. Another aspect is that emissivity of the sea at microwave frequencies is also very small and varies with the dielectric properties of seawater and the surface roughness. Since the dielectric constant varies with temperature, salinity, and frequency, the observations by a multichannel microwave radiometer must contain information not only about the sea surface temperature, but also about the ocean salinity and the sea state. Theoretically, 6 GHz is considered as the best frequency for SST because of the sensitivity of brightness temperature to the changes in SST peaks at this frequency with low sensitivity to both salinity as well as surface roughness (Fig. 1). A frequency of 10 GHz is also considered suitable for SST where there is adequate sensitivity for SST with the added advantage of better spatial resolution. Examples of passive microwave radiometer providing SST estimations are Nimbus SMMR, Oceansat-1 MSMR, TRMM TMI, Aqua AMSR, and NPP ATMS. A typical algorithm for retrieval of SST from microwave radiometer observations makes use of multichannel observations to correct for sea surface roughness, atmospheric water vapor, cloud liquid water, etc., and has the following form:

$$\begin{aligned} \mathrm{SST} = {} & A_1.T_{6H} + A_2.T_{6V} + A_3.T_{10H} + A_4.T_{10V} + A_5.T_{18H} \\ & + A_6.T_{18V} + A_7.T_{21H} + A_8.T_{21V} + A_9 \end{aligned}$$

where A_{1-9} are coefficients derived empirically, and T_{nH} and T_{nV} are H and V polarization brightness temperatures at frequency, $n = 6, 10, 18, 21$ GHz. In this relation, 6 GHz is the primary channel for SST retrieval, whereas 10 GHz provides the correction term for the wind-induced variable emissivity, and difference in the 18 and 21 GHz provides the correction term for the total column water vapor.

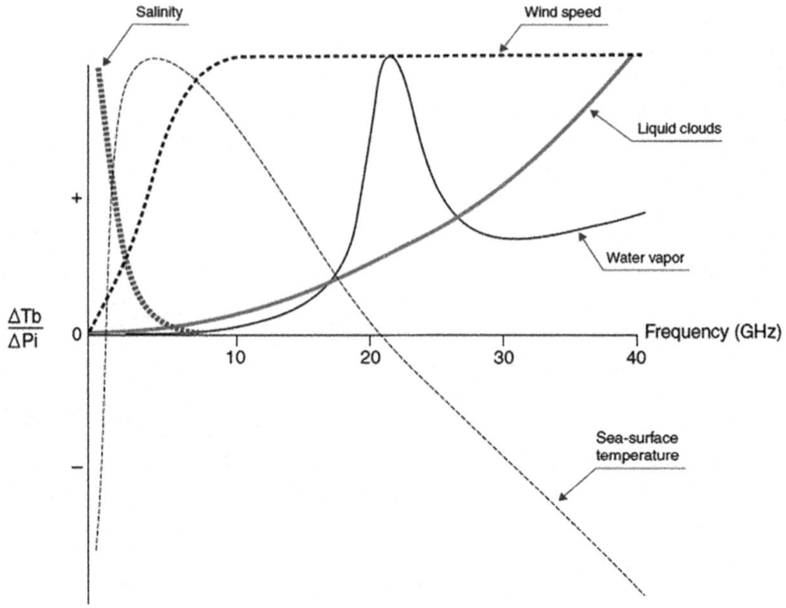

Fig. 1 Relative sensitivity of brightness temperature observation for various oceanic parameters as a function of microwave radiometer frequency (Source: Original figure by Thomas T. Wilheit, NASA/GSFC)

2.3 Accuracy, Precision, and Sampling

IR-based methods provides the best-quality and high spatial resolution SST with an accuracy better than 0.3 K. Typical resolution of SST from such sensors is ~1–4 km. However, SST measurement from IR sensors is limited to the clear sky conditions. This is a major drawback as far as getting SST from IR sensors is concerned, especially in the regions largely dominated by the clouds. For Indian Ocean region, specifically the Bay of Bengal, IR sensors are mainly useful during winter time when the sky conditions are largely clear.

In the current scenario, both polar orbiting and geostationary satellites have capabilities to measure SST. In fact, the currently existing geostationary satellites (GOES-E/W, Himawari-8/9, INSAT-3D/3DR, and Meteosat second generation series) are able to provide a global coverage of SST at a temporal resolution better than 30 min. MODIS which is onboard Aqua and Terra satellites has been continuously providing global coverage of SST at ~1 km resolution for more than 10 years. NOAA/AVHRR series of satellites have significantly contributed to the operational and research communities for more than 30 years as pathfinder AVHRR SST. Microwave sensors provide all weather SST measurements. However, the errors are large and the spatial resolution is coarse (~25 km) as compared to the IR sensors. Typical errors from microwave sensors range from 0.5 to 0.8 K. Microwave imager onboard tropical rainfall measuring mission (TRMM/TMI) provided almost

18 years of continuous data of SST in the tropics (±40° latitude belt). It was the most reliable source of SST from microwave sensor in all weather conditions (barring under high winds and rain conditions). Global maps of 3-day running-averaged SST from TMI at a resolution of 25 km are available from 1997 to 2014. Other instruments like AMSR onboard GCOM series and MWI onboard HY-2 series are also in operation providing microwave SST measurements ensuring the continuity.

2.4 *Applications of Sea Surface Temperature*

SST is a key "boundary forcing" to the atmosphere in the numerical weather prediction models and has a great influence on seasonal, interannual and to some extent on decadal predictions. Satellite-derived SST are assimilated in the ocean models for generating accurate ocean state forecasts. SST is the key variable in the air-sea interaction processes. High-resolution SST is quite useful in the determination of fine-scale horizontal thermal gradients or fronts (Fig. 2). These fine-scale structures can lead to the vertical movement of the biomass nutrients and, therefore, have a potential application in the fishery industry. Thermal fronts also modify the air-sea interaction processes significantly through heat flux exchange. This alteration in the air-sea interaction sometimes can even change the cyclone track [72]. Accurate and well-calibrated SST records are extremely useful for monitoring long-term temperature change and are pointers of climate change. SST fields help in detecting eddies and upwelling regions in the ocean, which are extremely useful for delineating potential fishing zones.

2.5 *New Frontier in SST Measurements*

There has been a tremendous improvement in the synergistic use of passive microwave and infrared sensors for providing continuous global high-resolution SST images for operational and research use. The Group of High Resolution SST (GHRSST [16]) is a merged product comprising of SST observations from several existing microwave and infrared sensors. Researchers involved in generating GHRSST are providing SST products on a near-real time basis for its use in the operational weather forecasting and in ocean process studies. The GHRSST program has resulted in the growth of data streams from all across the globe and has provided SST with common data format along with uncertainty estimates. All these efforts have led not only to the creation of long-term climate data records using existing satellite sensors, but also to the development of a procedure to provide a flawless integration of new satellite sensors.

Geostationary satellites can be quite crucial for providing synoptic measurements of SST at very high sampling frequencies. Efforts are on to increase the sampling rate up to 10 min interval so as to provide high frequency variability

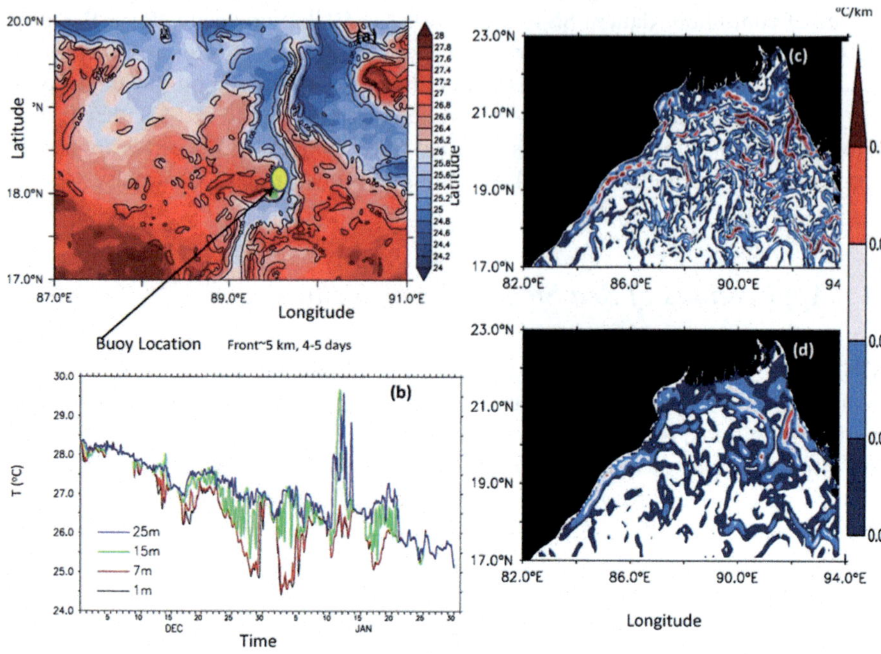

Fig. 2 Thermal fronts can be seen in the northern BoB from a composite satellite SST image (**a**), and time series of near-surface temperature at four different depths from a buoy at 18°N, 89.5°E (**b**). Horizontal SST gradient magnitude (°C/km) for January 15, 2013. The estimated temperature gradients based on 1 km resolution (**c**) and 10 km resolution (**d**) (Figure adopted from Wijesekera et al. [68], BAMS: doi: https://doi.org/10.1175/BAMS-D-14-00197.1)

of SST under cloud-free conditions. Himawari-8/9 of Japan Meteorological Agency, GOES-R of NOAA and Meteosat of Eumetsat are providing SST at high temporal sampling. GISAT of ISRO to be launched in a couple of years' time which will be giving SST from geostationary platform at 1 km resolution at 10 min interval.

It is very important to have continuity in the missions to get maximum benefits in operational oceanography services and to have long uninterrupted time series of data for climate studies. Operational continuity of microwave-based SST measurements is of utmost importance for all weather coverage. Future mission of ESA's microwat [54] is also being discussed. Efforts on ground truth collection are also needed to fine-tune the retrieval algorithms. Coastal observations from satellites have always been a challenge due to sudden transition of brightness temperature from land to sea; so, efforts must be towards building a methodology to address these concerns as coastal processes are of extreme importance. He et al. [35] have generated cloud-free daily SST product for West Florida Shelf at a 5-km spatial resolution by optimally combining microwave and infrared measurements.

3 Satellite Altimetry: A Versatile Tool for Ocean Applications

Satellite altimeter is undoubtedly one of the most versatile space-borne instruments for measuring ocean variables. With the primary application of altimetry in understanding the ocean dynamics by making use of sea surface height (SSH) information, it has come a long way where it is unthinkable of getting the estimate of global sea level rise without this instrument. Other ocean variables such as, significant wave height (SWH) and wind speed, are also retrieved from this instrument. These variables are contributing significantly to the operational oceanography.

3.1 History of Satellite Altimetry

Satellite altimeter has a very rich history. The first multipurpose microwave instrument onboard Skylab in 1974 and GEOS-3 (first dedicated altimeter mission) in the following year were more of technology demonstration. Seasat launched by NASA in the year 1978 in its 3-month lifetime demonstrated that altimeter could successfully detect mesoscale eddies. Launched in 1985, Geosat data was used to monitor eddy variability and also marine geoid. These altimeters were the first generation altimeters. Then came more sophisticated dual frequency altimeters (to take care of ionospheric effects) with onboard radiometer and improved orbit determination. First in this class was ERS-1 (1991) launched by European Space Agency, followed by US/French TOPEX/Poseidon (T/P) in 1992. Later a strategy was adopted [41] to observe ocean circulation with scales ranging from mesoscale to large-scale. It was recommended to have low-inclination altimetric mission having a high-accuracy altimeter (reference mission) on a non-sun-synchronous orbit for determining large-scale ocean currents and complementary higher inclination, sun-synchronous altimeter missions that can provide information on mesoscale eddies. This strategy was realized by NASA and CNES with the launch of Jason-1 in 2001 as a reference mission and Envisat launched in 2002 in a higher inclination. And then followed Jason-2/3 and Geosat follow-on missions. Another shift in altimetric measurements came with the launch of ISRO-CNES SARAL/AltiKa mission in the year 2013, which was a gap-filler between Envisat and Sentinel-3 [66]. The SARAL/AltiKa mission was launched at the behest of the international oceanographic community (Ocean Surface Topography-Science Team).

3.2 Measurement Principles

Altimeter is a nadir-viewing radar that transmits short pulses, typically of a few nanoseconds duration, and detects the return pulse along with the two-way travel time. The shape of the return pulse, known as "waveform," represents the time evolution of the reflected pulse from the footprint of the altimeter. As the name signifies, the primary goal of an altimeter mission is to measure the altitude of the sea surface from a reference ellipsoid. By measuring the two-way travel time of the radar pulse and knowing the speed of the electromagnetic wave, altimeter height above the sea surface called as "range" is computed. Making use of precise orbit determination, the height of the satellite above a reference ellipsoid is obtained. From these two measurements, one can then easily compute the sea surface height (SSH), height with respect to reference ellipsoid, by subtracting the range from the orbit of the satellite (Fig. 3). However, because of the slowing down of the radar pulse during its passage through ionosphere, and atmosphere, several corrections have to be applied. Apart from atmospheric corrections, one needs to correct for sea-state bias and skewness effects. A typical waveform from altimeter over the open ocean is shown in Fig. 4c (upper panel, an example from SARAL/AltiKa). Apart from SSH, one also gets significant wave height (SWH), related approximately inversely to the slope of the leading edge of the reflected pulse or the waveform. The third quantity of interest is ocean surface wind speed, which is empirically related to the maximum backscattered power. Over the ocean, observations are averaged over 1 s giving the along-track resolution of nearly 7 km (varies with ocean wave conditions) and cross-track separation of 40–300 km, depending upon the repeat cycle of the satellite. For altimeters, footprint is determined by the pulse-limited (duration of the pulse) geometry rather than beam-limited geometry. All these details and more on pulse compression method to achieve better range accuracy are described in the work by Chelton et al. [15].

3.3 Retrieval of Geophysical Parameters (Sea Surface Height, Significant Wave Height, and Wind Speed)

In the open ocean, the altimetric echo follows a standard shape, with steeply rising leading edge followed by a trailing edge with gradually diminishing power. This standard shape is in agreement with the theoretical Brown model [8] and hence can be modeled. Though details of the theoretical framework of the radar returns from the ocean surface are given by Brown [8], for the sake of brevity, it is once again briefed here.

Radar return pulse $W(t)$ is a convolution of three terms: (a) the flat sea surface response (FSSR); (b) the sea surface elevation probability distribution function (PDF); and (c) the radar system point target response (PTR) (transmitted pulse as affected by the receiver bandwidth). The first term (a) includes the effects of antenna beam width and the off-nadir pointing angle. The mean return waveform as a function of time t (generally measured in nanoseconds) is expressed as the following convolution:

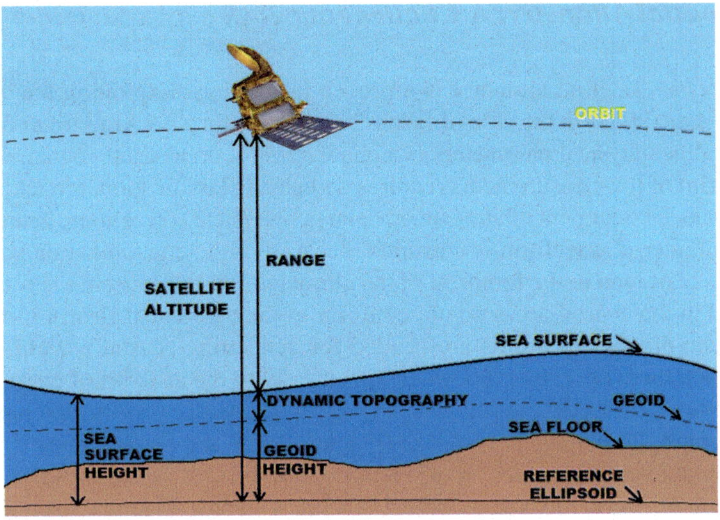

Fig. 3 Schematic representation of principles of satellite altimetry

$$W(t) = \mathrm{FSSR}(t) * \mathrm{PDF}(t) * \mathrm{PTR}(t)$$

Calculation of the convolution with the assumption that mispointing angle is less than 0.3 degrees gives the analytical expression

$$W(t) = (A/2)\exp(-v)\left[1 + \mathrm{erf}(u)\right]$$

where the auxiliary parameters u and v depend on the oceanic parameters of interest.

Geophysical parameters are obtained from altimeter data using retracking algorithms. In the open ocean, the waveforms are Brownian, and suitable algorithms exist for them [5]. Retrieval algorithm normally employed is physically based type and fitting algorithm is largely based on maximum likelihood estimator [58]. The algorithm retrieves mainly three parameters, namely the amplitude, the epoch (time passed since first return), and the backscatter coefficient. Least square method [47, 59] or Bayesian inference [61] are also employed to retrieve geophysical parameters. Accuracies of derived parameters from contemporary altimeters are ~0.3 m for SWH, ~3–4 cm for sea surface height anomaly (SSHA), and better than 2 m/s for wind speed. More complex waveforms normally encountered in coastal regions, continental waters, and over sea ice are retracked using empirical algorithms. Coastal altimetry, a new emerging domain under nadir-looking conventional altimetry, is described in detail in the next section.

3.4 Coastal Altimetry: A Challenging Task

In open ocean, satellite altimetry is a proven technology. Exploiting the high-rate altimeter data (20 Hz in the case of Jason-2 and 40 Hz for SARAL/Altika) for deriving coastal geophysical parameters is a challenging task. Coastal contamination in the footprint of the measurements requires dedicated classification, retracking strategy, and special treatment of atmospheric and geophysical corrections. In the coastal area, an altimetric waveform is corrupted because of the contamination caused by the presence of land in the footprint of the altimeter. For this reason, the waveforms measured in the coastal areas do not conform to the theoretical Brown model, and special data-processing efforts are needed for generating coastal waveform products. In fact, there are projects devoted specifically to the analysis of coastal waveforms, namely PISTACH [18] and COASTALT [30]. AltiKa was the first instrument to be operating at Ka-band (35-GHz) frequency with the bandwidth of 500 MHz that enabled a vertical resolution of 0.3 m instead of 0.5 m in Jason Ku-band [63]. A footprint of 3 dB in the case of AltiKa is 8 km as against 20 km in Jason altimeter. Along with this, high pulse repetition (4000 per second) results in better along-track sampling, enabling recovery of useful geophysical parameters near to the coast.

Normally, retracking algorithms appropriate for altimeter return echoes from open ocean that are based on the Brown model are not applicable for the coastal oceans. Hence, specific algorithms are to be devised for such echoes coming from coastal areas. Over the years, several retracking algorithms have been developed for specific surfaces, for example, Beta 5/9 algorithm [46] and the OCOG (offset center of gravity) technique [71]. These are also applicable for retracking of coastal waveforms. Guo et al. [32] proposed an improved threshold retracker, based on the leading edge detection and subwaveform extraction. Brown's model with the Gaussian peak model was developed by Halimi et al. [33] to model the contaminations caused by the land footprint in the form of Gaussian peak in the trailing edge of waveform. Launch of SARAL/AltiKa signifies a major leap in coastal altimetry owing to better signal-to-noise ratio, smaller footprint, and high along-track sampling. Altimeter footprint is the area of the surface over which the reflected power is accumulated over the designated number of gates for a single pulse. In the case of AltiKa, there are 128 gates, while for Jason-2 the number of gates are 104.

We will show a few examples to illustrate the usefulness of AltiKa instrument over Jason-2 for coastal applications. In Fig. 4a, SARAL/AltiKa ascending pass 223 (blue) and Jason-2 pass 155 (red) over the coast of Visakhapatnam (India) in the western Bay of Bengal are shown. AltiKa and Jason-2 waveforms as a function of distance from the coast near the Visakhapatnam region (east coast of India) are plotted over these tracks in Fig. 4b. Return power sampled in various gates (representing time elapsed since first return) is shown on y-axis. One can easily see that while waveforms in the case of Jason-2 get contaminated beyond 12 km shoreward, the same in the case of AltiKa show less contamination, and one can retrieve geophysical products up to 3–4 kms shoreward. SARAL/AltiKa waveforms in the open ocean and in the coastal region are shown in Fig. 4c. Another example of

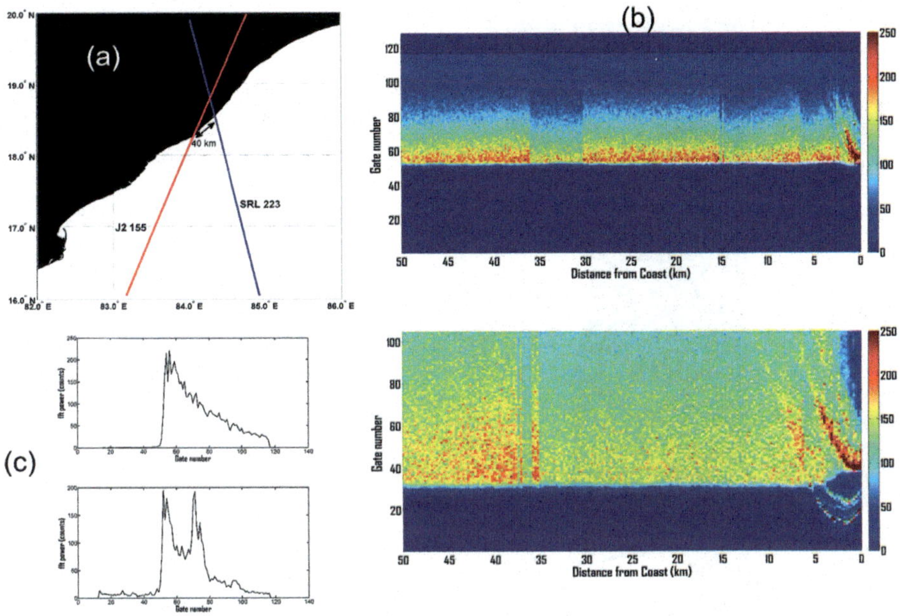

Fig. 4 (**a**) SARAL/AltiKa ascending pass 223 (*blue*) and Jason-2 pass 155 (*red*) passing along over the coast of Visakhapatnam (India) in the western Bay of Bengal (*upper*). (**b**) SARAL/AltiKa (*upper*) Ka-band waveforms and Jason-2 (*lower*) Ku-band waveforms are shown over these passes. X-axis denotes distance from the coast. (**c**) An example of typical waveform shape from open ocean (*upper*) and coastal region (*bottom*) from SARAL/AltiKa mission

cross-track surface current estimation from AltiKa and its comparison with currents measured by high-frequency (HF) radar is shown for the Chennai coast (east coast of India) in Fig. 5. One can see a good agreement of AltiKa current with HF radar currents.

3.5 Oceanographic Applications of Altimeter-Derived Parameters

Operational Oceanography is now well established, thanks to the efforts by various nations under Global Ocean Data Assimilation Experiment (GODAE) program, initiated in 1997, to understand the modeling and ocean forecasting needs. Altimetric measurements of sea level and significant wave height are the backbone of Operational Oceanography. The need to maintain the continuity of altimetric system for ocean forecasting is now firmly established. Altimeter-derived sea level and SWH are routinely assimilated in numerical models for generating ocean state forecasts. Forecasting the ocean state with 5–7 days lead time has several applications in the marine

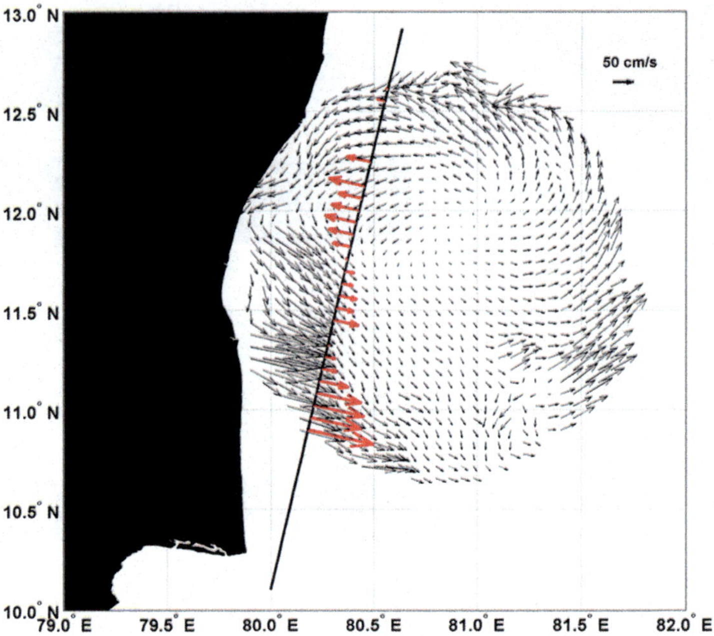

Fig. 5 A field of HF radar currents (in *black color*) near Chennai region on September 09, 2013. SARAL/AltiKa-derived across-track geostrophic currents (*red color*) computed for the track 868 for the same date are overplotted

fisheries, navigation, naval operations, oil-spill monitoring, etc. As the sea level, variations represent the integrated effect of the water column density variations; these data are now routinely used for hurricane forecasting [31]. Of course, this application requires merging of data from several altimeters. Altimeter-derived sea level anomaly are widely being used by researchers to monitor the progress of El Nino [38], as these events affect the climate and have impact on the economic conditions of nations. High-resolution altimeter data are being used to monitor inland river levels which may be useful for flood forecasting. Altimeter observations play a very important role in ice-sheet mass balance studies. Sea level rise is one of the most severe manifestations of the present-day global warming. Tide gauge and altimetric data from 1992 onward have revealed that global mean sea level (GMSL) has risen by 0.19 m between 1901 and 2010. The rate of increase in GMSL during 1993–2010 was 3.2 mm/year [64]. To capture this kind of small change, altimetric system needs to have a "mm" level control on the system drift. This calls for the continuity of mission, homogenization, and intercalibration of different altimeters to minimize the bias. Challenges associated with the accurate estimate of global sea level rise using satellite altimetry are provided in detail by Willis et al. [69].

3.6 GNSS-R and Swath Altimetry

Ocean Surface Topography Science Team (OSTST) and CEOS virtual constellation team consolidates the newer demands of global researchers, formulates the white paper, and coordinates the altimetric missions of various space agencies to maximize the benefits. Conventional altimeters which are nadir-looking measure the sea level only along the subsatellite tracks. Pascual et al. [52] have shown that four altimeters in a constellation resolve the mesoscale features in a more realistic manner. One way to enhance the coverage of nadir-looking altimeter is to have several altimeters in a constellation. A community white paper by Wilson et al. [70] highlights the requirement of multiple altimeters. Two new concepts, GNSS-R and Swath altimetry are described below.

Global Navigational Satellite System Reflectometry (GNSS-R) which works on the concept of bistatic radar is another emerging technology which can provide data on sea level, wind speed, and SWH on a large swath. Signals emitted by GNSS satellites (GPS, Galileo, and IRNSS) and reflected by ocean surface are received by low-orbiting satellites. GNSS signal consists of ranging codes known as C/A codes which belong to the family of pseudo-random noise (PRN) codes. GNSS signal is reflected from the earth, and the area which significantly contributes to the reflected signals is known as the glistening zone. These signals have a variety of delays (due to position) and Doppler frequency shifts (due to relative velocity). These shifts are mapped in receivers through a two-dimensional Delay Doppler Map (DDM). This DDM is then used to derive various parameters by using either theoretical models like that of the Zavorotny and Voronovich [73] scattering model or empirical relations [27]. This technology is still under demonstration phase.

The next available technology is the wide-swath altimeter [25], which will extend the observational capability of altimetry to the cross-track direction. It is expected that interferometric synthetic aperture radar processing of the returned signal, averaged over 1 square km, can give better than 2 cm height precision. The upcoming Surface Water Ocean Topography (SWOT) mission, proposed for launch in a few years is based on this technology. With ~130 km swath and an order-of-magnitude finer resolution, SWOT will represent a paradigm shift in the measurement capability. Along with oceanic mesoscale, submesoscale, and coastal observations, it would also contribute significantly to the land hydrology [26].

4 Satellite Scatteromerty: Measuring the Ocean Surface Winds

Ocean surface vector wind, being one of the major parameters of importance for forecasting of weather and ocean state, needs regular monitoring with good accuracy. A scatterometer is designed to measure ocean surface vector winds by utilizing the scattering mechanism of the incident microwave signal due to the surface waves [53]. The primary function of a scatterometer is to utilize backscatter dependence on the radar azimuth to retrieve ocean surface wind vectors over the global oceans with 1–2 days interval. Surface winds are the major sources of momentum for the generation of surface waves and basin-scale ocean currents. Air-sea fluxes of heat, moisture, and gases are modulated by the action of winds. In this way, winds influence the regional as well as global climate. Thus, scatterometer, being an instrument to measure ocean surface winds in all weather conditions by using microwave signals, is one of the important space-borne sensors for the routine monitoring of the earth system processes.

4.1 Past, Present, and Future Scatterometers

Several scatterometers have been flown in the space by NASA, ESA, JAXA, and ISRO. Seasat launched by NASA in the year 1978 was the first operational scatterometer which operated at Ku-band (14 GHz) frequency and employed fan-beam system. Seasat provided global wind product at 50 km resolution on a swath of 750 km on each side. Subsequently, ESA launched ERS-1 Active Microwave Instrument – scatterometer at C-band (5.3 GHz), and NSCAT (Ku-band) was launched by NASA, once again with the fan-beam concept. Then came the generational shift in the scatterometry concept, when first pencil-beam scanning scatterometer "SeaWinds" onboard QuikSCAT was put in the orbit by NASA in June 1999. The operating frequency for this QuikSCAT was 13.6 GHz with large swath of 1800 km. QuikSCAT provided very useful wind product for nearly 10 years. Advanced SCATterometer (ASCAT) onboard Metop-A launched by EUMETSAT in 2006 operated at C-band. In the year 2009, ISRO launched OSCAT onboard Oceansat-2 satellite. It provided very useful data until March 2014. ISRO launched SCATSAT-1 in September 2016, which was a repeat mission of OSCAT with the same specifications. Also, there is a plan by ISRO to have scatterometer onboard Oceansat-3.

4.2 Basic Measurement Techniques: em Interaction with Roughness

A scatterometer is a side-looking radar system that transmits and receives microwave (electromagnetic) pulses. When the electromagnetic radiation transmitted from a scatterometer impinges on the ocean surface, most of the incident radiation gets scattered in different directions. Depending upon the roughness of the ocean surface, a portion of the incident radiation gets reflected towards the scatterometer antenna. This is called the phenomenon of backscattering. The backscattered power measured by the scatterometer is proportional to the surface roughness caused by oceanic winds. If the winds with higher magnitude blow over the ocean surface, the surface roughness will be more and thus a scatterometer will receive more backscattered power and vice versa. However, such proportionality is not uniform throughout the possible wind speed regimes.

4.3 Retrieval of Ocean Surface Winds from Backscattering

The backscattered power intercepted by a scatterometer is measured in terms of backscattering coefficient or sigma-naught (σ^0). As mentioned earlier, the backscattered power and hence the σ^0 values are proportional to the wind speed, but the problem arises when researchers try to retrieve winds from σ^0. One can refer to the work by Ulaby et al. [65] for details on this aspect. However, for brevity, we briefly describe it here. Difficulty in retrieving winds emerges simply because scatterometer measures only the σ^0 that has the influence of the wind stress, but not the winds. Thus, the retrieval of the winds from the scatterometer measurements is basically an inverse problem where we need to find a suitable forward model for the σ^0 dependent on winds and then we need to invert that model to derive winds. It has been found that σ^0 is adequately described by the forward model:

$$\sigma^0 = \mathrm{a}V^{\gamma}\left(1+\mathrm{bCos}\psi+\mathrm{cCos}2\psi\right)$$

where a, b, c, and γ are constant for a given incidence angle, polarization and frequency, V is the wind speed, and ψ is the angle between the wind vector and the scatterometer look direction.

Such a forward model is known as geophysical model function (GMF) in scatterometer terminology. A GMF is developed by fitting collocated true winds (e.g., winds from in-situ observations like moored buoys, ships, etc.) and measured σ^0 values. In most of the cases, the GMFs are developed empirically, and they depend on the wind speed, direction, scatterometer incidence and azimuth angle (the direction along the track of the instrument) and polarization. Because of the wind direction dependence, a GMF exhibits biharmonic behavior over the various direction zones. In practice, the GMF is generally developed post-facto using collocated wind observations mostly from NWP model and the scatterometer observations over a period of several months covering the full dynamic range of wind vector. The GMF is essential for deriving the wind vector from scatterometer observations made by the same or similar scatterometer system. The GMF developed for one scatterometer system, in principle, can only be applied to the same system due to inherent characteristics embedded empirically in the derived GMF. However, its use with another system is possible provided the parameters of that system are kept unchanged or least deviated, which then can be further fine-tuned.

The exact behavior of radar backscatter varies with scatterometer operating frequency, polarization, radar azimuth and the incidence angle. It has been observed over decades using earlier satellite missions and also based on theoretical models that radar backscatter depends upon wind speed with a power law while it depends biharmonically on wind direction. This harmonic nature of radar backscatter on wind direction leads to multiple possibility of wind vectors yielding the same radar backscatter value and thus causes ambiguity in wind direction determination. The radar backscatter decreases with incidence angle for a given wind speed and polarization. Moreover, radar backscatter from the ocean surface is more in vertical polarization than in the horizontal polarization.

The basic technique generally employed for extracting wind speed and direction from oceanic radar backscatter measurements made by space-borne microwave scatterometer makes use of the dominant dependence of radar backscatter on polarization and on wind speed and direction for a fixed observational geometry of scatterometer at the data location on ocean surface. The observational geometry varies across the swath for both types, viz. fan-beam and pencil-beam scatterometer systems. In the case of fan-beam scatterometer, the azimuth geometry remains unchanged with the incidence angle varying across the swath, while in case of pencil-beam scatterometer the incidence angle remains constant with azimuth angle varying across the swath. However, the constant parameters (depending upon the scatterometer type) change slightly due to the earth's curvature and the satellite orbit inclination and attitude. An example of scanning pencil-beam viewing geometry of SCATSAT-1 is shown in Fig. 6. SCATSAT-1 carries a scanning pencil-beam Ku-band scatterometer with a swath of 1800 km and wind vector cell size of 25 km × 25 km.

Assuming other parameters as constant and the dominant dependency of radar backscatter on ocean surface wind vector (speed and direction), extraction of wind speed and direction is carried out by comparing the measured radar backscatter with those simulated using suitable GMF for assumed wind speed and direction varied in its entire range valid for the GMF being used [28]. This process yields multiple solutions of wind vector among which one solution corresponds to true wind vector while others are ambiguities. The wind speed values of these vector solutions have small differences while the direction values are quite different. These solutions are prioritized according to the deviation of measured radar backscatter from the simulated values with the vector solution having minimum deviation treated as highest priority solution [29]. Under noise-free conditions, the highest priority vector solution always identifies the correct (true) wind vector, while under moderately noisy conditions, the highest priority solutions identify the correct wind vectors in about half of the data cases considered. Such performance of the algorithm is heavily dependent on the noise present in the radar backscatter data. The characteristic of these prioritized solutions is such that the majority of correct wind vector cases can be identified between the first two highest priority solutions. Moreover, in most of the cases, the directions of the first two highest priority solutions are mostly opposite to each other. Thus, when the wind vectors are retrieved from scatterometer data over the swath, about half of the directions may be found in opposite direction to the overall wind directional flow in the data region. These directional ambiguities are

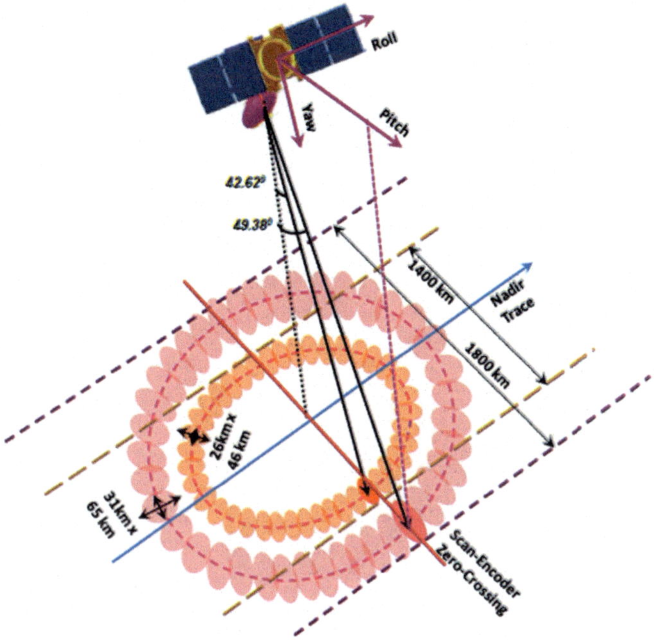

Fig. 6 SCATSAT viewing geometry (Source: SAC/SCATSAT/PDR/01)

filtered out by using another process known as directional ambiguity removal process.

4.4 Accuracy, Swath, and Resolution

The accuracies of the present-day scatterometer-derived winds are well within the research requirements. The mission requirements for a standard scatterometer-derived wind are less than 2 m/s RMSE in the wind speed and less than 20° in the wind direction. Once a scatterometer is launched, the data products that come from the scatterometer undergo rigorous calibration-validation (CAL-VAL) phase. For example, after the successful launch of Oceansat-2 Scatterometer (OSCAT), the initial 9 months of data from it was used for extensive CAL-VAL [10, 11]. The ocean surface winds derived from OSCAT were validated against observations from global moored buoys, winds available from NWP models, and other contemporary scatterometers [43]. Such extensive validation is required to establish the fulfillment of the mission-specific requirements [24].

The width of the sub-satellite track is defined as the satellite swath. The width of such swaths determines the coverage over the global surface. Depending upon the incident beam geometry, the swath varies for different scatterometers. For

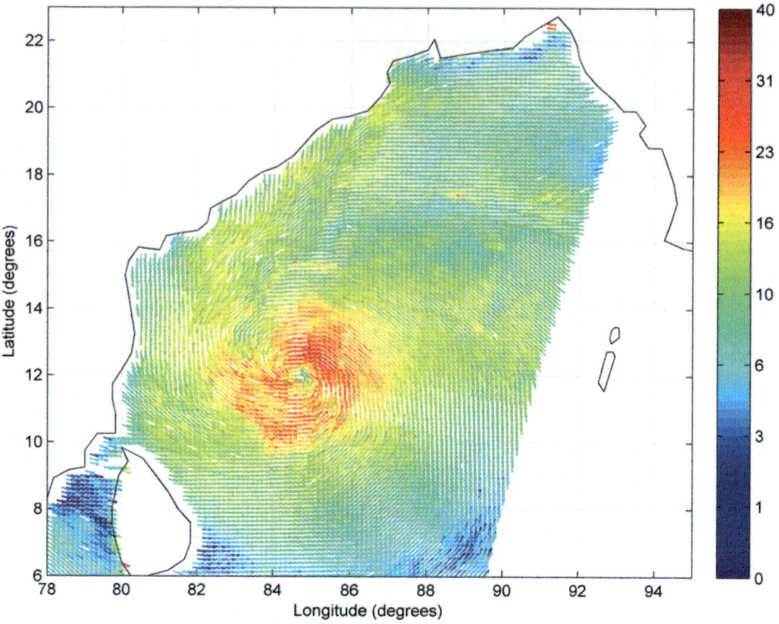

Fig. 7 OSCAT scatterometer derived wind fields on December 28, 2011 for the cyclone Thane

instance, QuikSCAT scatterometer launched by NASA in June 1999 had large swath of 1800 km for the outer beam which is quite similar for the case of OSCAT launched by ISRO in 2009. Large swaths of these scatterometers provided synoptic wind fields which helped in the studies of cyclogenesis and cyclone track prediction. Figure 7 shows one example of OSCAT-derived winds for the case of cyclone Thane. Advanced SCATterometer (ASCAT) onboard Metop-A launched by EUMETSAT in 2006 had two swaths each with 550 km separated by a nadir gap of 700 km.

The resolution of a scatterometer depends on various factors. The smallest element from which wind information from a scatterometer is obtained is called wind vector cell (WVC). The signals from multiple WVC are averaged to remove noises, and that leads to nominal resolution for operational product from a scatterometer. For example, the operational horizontal resolutions of the QuikSCAT, ASCAT, and OSCAT were 25 km, 25 km, and 50 km, respectively, though there were developmental finer resolution versions available.

4.5 Ocean and Ice Applications of Scatterometry

Ocean general circulation models (OGCMs) are forced by air-sea fluxes (wind stress, heat fluxes, and freshwater fluxes) at the ocean-atmosphere interface. Out of these surface boundary forcings, wind stress plays the leading role, particularly in the tropics, where the ocean circulation is primarily wind-driven. Hence, an accurate wind forcing is essential for simulating realistic circulation features [12, 13]. However, the surface wind vectors retrieved from a scatterometer are irregular in both space and time due to the limited beamwidth of the scatterometer geometry. Such scattered observations from scatterometers are analyzed to produce synoptic gridded wind vectors (analyzed wind vectors) regular in spatiotemporal coordinates [14] that are used to provide forcing to the ocean circulation as well as ocean wave models. Hence, the major application of scatterometer observation for oceanography is to provide forcing to the numerical ocean models. When it comes to applicability of scatterometer winds in ocean state forecasts, these winds are assimilated in the numerical weather prediction (NWP) models for providing the necessary forecasted surface boundary forcings for OGCM at more frequent time intervals (daily or even 6-hourly). Scatterometer-derived winds are also used to compute ocean surface currents along with altimeter observations. The phenomena of land and sea breeze along the coasts can also be studied using scatterometer data. Apart from this, scatterometer observations are utilized to monitor the extent and variability of sea ice. Here, the geophysical product that finds its application is the backscattering coefficient itself rather the ocean surface winds. Also, the scatterometer observations help in detecting large icebergs in the polar oceans.

4.6 New Concept in Scatterometry

The importance of scatterometer in met-ocean studies and operational forecasting purposes is now already established. There are several new upcoming concepts in scatterometer apart from the conventional configurations. A rotating fan-beam scatterometer named as RFSCAT will be flown in the Chinese-French Oceanic Satellite (CFOSAT). Also, engineering efforts are being engaged to develop scatterometer processor that will be doing the retrieval onboard. Efforts are also envisaged to measure the ocean surface currents along with ocean surface winds from a single scatterometer.

5 Synthetic Aperture Radar: Exploring Fine-Scale Processes

5.1 Concept and Principles of SAR Technology

Similar to the scatterometer systems, Synthetic Aperture Radar (SAR) is also an active radar, with a side-looking capability to observe various ocean, land, and atmosphere parameters, day and night, as well as practically in all weather conditions. In the lower frequency range of operations, they are also able to penetrate through clouds and light rains. General side-looking imaging radars work on the principle similar to the scatterometer with imaging capability, whereas Synthetic Aperture Radar provides the advantage of obtaining high-resolution images (~1 m) of the land and ocean surfaces. The basic difference between scatterometers, discussed in last section, and SAR is that SAR is able to provide high-resolution images of the targets, whereas scatterometers provide coarser resolution backscatter data.

The minimum area, which can be differentiated from the neighboring ones is called the resolution of the system. Basically, there are two types of resolutions, one in the direction of the spacecraft movement (azimuth resolution) and the other in the transverse direction (range resolution). Range resolution basically depends upon the width of the pulse. Radars achieve high resolution in range direction by transmitting a short pulse. Along-track or azimuth resolution of the radar is defined as βR, where β is the beamwidth of the antenna and R is the range of the target. The beamwidth is inversely proportional to the length of the antenna. In SAR system, the length of the aperture is synthesized to achieve high resolution.

Consider two objects A and B over the ground along the range direction (Fig. 8a). Assume that the satellite height is H, the range distance from the satellite to A and B are R_1 and R_2, respectively, the look angle is θ, and the pulse separation time is τ. If A' is the projection of object A along the slant range, to resolve the two objects along the slant range direction, the minimum distance traveled by the incident pulses should be (R_2–R_1), that is, the slant range resolution should match the criterion given by

$$2R_2 / C - 2R_1 / C >= \tau, \text{where } C \text{ is speed of light}$$
$$\text{or,} \left(R_2 - R_1\right) >= C^* \tau / 2$$

Along the ground range, the projection of (R_2–R_1) will be (R_2–R_1)*sin θ. Hence, the ground range resolution of a SAR system will be

$$\left(R_2 - R_1\right) >= C^* \tau / \left(2^* \sin\theta\right)$$

Next, we consider azimuth (along-track) resolution (Fig. 8b). We focus on a smallest element over the satellite swath. Suppose a signal is emitted from the radar and it reflects back from different targets. The signal from the point opposite the antenna reaches first the center of the antenna and later the ends of the antenna, that

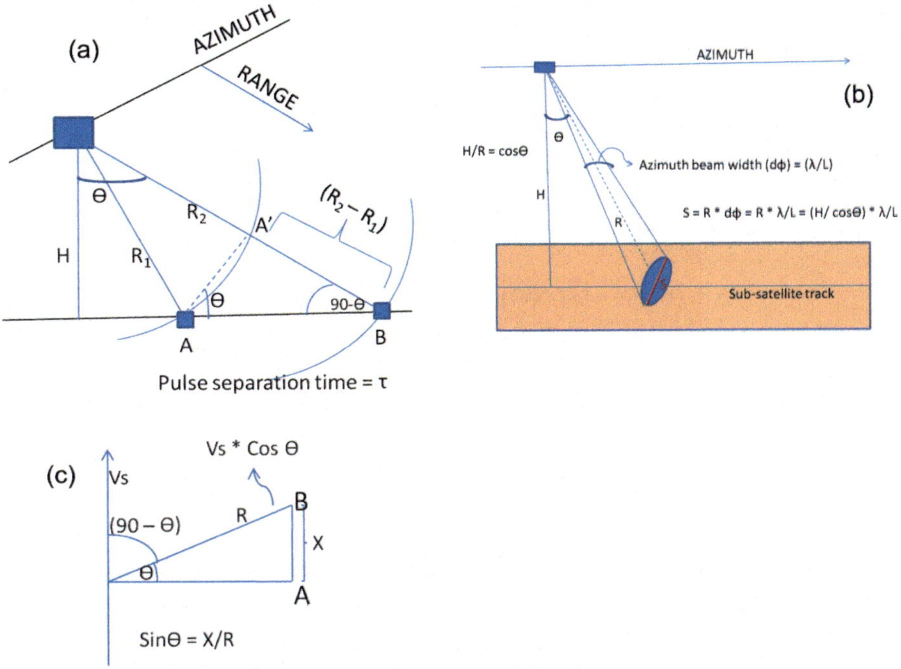

Fig. 8 Synthetic aperture radar (**a**) range (across-track), (**b**) azimuth (along-track) resolution measurement principles, (**c**) concept of Doppler shift

is, small phase change will occur along the aperture, but phase distribution will be symmetrical about the center, and the overall summation of signal across the aperture will be constructive. But, if the target is toward one end of antenna, the phase change around the center of antenna will not be symmetric, and hence destructive interference will occur, and when the signal is integrated, it will not contribute to the signal. Similar things will happen for targets between these two points. The larger the aperture of antenna, the closer the point from the center will try to create destructive interference and will not contribute to received signal; hence, resolution will be higher. SAR uses the same principle, but instead of using larger antenna, it uses smaller antenna, and signal is integrated over a long period of time.

If R is the range distance to the central point of the element and θ is the look angle, then from geometry we get H/R = cos θ. The azimuth beam width β is defined as

$\beta = \lambda/L$, where λ is the radar wavelength and L is the antenna length. Thus, the radar azimuth resolution (S) can be computed as

$$S = \beta R = \lambda / L \cdot R = \left(H / \cos\theta\right)^{*} \lambda / L.$$

At this stage, let us consider an interesting example. The following are the provided specifications of RISAT-1 SAR:

$\lambda = 5.6$ cm = 0.056 m (f = 5.35 GHz)
$H = 580$ km = 580,000 m
$L = 6$ m
$\theta = 25^0$ (say)

For this configuration, the azimuth resolution can be computed as

$$S = (0.056 * 580000)/(6 * \cos(25)) = 5972\,\text{m}$$
$$= 5.9\,\text{km which is too large for a sensor like SAR.}$$

RISAT-1 SAR is having various mode-specific resolutions, ranging from 2 to 50 m. For instance, if $S = 10$ m, then the required antenna length can be computed as

$$L = \lambda * H/(Ys * \cos\theta) = 0.056 * 580000/(10 * \cos(25)) \approx 35\,\text{km}$$

It is practically impossible to mount such a big antenna onboard a SAR system. To mitigate this problem, SAR continues to look at a particular object for a sufficient dwell time, and during that time the distance traveled by the sensor is used as a synthetic aperture length to process the SAR signal, and thus the fine resolution is achieved along the azimuth.

Now, consider SAR system is moving with the speed Vs and along the azimuth there are two nearby objects A and B separated by a distance X (very small in magnitude). If R is the range distance and θ is the look angle, then from geometry (Fig. 8c) we get

$\sin\theta = X / R$. Now if we recall basic Physics, we can define Doppler frequency as

$f_o = (1 + v/c) f_s$, where the subscript "o" and "s" stand for observer and source frequency, respectively, v is the relative speed between them, and c is the speed of light. So, the Doppler shift is

$$df = f_o - f_s = (v / c) f_s = v / \lambda,\ \ \lambda(= f_s / c, \text{being the wavelength of the source signal}).$$

Hence, the total Doppler shift (approaching + receding) = $\Delta f = 2 * df = 2 * v/\lambda$. For side-looking SAR, $v = Vs * \cos(90 - \theta) = Vs * X/R$

So, $\Delta f = 2 * Vs * X/(\lambda * R)$

Now, for a small shift, $\delta(\Delta f) = \{ 2 * Vs/(\lambda * R) \} * \delta X$.

SAR integration time, Tint = $1/\delta(\Delta f)$ = Footprint/sensor velocity = $(\lambda/L) * R/Vs$.

On using the above two expressions, we get $\delta X = L/2$.

Thus, the azimuth resolution of SAR is just half the antenna length. However, to achieve much finer resolution, the antenna length cannot be simply reduced because the sampling bandwidth (or the pulse repetition frequency or PRF) must be greater or equal to Doppler bandwidth, that is,

$$\text{PRF} >= Vs / \delta X \text{ which implies PRF} >= 2Vs / L.$$

These computations are well valid for the land-based targets where Doppler shift happens because of the relative velocity between the fixed targets and the SAR sensor. In case of moving targets, for example, ocean surfaces, there will be an additional velocity component from the motion of the targets.

For such a situation, SAR data processing becomes more complicated.

5.2 Ocean Surface Imaging

The high resolution by SAR relies on the precise measurement of phase and Doppler and signal processing. The SAR signal processing has the assumption that the Doppler shift of the signal is due to the movement of SAR system, whereas the inherent motion of the object in consideration also produces a Doppler shift and affects its appearances in SAR image. The backscattered signal received by a SAR receiver is due to the interaction of the transmitted signal whose characteristics are determined by the radar's frequency, polarization, viewing geometry, and the target surface whose characteristics depend on roughness features, electrical properties, and material composition. Over the ocean surface, SAR energy is primarily scattered by the presence of small-scale wind-induced capillary waves. For the ocean imaging, the surface is always in motion, and the mean wave structure will include a variety of motions with components along the line-of-sight to the radar. These motions will induce Doppler frequency shifts on the backscattered signal. These shifts, and the resulting misregistration of scene scatterers, produce a smearing or blurring in the azimuth direction. These shifts tend to be different for different phases of the dominant (long) waves, and the magnitude of the effect depends primarily on significant wave height and other parameters. This phenomenon is called velocity bunching. It is a limiting factor in a SAR's ability to image ocean wave fields

Similar effect is also observed from other moving objects such as ships, which are displaced from their actual position, and trains that appear to be moved from the railway tracks. The ocean features commonly seen on SAR imagery include surface waves, mesoscale ocean circulation features such as eddies and currents, oil slicks, and surface manifestations of internal waves, and subsurface currents over shallow shoals. In addition to the wind speed, one can also get information on the patterns and structures of winds within the atmospheric boundary layer.

The interactions of short and long waves affect the radar scattering by a process known as the two-scale scattering theory. There are three primary mechanisms in which long waves modify the Bragg waves to effect SAR imaging: the hydrodynamic, tilt, and velocity bunching [1, 2, 4, 21, 34]. The hydrodynamic modulation occurs due to the modulation of the energy of short ripples through interaction between ripples and long waves. The small-scale waves ride upon the large-scale waves somewhat nonuniformly. Due to the different directions of the orbital velocities of long waves along the wave, short waves pile up at the crest and spread them out in trough. Tilt modulation is purely due to the geometric effect that Bragg scattering waves are seen by radar at different local incidence angles depending upon the slope of long-scale waves. This effect is independent of hydrodynamic interaction and will occur even if ripples are uniformly distributed over the longer waves.

In the case of actual ocean surface, as long waves grow steeper, the radial velocity components increase, resulting in more random azimuth displacements and smearing in the imagery. This effect reduces the azimuth resolution and thus limits the detectable ocean wavelengths. The azimuth shift is estimated by the distance between the radar and the surface and its velocity.

5.3 *Retrieval of Oceanographic Parameters*

High-resolution surface images from SAR over the oceans contain signatures of the surface features. Hence, it is possible to retrieve information about those features by interpreting or processing the SAR images. One of the important applications of SAR is the retrieval of ocean surface wave information. In general, a wave field consists of a number of waves with different wavelengths and directions. The best way to extract wave information from SAR image is by analyzing the power spectrum. All dominant wave peaks can be seen in the power spectrum, which is obtained by taking the square of 2D Fourier transform of image data. However, due to system response and noise, different corrections have to be applied to obtain noise-free image spectrum, but not ocean wave spectrum.

The ocean wave spectrum S(K) results after applying modulation transfer function, consisting of tilt, hydrodynamic modulation and velocity bunching, to the SAR image spectrum. The linear and nonlinear schemes, depending upon the ocean conditions, have been generally used for the inversion of SAR image spectra to ocean wave spectra [21, 38].

The wave height can be estimated by

$$\mathrm{SWH} = 4.0. \times \mathrm{Energy}$$

where $\mathrm{Energy} = \int S(K)\,dK$

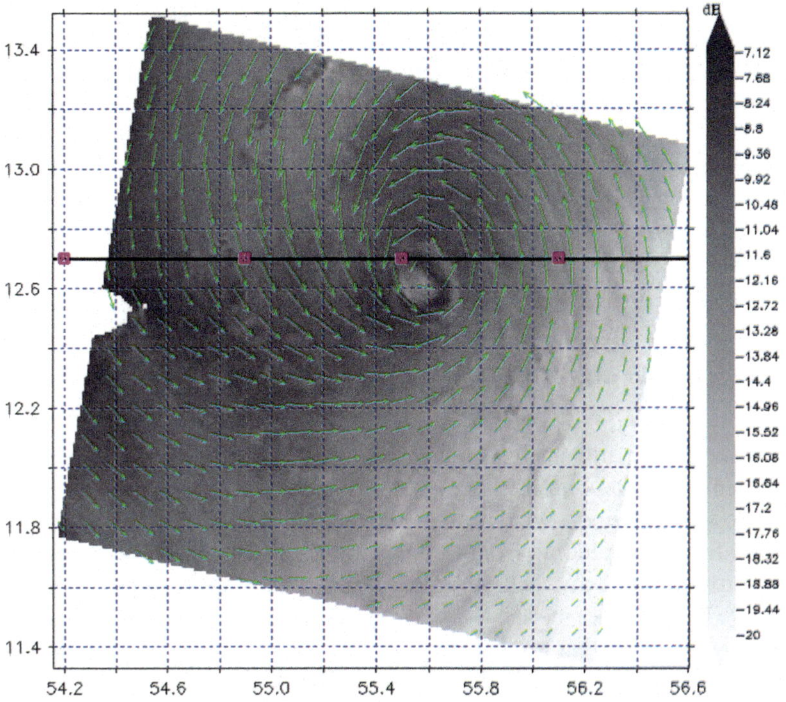

Fig. 9 RISAT-1 SAR image (VV-Pol) of cyclone Megh acquired on November 08, 2015. Wind direction derived from the SAR image (*green arrows*) overlaid on the SAR image. Cyclone track data (November 7–8, 2015) are from Joint Typhoon Warning Centre, shown by the *black line*

SAR is able to retrieve high-resolution and wide-coverage wind fields for applications where knowledge of the wind field is crucial. With the SAR-derived wind fields, one is now able to position atmospheric fronts and lows to a high accuracy. The high-resolution wind fields help in many applications like weather prediction, cyclone studies (Fig. 9), climate research, risk management, and commercial applications like energy production, ship routing, and structure design to operate in coastal environment. SAR-derived wind fields are also useful in wave retrieval by providing a first guess wind wave spectrum.

There are different types of techniques for wind retrieval. One of them is scatterometry-based approach. This is based on the idea that as the wind blows across the surface, it creates surface roughness commonly aligned with the wind. Consequently, the radar backscatter arising from this roughened surface is related to the wind speed and direction [37]. The dominant mechanism for scatterometer and SAR incident angles is Bragg's scattering, which means that the dominant return is proportional to the roughness of the ocean surface on the scale of the radar wavelength.

To retrieve winds from SAR imagery, image calibration is required to convert the digital number values of the imagery into the backscattered power. Calibration process requires the information of the external calibration constant and the local incidence angles. The calibrated SAR image is then inverted to obtain the wind speed using a Geophysical Model Function (GMF) and using auxiliary information of wind direction from NWP models or based on the information embedded in the data itself seen as wind streaks. Most of the GMF developed presently are based on VV polarization SAR images. However, the wind retrievals from HH polarized images are performed using an azimuth and incident angle-dependent parameterization for the effective polarization ratio [50]. SAR is also capable of estimating high-magnitude cyclonic winds with good accuracy [36].

5.4 Oceanographic Applications of SAR

SAR provides a two-dimensional image of the sea surface. Surface waves can be clearly seen in SAR images. Since the surface waves are formed primarily in response to surface winds, winds can be derived by SAR. Detection of upper layer circulation features including fronts, eddies, upwelling, internal waves, tidal circulation, bottom topography, and ship speeds have been demonstrated using SAR data. Due to lack of space-borne data between SEASAT in 1978 and ERS-1 in 1991, the studies were somewhat limited. Most of the studies have been obtained through airborne system; however, since the launch of ERS-1, SAR data has been regularly available and a large number of demonstrative studies have been conducted.

Ocean surface currents can also be retrieved from SAR imagery. However, this requires along-track interferometric configuration of the SAR antenna. This configuration has two antennas. One transmits and receives in usual way while the other antenna is used for only receiving. Interferometric configuration is not common for all the available SAR sensors. From the SAR systems without interferometric capability surface currents can be measured but only along the range direction (single component only) by using Doppler shifts. However, one-component currents thus retrieved have limited usability.

Monitoring of coastal bathymetry is vital for the exploitation of living and non-living resources, operations on engineering structures, and ocean circulation studies. The estimation of shallow water bathymetry depends upon the refraction of deepwater wavelength and wave direction in the shallow region. With the propagation of long waves toward the shallow region, waves start feeling bottom, and their wavelength as well as direction changes [43, 60]. Another method of deriving bathymetry is more complex, however it provides better estimates. In this approach, imaging mechanisms, consisting of various interaction processes, such as depth-current interaction, current-wave interaction, and wave-radar interaction, are being used. To estimate the bathymetry, the interaction mechanisms discussed above are inverted using a data-assimilation approach in conjunction with SAR data and a limited amount of in-situ data. [3, 9, 67].

It might be surprising that SAR, being a surface-viewing sensor, is also able to detect processes which take place within the sea and particularly at the thermocline.

Internal waves are waves that travel underneath the ocean surface. They are generated and propagated along the interface between waters of different densities. Due to large amplitude and orbital velocities, manifestations of these waves can be seen on the surface. The signature on the surface of ocean due to internal waves is related to the convergence of surface velocities at the surface above the slope behind the wave crest. Characteristics of the internal waves from a SAR image can be estimated using Fast Fourier Transform Technique (FFT). Pollution of the sea surface by mineral or petroleum oil is a major environmental problem. Despite the International Convention for Prevention of Pollution from ships, large quantities of mineral oil are still being discharged from ships in these special areas. SAR images are also very useful in the detection of oil spills.

The strong signals from targets like ships make SAR systems particularly useful for detecting vessels at sea. Ships are detected by three mechanisms: by observing radar backscatter directly from the ships; by detecting the wake of ships, and by identifying surface slicks resulting from ship engines. Other phenomena revealed by SAR include the detection of current patterns, eddies, and gyres, by their influence on surface waves. In addition to ocean features, SAR imagery is also being used to detect and identify various features such as ice type, ice edge, icebergs, and ice islands. SAR-derived wave spectra have the capability of assimilation in wave prediction models; however, operational use of such spectra or wind is limited by the lower repetivity and smaller coverage over the oceans. The Advanced SAR (ASAR) onboard EnviSAT as well as Sentinel-1 has an additional wave-mode acquisition dedicated for the retrieval of waves and winds only.

5.5 *Future Advancements in SAR*

For the last four decades, high-resolution imageries over the ocean surfaces captured by various tandem SAR missions have been providing resourceful information to the research community. At present scenario SAR is capable of working in monostatic mode, which suffers from receiving a major portion of the backscattered signals. To avoid this, the idea of bistatic SAR has already been conceptualized. Efforts are being dedicated presently over the globe to implement such systems practically. Also, several SAR constellations are being planned. ISRO has planned a follow-on mission of RISAT-1 as RISAT-1A. NASA and ISRO have also planned a joint dual frequency (L-band and S-band) SAR mission, to be launched in 2021, known as NISAR.

6 Remote Sensing of Ocean Salinity: Filling the Missing Gap in Ocean Observation

Sea surface salinity (SSS) variations are key indicators of the hydrological cycle encompassing evaporation, precipitation, freezing/melting of ice, and river run-off. There have been several studies that highlight the importance of SSS for ocean circulation and climate change. A special section on ocean salinity in the *Journal of Geophysical Research* [45] highlights the importance of this parameter in a wide variety of ocean studies.

Ocean salinity can be measured accurately with ships, buoys, and Argo floats at different depths in the ocean, but such measurements are very sparse. Although with the availability of the Argo floats, the salinity observations have considerably increased, still satellite-based observations with better spatial and temporal coverage hold a very good promise. Ocean average surface salinity is about 35 psu with a range of 32–37 psu, however the regions which are strongly affected by river water, salinity can go down to as low as 26–27 psu. Salinity retrieval from space is relatively a new concept. Soil Moisture and Ocean Salinity (SMOS) and Aquarius were the dedicated space-borne salinity missions that paved the way for a new era in ocean remote sensing.

6.1 *Satellite Instruments for Salinity*

The satellite, Soil Moisture and Ocean Salinity (SMOS), for the estimation of ocean surface salinity was launched in November 2009 by ESA. SMOS makes observations at 1.4 GHz using Microwave Imaging Radiometer with Aperture Synthesis (MIRAS) instrument for the surface salinity observation over ocean and surface soil moisture over land. SMOS used a dual-polarized L-band radiometer and adopted the 2D aperture synthesis technique to achieve a ground resolution better than 50 km without putting a large antenna into the orbit [40]. Subsequently, NASA launched Aquarius mission in June 2011 that carried three radiometers and a scatterometer having swath of 390 km. Aquarius provided salinity using 1.4 GHz passive microwave measurements with an accuracy of 0.2 psu on a monthly scale. Measurement of the ocean surface salinity from Aquarius was based on a real aperture 3-beam push-broom design. Aquarius was a dedicated surface salinity mission with enhanced capability in terms of better signal-to-noise ratio. Unfortunately, it suffered from failure in the power supply, and the mission ended in June 2015. The Soil Moisture Active and Passive (SMAP) mission by NASA (in collaboration with JAXA) was launched in January 2015. Although the primary objective of SMAP was the estimation of soil moisture over land, its 1.4 GHz passive radiometer had the potential for the estimation of ocean surface salinity. The SMAP [7, 22] mission, with both active and passive instruments, provides SSS at a spatial resolution of 40 km over a wide swath of ~1000 km, which is a clear advantage over coarser resolution (100–150 km) and narrower swath (390 km) from Aquarius. The L-band synthetic aperture radar (active sensor) onboard SMAP stopped functioning during July 2015, but salinity data continues to be derived from this instrument.

6.2 Measurement Principles and Challenges for Salinity Retrieval from Space

Theoretical basis for ocean salinity retrieval from passive microwave radiometric measurements is to exploit the sensitivity of emission to ocean salinity through its effect on dielectric constant of water. Dielectric constant of water decreases with the increase in salt content. Figure 1 shows the relative sensitivity of microwave radiometer frequencies for various oceanic parameters which shows that the sensitivity of microwave brightness temperature for sea surface salinity is maximum towards the lower frequency. 1.4 GHz is considered the best frequency for salinity retrieval as below this frequency there is significant radio frequency interference due to man-made RF transmitters. This frequency is least sensitive to SST, surface roughness, atmospheric water vapor, and liquid water content. Hence, salinity can be retrieved primarily from passive radiometer at L-band microwave frequency, with scatterometer or synthetic aperture radars used for correcting the surface roughness effect. For salinity retrieval, normally mono-frequency is preferred. The salinity retrieval algorithm is normally based on an iterative convergence approach which minimizes the difference between the satellite radiometer-measured brightness temperature and those generated from forward radiative transfer model. Forward modeling is performed for ocean surface emissivity which depends on the sea state, SST, viewing angle, and polarization. RT model also includes the atmospheric effects, galactic radiation contamination, and the sun-glint effect.

One of the very important points in salinity retrieval is that although the sensitivity is very small one has to take into account the effect due to sea surface roughness, SST, foam, sun-glint, rainfall, ionospheric effects, and galactic background impact on the brightness temperature. Another very important point to be remembered is the low sensitivity of brightness temperature to salinity (0.75 K at 30 °C, 0.5 K at 20 °C, and 0.25 K at 0 °C) that puts a stringent requirement on the radiometer to have a very high signal-to-noise ratio. Additional requirements are multiangular and multipolarization measurements. Low sensitivity of brightness temperature to salinity requires more energy to be gathered so that it is above the noise level, and hence footprint of the radiometer needs to be large. The active sensor scatterometer, onboard Aquarius, and the synthetic aperture radar, onboard SMAP, were used for correcting surface roughness effect on the brightness temperature. As mentioned earlier, SAR sensor onboard SMAP has stopped working. SMAP instrument [7] employs a single horn, with dual-polarization and dual-frequency capability (radiometer at 1.41 GHz and radar at 1.26 GHz). The SMAP radiometer provides a real aperture resolution in which the dimensions of the 3 dB antenna footprint projected on the surface meets the 40-km spatial resolution requirement. The radiometer measures four Stokes parameters at 1.41 GHz to provide a capability to correct for possible Faraday rotation caused by the ionosphere. The chosen 6-AM/6-PM sun-synchronous orbit configuration also minimizes such Faraday rotation.

6.3 Accuracy and Spatiotemporal Sampling

Global observation of ocean salinity with an accuracy of 0.1 psu, every 10 days at 200 km spatial resolution, was envisaged under GODAE. The passive microwave radiometers have a major limitation in resolving small-scale salinity gradients due to their coarser spatial resolution (~40–100 km). The SMOS instrument uses a synthetic aperture antenna that yields multiangular brightness temperature mapping at about 40-km resolution. The Aquarius mission had a revisit of 7 days and provided global maps of SSS with accuracy of 0.2 psu at 150 km resolution on a monthly timescale. The SMAP has a higher spatial resolution of 40 km and a wider swath of 1000 km that enable global coverage in 2–3 days. The finer spatial resolution of the SMAP makes it noisier than Aquarius, but the noise is reduced by larger temporal averaging. Therefore, SMAP provides only 8-day running mean and monthly SSS product.

In the Indian Ocean, Ratheesh et al. [56] validated Aquarius SSS with the Argo data for the period 2011–2012. The coefficient of determination between SSS and reference measurements was found to be 0.84, and root mean square difference (RMSD) was about 0.45 psu. A similar analysis was performed by Ratheesh et al. [55] with daily level 3 product of SMOS SSS on a grid of 0.25 × 0.25 deg. Limited validation showed SMOS SSS accuracy of 0.36 psu and 0.34 psu at two RAMA buoys in the Indian Ocean. Drucker and Riser [17] found that Aquarius level 2 salinity differs from Argo salinity by +0.018+/−0.42 psu on a global scale.

Aquarius was declared nonfunctional in May 2015, leaving, once again, a huge void in salinity measurements over the global oceans. Since April 2015, SSS data from NASA's Soil Moisture Active Passive sensor is made available at http://www.remss.com/missions/smap [49]. Figure 10 shows a plot of sea surface salinity from SMAP averaged for August 21–September 18, 2015 in the Bay of Bengal. During this time, in-situ salinity measurements from the thermosalinograph were available from the US R/V Roger Revelle under the joint ASIRI-OMM program [68]. These observations along the ship track are overlaid on SMAP salinity (Fig. 10). Figure 10 indicates that the quality of salinity from SMAP is promising.

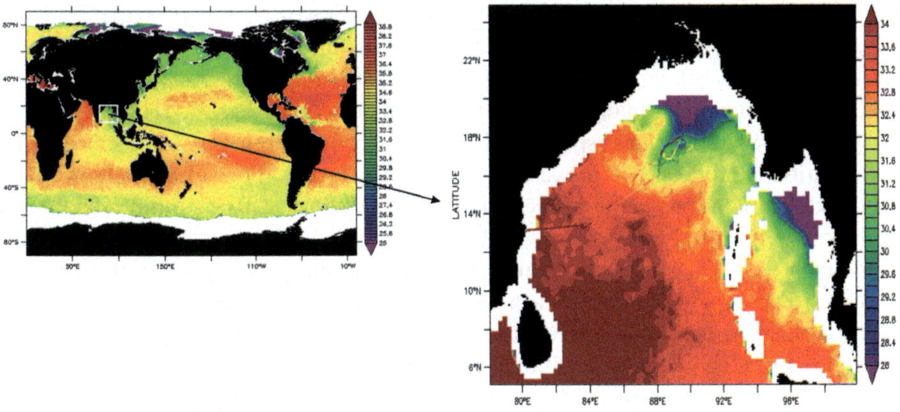

Fig. 10 Averaged (August 21–September 15, 2015) sea surface salinity (psu) from Soil Moisture Active and Passive Sensor. Surface salinity data from thermosalinograph of R/V RogerRevelle cruise in the Bay of Bengal for the same period are overlaid on SMAP salinity (blown-up image) (TSG data courtesy: Prof Jonathan D. Nash, Oregon State University. SMAP data source: Beta version release by Remote Sensing Systems – https://remss.com)

6.4 Applications of Satellite-Derived Salinity

Ocean surface salinity is increasingly being recognized as a key parameter in ocean-atmosphere interaction. Apart from taking part in this interaction, surface salinity plays a vital role in oceanography in the stand-alone mode also by influencing ocean dynamics and thermodynamics [20]. In the tropical Pacific, an in-depth understanding of the dominant El Nino Southern Oscillation (ENSO) event using numerical ocean circulation models is just not possible without a faithful representation of SSS in these models [39, 48, 51]. Knowledge of the global salt distribution and its variations is critical to understanding the role of oceans in the climate system. It is well established that ocean circulation, air-sea heat exchanges, and heat transport play important roles in regulating the climate. Three-dimensional ocean flow which is famously known as "thermohaline circulation" is largely governed by salinity variations due to evaporation and ice melting/freezing. Ocean surface salinity is linked to evaporation minus precipitation. Hence, measuring salinity will be helpful to constrain the estimation of E-P and will help in better estimate of air-sea flux. Using 50 years' of observed global surface salinity changes and climate model simulations, there is evidence of intensified global water cycle at a rate of 8 +/– 5% per degree of surface temperature warming [19].

The hydrological cycles in the ocean and atmosphere are intimately linked. This linkage is even much stronger in the river-dominated Bay of Bengal due to heavy river discharge. The stability caused by freshening isolates the upper layer of northern Bay of Bengal from its interior, which results in a strong barrier layer. Observations suggest that storm-induced vertical mixing is limited to the upper warm layer which favors intense tropical cyclones in this region. Salinity from space will help unravel many unanswered processes at play in such regions.

Another very important aspect of global salinity measurements from space is its assimilation in models to improve the ocean state estimation for ocean process studies. Model simulations in the tropical Indian Ocean show surface salinity variability in the range from 0.2 to 1.5 psu, with larger values in regions with strong seasonal transitions of surface currents (south of India) and along the coast in the Bay of Bengal [62]. Assimilation of surface salinity from Aquarius (NASA Salinity Mission) and surface temperature from AVHRR [12, 13]) has shown positive impact on Indian Ocean equatorial jet simulation.

7 End Remarks

This chapter summarized the development of satellite oceanography over the years. It is now well established that satellite observations are integral components of ocean research and applications. Immense work has been carried out using the data from various satellites working in different regions of the *em* spectrum. Still there are several challenges in terms of sensor resolution, retrieval accuracy, and continuity of the missions. Exploiting the full potential of altimetry in the coastal region is an active area of research. Making use of opportunity signals from global navigational systems for extracting useful ocean information is another emerging field. Nadir altimeters are in operation for so many years now, new concept of swath altimetry (SWOT mission) will revolutionize the science of altimetry. Improvement in spatial resolution with microwave instruments requires advanced signal processing. With a lot of data coming now from in situ platforms, one needs to understand how best one can synergistically combine the in-situ and satellite data to address many unresolved problems of ocean research. There is also a need to adopt sophisticated data-mining approach to handle vast amount of ocean data obtained from satellite platforms. Another important aspect is to have coherent and coordinated efforts by different space agencies to plan out satellite constellation and maintain the continuity of the missions. In this regard, multimission science teams and virtual constellation teams under the umbrella of Committee on Earth Observation Satellites (CEOS) are working together to define data policy and evolve user requirements.

Acknowledgments The content presented in this chapter is the result of help provided by many of the authors' colleagues at the Space Applications Centre, Ahmedabad. In particular, the authors would like to express their sincere gratitude to Dr Sujit Basu, Dr Pradeep Thapliyal, Dr Neeraj Agarwal, Sh Aditya Chaudhary and Dr Abhisek Chakraborty. In-situ salinity data used in Fig. 10 are from OMM-ASIRI ship cruise (R/V Roger Revelle) in the Bay of Bengal. Ocean Mixing and

Monsoon (OMM) is a multi-institutional project funded by MoES. The authors thank Dr R. Venkatesan of National Institute of Ocean Technology (NIOT), Chennai, India for giving them the opportunity to contribute this chapter. They are grateful to the reviewer who provided thoughtful comments and suggestions. HF Radar and buoy data were obtained from Indian National Centre for Ocean Information Services (INCOIS), Hyderabad, India.

References

1. Alpers W (1983a) Imaging ocean surface waves by synthetic aperture radar: a review. Satellite Microwave Remote Sensing. Hg. Allan, T. D, England: Ellis Horwood, Chapter 6, pp 107–119
2. Alpers W (1983b) Monte Carlo simulations for studying the relationship between ocean wave and synthetic aperture radar image spectra. J Geophys Res Oceans 88(C3):1745–1759
3. Alpers W, Hennings I (1984) A theory of the imaging mechanism of underwater bottom topography. J Geophys Res 89:1029–10546
4. Alpers W, Ross DB, Rufenach CL (1981) On detectability of ocean surface waves by real and synthetic aperture radar. J Geophys Res 86:6481–6498
5. Amarouche L, Thibaut P, Zanife O-Z, Dumont P, Vincent J-P, Steunou N (2004) Improving the Jason-1 ground retracking to better account for attitude effects. Mar Geod 27:171–197
6. Bhowmick SA, Kumar R, Kiran Kumar AS (2014) Cross calibration of the OceanSAT-2 scatterometer with QuikSCAT scatterometer using natural terrestrial targets. IEEE TGRS 52(6). https://doi.org/10.1109/TGRS.2013.2272738
7. Brown ME, Escobar V, Moran S, Entekhabi D, O'Neill PE, Njoku EG, Doorn B, Jared K (2013) NASA's Soil Moisture Active Passive (SMAP) mission and opportunities for applications users. Bull American Met Soc 94(8):1125–1128
8. Brown GS (1977) The average impulse response of a rough surface and its applications. IEEE Trans Antennas Propag AP-25:67–74
9. Calkoen CJ et al. (1993) The imaging of sea bottom topography with polarimetric P-, L-, and C-band SAR. Report BCRS project 2.
10. Chakraborty A, Deb SK, Sikhakolli R, Gohil BS, Kumar R (2013a) Intercomparison of OSCAT winds with numerical model generated winds. IEEE Geosci Remote Sens Letts 10:260–262
11. Chakraborty A, Kumar R, Stoffelen A (2013b) Validation of ocean surface winds from the OCEANSAT-2 scatterometer using triple collocation. Rem Sens Letts 4(1):85–94
12. Chakraborty A, Sharma R, Kumar R, Basu S (2014a) A SEEK filter assimilation of sea surface salinity from Aquarius in an OGCM: Implication for surface dynamics and thermohaline structure. J Geophys Res (Oceans) 119:4777–4796
13. Chakraborty A, Sharma R, Kumar R, Basu S (2014b) An OGCM assessment of blended OSCAT winds. J Geophys Res (Oceans) 119:173–186
14. Chakraborty A, Kumar R (2013) Generation and validation of analysed wind vectors over the global oceans. Remote Sens Lett 4(2):114–122
15. Chelton DB, Ries JC, Haines BJ, Fu L-L, Callahan PS (2001) Satellite altimetry. In: Fu L-L, Cazenave A (eds) Satellite altimetry and Earth sciences: a handbook of techniques and applications. Academic, San Diego, pp 1–132
16. Donlon CJ et al (2007) The GODAE high resolution sea surface temperature pilot project (GHRSST-PP). Bull Am Meteorol Soc 88:1197–1213
17. Drucker R, Riser S (2014) Validation of Aquarius sea surface salinity with Argo: analysis of error due to depth of measurement and vertical salinity stratification. J Geophys Res (Oceans) 119:4626–4637
18. Dufau C, Mercier F, Bouffard J, Bronner E, Lombard A, Picot N, CLS PISTACH Team (2009) Improved Jason-2 altimetry products for coastal ocean: PISTACH. Geophysical Research Abstracts, vol 11, EGU2009-8980, EGU General Assembly

19. Durack PJ, Wijffels SE, Mater RJ (2012) Ocean salinities reveal strong global water cycle intensification during 1950 to 2000. Science 336:455–458
20. Durand F, Gourdeau L, Delcroix T (2003) Can we improve the representation of modeled ocean mixed layer by assimilating surface-only satellite data? A case study for the tropical Pacific during the 1997–1998 El Nino? J Geophys Res 108(C6):3200. https://doi.org/10.1029/2002JC001603
21. Elachi C, Brown WE (1977) Models of radar imaging of the ocean surface waves. IEEE Trans Antennas Propag AP-25:84–95
22. Entekhabi D et al (2010) The Soil Moisture Active Passive (SMAP) Mission. Proc of the IEEE 98.5 (2010):704–716 Copyright 2010 IEEE.
23. Flament P, Firing J, Sawyer M, Trefois C (1994) Amplitude and horizontal structure of a large sea surface warming event during the Coastal Ocean Dynamics Experiment. J Phys Oceanogr 24:124–139
24. Freilich MH, Dunbar RS (1999) The accuracy of the NSCAT-1 vector winds: comparisons with National Data Buoy Center buoys. J Geophys Res 104(C5):11231–11246
25. Fu L-L, Alsdorf D, Rodriguez E, Morrow R, Mognard N, Lambin J, Vaze P, Lafon T (2010) The SWOT (Surface Water and Ocean Topography) Mission: Spaceborne Radar Interferometry for Oceanographic and Hydrological Applications. In: Hall J, Harrison DE, Stammer D (eds) Proceedings of the "OceanObs'09: Sustained Ocean Observations and Information for Society" Conference (Vol. 2), Venice, Italy, 21–25 September 2009, ESA Publication WPP-306
26. Fu L-L, Douglas A, Rodriguez E, Morrow R, Mognard N, Lambin J, Vaze P, Lafon T (2009) The SWOT (Surface Water and Ocean Topography) Mission: Spaceborne Radar Interferometry for Oceanographic and Hydrological Applications. In Proceedings of OceanObs'09, Venice, Italy, Sept. 21–25, 2009
27. Gleason ST, Adjrad M (2005) An Attempt to Sense Ocean Winds and Waves Empirically Using Bistatic GNSS Reflections in Low Earth Orbit. In Proceedings of the IEEE International Geosci and Rem Sens Symposium , Seoul, South Korea, 25–29 July 2005
28. Gohil BS, Sarkar A, Agarwal VK (2008) A new algorithm for wind vector retrieval from scatterometer. IEEE Geosci Rem Sens Letts 5:387–391
29. Gohil BS, Sharma P, Shikhakoli R, Sarkar A (2010) Directional stability and conservation of scattering (DiSCS) based directional-ambiguity removal algorithm for improving wind-fields from scatterometer: a QuikSCAT example. IEEE Geosci Rem Sens Letts 7:592–595
30. Gommenginger C et al (2011) Retracking altimeter waveforms near the coasts. In: Vignudelli S, Kostianoy A, Cipollini P, Benveniste J (eds) Coastal altimetry. Springer, Berlin Heidelberg, pp 61–102
31. Goni G et al (2010) The Ocean Observing System for Tropical Cyclone Intensification Forecasts and Studies. In: Hall J, Harrison DE, Stammer D (eds) Proceedings of the "OceanObs'09: Sustained Ocean Observations and Information for Society" Conference (Vol. 2), Venice, Italy, 21-25 September 2009, ESA Publication WPP-306
32. Guo JY, Hwang CW, Chang XT, Liu YT (2006) Improved threshold retracker for satellite altimeter waveform retracking over coastal sea. Prog Nat Sci 16:732–738
33. Halimi A, Mailhes C, Tourneret J-Y, Thibaut P, Boy F (2013) Parameter estimation for peaky altimetric waveforms. IEEE Trans Geosci Rem Sens 51:1568–1577
34. Hasselmann K et al (1985) Theory of synthetic aperture radar ocean imaging: a MARSEN view. J Geophys Res (Oceans) 90(C3):4659–4686
35. He R, Weisberg RH, Zhang H, Muller-Karger FE, Helber RW (2003) A cloud-free, satellite-derived, sea surface temperature analysis for the West Florida Shelf. Geophys Res Lett 30(15):1811. https://doi.org/10.1029/2003GL017673
36. Hersbach H, Stoffelen A, Haan De S (2007) An improved C-band scatterometer ocean geophysical model function: CMOD5. J Geophys Res 112:1–18. https://doi.org/10.1029/2006JC003743
37. Horstmann J, Koch W, Lehner S, Tonboe R (2000) Wind retrieval over the ocean using synthetic aperture radar with C-band HH polarization. IEEE Trans Geosci Remote Sens 38:2122–2131
38. Jackson CR, Apel JR (2004) Synthetic aperture radar marine user's manual. US Department of Commerce, National Oceanic and Atmospheric Administration, National Environmental Satellite, Data, and Information Serve, Office of Research and Applications

39. Ji M, Reynolds RW, Behringer DW (2000) Use of Topex/Poseidon sea level data for ocean analyses and ENSO prediction: some early results. J Clim 13:216–231
40. Kerr YH, Waldteufel P, Wigneron J-P, Martinuzzi J-M, Font J, Berger M (2001) Soil moisture retrieval from space: The Soil Moisture and Ocean Salinity (SMOS) mission. IEEE Trans Geosci Rem Sens 39:1729–1735
41. Koblinsky C, Gaspar P, Lagerloef G (eds) (1992) The future of spaceborne altimetry – oceans and climate change: a long-term strategy. Joint Oceanographic Institutions, Inc., 85 pp
42. Krishna Rao P, Smith WL, Koffler R (1972) Global sea-surface temperature distribution determined from an environmental satellite. Mon Weather Rev 100(1):10–14
43. Kumar R, Sarkar A, Pandey PC (1999) Estimation of ocean depths off Goa coast using ERS-1 Synthetic Aperture Radar. Cont Shelf Res 19(2):171–181
44. Kumar R, Chakraborty A, Parekh A, Shikakolli R, Gohil BS, Kiran Kumar AS (2013) Evaluation of Oceansat-2 derived ocean surface winds using observations from global buoys and other scatterometers. IEEE TGRS 51(5):2571–2576
45. Lagerloef GSE (2002) Introduction to the special section: the role of surface salinity on upper ocean dynamics, air sea interaction and climate. J Geophys Res 107(C12):8000. https://doi.org/10.1029/2002JC001669
46. Martin TV, Zwally HL, Brenner AC, Bindschadler RA (1983) Analysis and retracking of continental ice sheet radar altimeter waveform. J Geophys Res 88:1608–1616
47. Maus S, Green CM, Fairhead JD (1998) Improved ocean-geoid resolution from retracked ERS-1 satellite altimeter waveforms. Geophys J Int 134:243–253
48. McPhaden MJ, Delcroix T, Hanawa K, Kuroda Y, Meyers G, Picaut J, Swenson M (2001) The El Nino/Southern Osclillation (ENSO) observing system, in Observing the Ocean in the 21st Century. In Aust. Bur. of Meteorol., Melbourne, Victoria, Australia, 17 pp
49. Meissner T, Wentz FJ, Scott J (2015) Remote Sensing Systems SMAP level 3 Ocean Surface Salinities [Running 8-day, Monthly] on 0.25 deg grid, version 1.0 (BETA). www.remss.com/missions/smap
50. Mouche AA, Hauser D, Daloze JF, Guerin C (2005) Dual polarization measurements at C-band over the oceans: results from airborne radar observations and comparison with ENVISAT ASAR data. IEEE Trans Geosci Remote Sensing 43:753–769
51. Murtugudde R, Busalacchi AJ (1998) Salinity effects in a tropical ocean model. J Geophys Res 103:3283–3300
52. Pascual A, Faugère Y, Larnicol G, LeTraon P-T (2006) Improved description of the ocean mesoscalr variability by combining four satellite altimeters. Geophys Res Letts 33(2):13–16
53. Plant WJ (1997) A model for microwave Doppler sea return at high incidence angles: Bragg scattering from bound, tilted waves. J Geophys Res 102:21131–21146
54. Prigent C, Aires F, Bernardo F, Orlhac J-C, Goutoule J-M, Roquet H, Donlon C (2013) Analysis of the potential and limitations of microwave radiometry for the retrieval of sea surface temperature: definition of MICROWAT, a new mission concept. J Geophys Res Oceans 118:3074–3086
55. Ratheesh S, Mankad B, Basu S, Sharma R (2013) Assessment of satellite-derived sea surface salinity in the Indian Ocean. IEEE Geosci Rem Sens Lett 10(3):428–431. https://doi.org/10.1109/LGRS.2012.2207943
56. Ratheesh S, Sharma R, Sikhakolli R, Kumar R, Basu S (2014) Assessing sea surface salinity derived by Aquarius in the Indian Ocean. IEEE Geosci Rem Sens Letts 11:719722
57. Robinson IS (2005) Measuring the oceans from space. Praxis Publishing, Chinchester, 655 pp
58. Rodriguez E (1988) Altimetry for non-Gaussian oceans: Height biases and estimation of parameters. J Geophys Res 93:14107–14120
59. Sandwell DT, Smith WHF (2005) Retracking ERS-1 altimeter waveforms for optimal gravity field recovery. J Geophys Res 163:79–89
60. Schuchman RA, Lyzenga DR, Meadows GA (1985) Synthetic aperture radar imaging of ocean-bottom topography via tidal-current interactions: theory and observations. Int J Rem Sens 6:1179–1200

61. Severini J, Mailhes C, Tourneret J-Y, Thibaut P (2008) Bayesian estimation of altimeter echo parameters. In Proc. IEEE IGARSS, Boston, MA 3, pp III-238–III-241
62. Sharma R, Agarwal N, Momin IM, Basu S, Agarwal VK (2010) Simulated sea surface salinity variability in the Tropical Indian Ocean. J Clim 23:6542–6554
63. Steunou N, Desjonqueres JD, Picot N, Sengenes P, Noubel J, Poisson JC (2015) AltiKa Altimeter: instrument description and in flight performance. Mar Geod 38(S1):22–42
64. Stocker TF et al (2013) Technical Summary. In: Climate Change 2013: The Physical Science Basis. Contribution of Working Group I to the Fifth Assessment Report of the Intergovernmental Panel on Climate Change [Stocker, T.F., D. Qin, G.-K.Plattner, M.Tignor, S.K. Allen, J. Boschung, A. Nauels, Y. Xia, V. Bex andP.M. Midgley (eds.)]. Cambridge University Press, Cambridge, UK and New York, pp 33–115, doi:https://doi.org/10.1017/CBO9781107415324.005
65. Ulaby FT, Moore RK, Fung AK (1986) Microwave remote sensing: active and passive, Voll III: Volume scattering and emission theory, advanced systems and applications. Artech House Inc., Dedham, 1100 pp
66. Vincent P, Steunou N, Caubet E, Phalippou L, Rey L, Thouvenot E, Verron J (2006) AltiKa: A Ka-band altime-try payload and system for operational altimetry during the GMES period. Sensors 6:208–234
67. Vogelzang J (1989) A comparison of the hydrodynamic modulation in some existing models. Int J Rem Sens 10(9):1503–1518
68. Wijesekera HW et al (2016) ASIRI: an ocean atmosphere initiative for Bay of Bengal. Bull Amer Meteor Soc 97:1859–1884. https://doi.org/10.1175/BAMD-D-14-00197.1
69. Willis JK, Chambers DP, Kuo C-Y, Shum CK (2010) Global sea level rise: recent progress and challenges for the decade to come. Oceanography 23(4):26–35. https://doi.org/10.5670/oceanog.2010.03
70. Wilson S and Co-authors (2010) The Ocean Surface Topography Constellation: The Next 15 Years in Satellite Altimetry. In: Hall J, Harrison DE, Stammer D (eds) Proceedings of OceanObs'09: Sustained Ocean Observations and Information for Society (Vol. 2), Venice, Italy, 21–25 September 2009, ESA Publication WPP-306, doi:https://doi.org/10.5270/OceanObs09.cwp.92
71. Wingham DJ, Rapley CG, Griffiths HJ (1986) New technique in satellite altimeter tracking systems. IEEE-IGARSS 1:185–190
72. Yu L, McPhaden MJ (2011) Ocean preconditioning of cyclone Nargis in the bay of bengal: interaction between rossby waves, surface fresh waters, and sea surface temperatures. J Phys Oceanogr 41(9):1741–1755
73. Zavorotny VU, Voronovich AG (2000) Scattering of GPS signals from the ocean with wind remote sensing application. IEEE Trans Geosci Rem Sens 38:951–964

Near Real-Time Underwater Passive Acoustic Monitoring of Natural and Anthropogenic Sounds

Mark F. Baumgartner, Kathleen M. Stafford, and G. Latha

Abstract Passive acoustic monitoring takes advantage of the relative opacity of the ocean to sound. Traditionally, long-term monitoring has employed archival instruments from which data are accessed only when the recording instrument is retrieved. Recent advances in low-power instrumentation and computational speed allow passive acoustic data to be collected, processed and relayed to shore in near real time from fixed and mobile platforms deployed at the sea floor, in the water column or on the ocean's surface. Measurements of ambient noise provide insight into natural sound sources, such as rainfall, earthquakes or marine animals, as well as anthropogenic sound sources, such as shipping or resource extraction. Near real-time passive acoustic measurements allow scientists and agencies to monitor shipping, observe underwater seismicity and detect the presence of critically endangered large whales. The development and use of real-time passive acoustic monitoring systems will grow in coming decades to help better manage increasing industrialization of the oceans. This chapter reviews the capabilities of real-time passive acoustic monitoring to address civilian scientific needs. The currently available suite of instrumentation and platforms used for passive acoustic monitoring are discussed along with the wide variety of measurements that can be made with this technology. Finally, examples of how real-time passive acoustic monitoring has improved our understanding of the ocean are presented.

M.F. Baumgartner (✉)
Biology Department, Woods Hole Oceanographic Institution, Woods Hole, MA, USA
e-mail: mbaumgartner@whoi.edu

K.M. Stafford
Applied Physics Laboratory, University of Washington, Seattle, WA, USA

G. Latha
Ocean Acoustics & Modeling, National Institute of Ocean Technology, Chennai, India

R. Venkatesan et al. (eds.), *Observing the Oceans in Real Time*, Springer Oceanography, DOI 10.1007/978-3-319-66493-4_10

1 Introduction

Sound in the ocean has been used to study a wide variety of phenomena ranging from rainfall in the deep ocean to the migration patterns of great whales. Researchers can study the acoustics of both abiotic and biotic sources because of the relative transparency of the ocean to sound. Unlike light or chemical signals that dissipate within tens to hundreds of meters underwater, sound waves can travel for tens to hundreds, even thousands of kilometers with little attenuation. The degree to which sound propagates through the ocean depends on its frequency; a high-frequency echosounder on a fishing boat can provide information on schooling fish a few hundred meters away, while the low-frequency sounds produced by the propeller of a large cargo ship may be detected hundreds of kilometers away. Both the production and reception of underwater sound have been used in military applications for over a century; however, the widespread civilian use of passive acoustic monitoring is a relatively recent development that rapidly has become a common tool to study properties of the world's oceans. Passive acoustic monitoring involves instrumentation that listens for sounds, but does not produce any sound itself (in contrast to active acoustic instrumentation, which produces sound and records or processes the echo of the sound).

Traditionally, passive acoustic monitoring has been conducted with autonomous archival instruments that record audio in-situ; however, information derived from these recordings is unavailable until the instrument is recovered and the data are offloaded and analyzed. Critical ephemeral events, such as an earthquake or the presence of a vocalizing endangered species, cannot be detected until long after the event has occurred. For many research applications, such delays in data access and analysis are perfectly acceptable, but most passive acoustic recording systems are unhelpful for applications involving real-time response or where storage or recovery of audio recordings is not feasible. In such cases, access to the audio, or measurements derived from the audio, is needed in real time or near real time. Meeting this requirement is challenging, however, because audio is often sampled at much higher sampling rates than the transfer rates of most available communication systems. For example, the detection of dolphin whistles requires audio sample rates of at least 5 kHz (often much higher) or 10,000 bytes per second for a 16-bit audio system, and Iridium satellite communications currently support data transfer rates of only 300 bytes per second. Audio is therefore difficult to transfer from sea to shore in real time except over very high bandwidth communication channels (e.g., cables). Without the appropriate communication infrastructure, audio must be either sent in short noncontinuous snippets or processed in-situ to derive and relay measurements of interest. Fortunately, advances in instrumentation are now allowing the collection and processing of audio in-situ, enabling near real-time access to passive acoustic measurements.

The capability to detect the presence of sound sources via passive acoustic monitoring in real time (i.e., at the time of detection) or in near real time (e.g., within minutes or hours of a detection) is growing within the scientific community, and it is becoming more widely available on a variety of manned and unmanned platforms for both short-term and long-term monitoring applications. One of the tremendous strengths of combining passive acoustic monitoring with unmanned autonomous platforms is persistence: the ocean can be acoustically monitored in real time continuously over long timescales. This persistence is revolutionary. Monitoring marine mammals, for example, has been traditionally conducted by human observers from ships or aircraft, and relies on the fact that air-breathing marine mammals must return to the surface where they can be visually detected. However, visual detection is severely limited by conditions of low light, fog, rain, and high winds-observations can only be made during daytime in quite good weather conditions. Real-time passive acoustic monitoring relies on animals making sound, but it can be done continuously regardless of weather or light conditions, and with the use of long-endurance autonomous platforms, it can be done in remote locations that are difficult to access by human-occupied platforms. This approach is significantly less expensive than large-scale ship or aircraft operations. Most importantly, a real-time capability allows immediate action in response to detection events, which can support improved science and conservation efforts. In the case of marine mammal monitoring, real-time detections can trigger immediate responsive changes in industrial activities such as shipping, fishing, or seismic exploration to reduce impacts of these activities on marine animals, or it can alert scientists to locations where they can find study animals for follow-up research using photo-identification, biopsy, or tagging.

In this chapter, we review the capabilities of real-time passive acoustic monitoring to address civilian scientific needs. The chapter discusses the suite of instrumentation and platforms used for passive acoustic monitoring, the wide variety of measurements that can be made with this technology, and finally presents examples of how real-time passive acoustic monitoring has improved our understanding of the ocean.

2 Instruments

The instrument that is central to all passive acoustic monitoring is the hydrophone. Hydrophones work by converting acoustic energy from the water into electrical energy using a piezoelectric transducer that measures pressure changes produced by a sound wave. Hydrophones can be omnidirectional, sensing sounds from all directions around the instrument, or directional, providing bearings to a sound. When used in arrays, techniques such as beamforming [19], normal mode backpropagation [53], or hyperbolic fixing via time difference of arrival [10, 110] can be used to estimate bearings, ranges, and (or) locations of sounds. Directional hydrophones can have multiple pressure vector sensors and a single hydrophone that together provide a bearing to a sound [35].

3 Platforms

There are a wide variety of instrument configurations and platforms from which real-time passive acoustic monitoring can be accomplished. Platforms can be fixed, mobile, surface-bound, bottom-mounted, or profiled throughout the water column, and instrument data can be transferred to a ship or shore via a cable, radio, or satellite. Each platform has particular space and timescales over which it can operate (Fig. 1), and it is critical to match these scales to those of the motivating research questions or monitoring needs. One of the most important aspects of a platform is the noise generated by the operation of the platform itself. Buoyancy-driven platforms, such as ocean gliders and profiling floats, produce almost no self-noise and minimal low-frequency flow noise while passively sinking or floating through the water column. In contrast, oceanographic ships produce an abundance of broadband noise and induce flow noise that makes low-frequency monitoring impossible; for

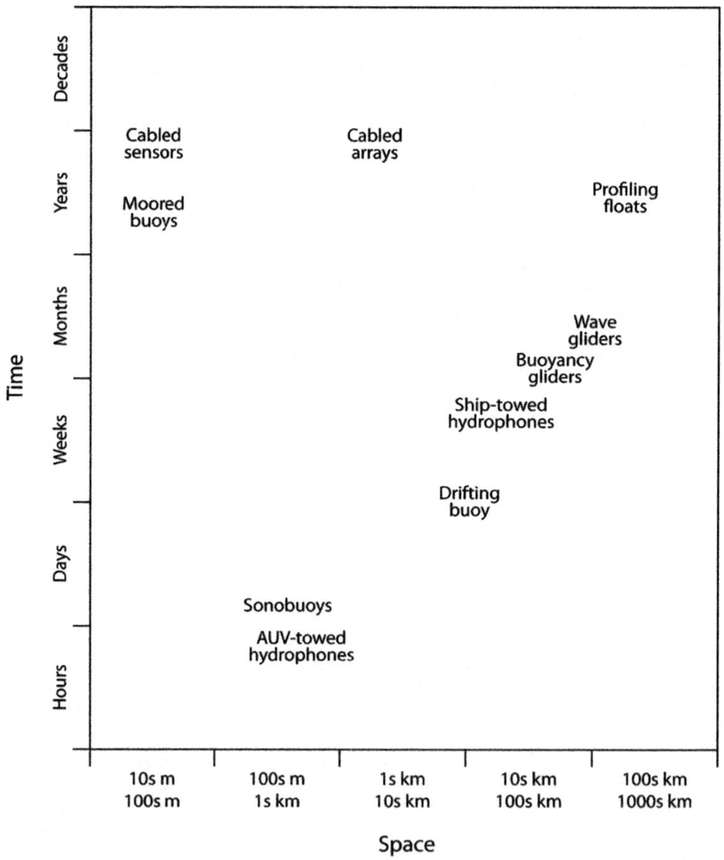

Fig. 1 Space and time scales of near real-time passive acoustic monitoring platforms

mid-frequency and high-frequency monitoring, the noise can only be mitigated with towed hydrophone arrays that can beamform to directionally monitor received sound at bearings that exclude the ship.

3.1 Fixed Platforms

Fixed platforms, such as moored buoys, cabled sensors, and cabled arrays, can offer extremely low self-noise when the hydrophones are mounted at or near the seafloor. Cabled sensors and arrays (i.e., those connected directly to shore) have the tremendous advantage of unlimited power and data storage, since the cable provides both power and a high bandwidth data communication channel. This allows for nearly continuous operation over very long timescales (years; although maintenance may still be required for biofouling, cable breaks, or instrument fatigue), and audio can be delivered directly to shore where it can be processed and stored in real time. Examples of cabled systems include (1) the Integrated Undersea Surveillance System that has been used to detect underwater earthquakes, volcanic eruptions, and track baleen whales [26, 97], (2) the Comprehensive Test Ban Treaty Organization (CTBTO) hydroacoustic stations that are designed to monitor illegal nuclear tests but are also capable of detecting Antarctic iceberg calving events [16], and (3) deep-sea observatories that can be used to monitor marine animals, shipping, and other anthropogenic activities [2].

In contrast to cabled acoustic sensors, autonomous moored sensors that transmit acoustic measurements in near real time must have a surface buoy to allow communications with shore via radio, cellular, or satellite communication systems (Fig. 2a). Like all autonomous platforms, the endurance of a moored buoy is limited by power, since it relies on batteries, and the bandwidth of the communication systems is quite low compared to a cabled sensor. Consequently, moored buoys are limited to transmitting measurements derived from audio collected in-situ, or to sending short snippets of audio recorded during detection events. Self-noise on moored systems with a surface expression can be substantial. Typical mooring components such as chains and shackles are exceedingly noisy if not properly treated (e.g., chains can be quieted by encasing them in urethane), and wire rope has the propensity to strum in strong currents, even when faired. Motion of the surface buoy caused by waves can create sloshing noise at the surface as water directly impacts the buoy, as well as impart motion in the mooring components, which may also cause noise. This motion can be dampened with the use of a compliant "stretch hose" [78] and subsurface flotation to isolate the lower part of the mooring (between the flotation and the seafloor where the passive acoustic instrument is located) from the motion of the surface buoy.

The challenge of delivering data from the submerged acoustic instrument to the surface is not trivial. If the instrument is close to the surface, electromagnetic (EM) cables can be used to transfer data from the instrument to an electronics package in the buoy for immediate or delayed transfer via radio, cellular, or satellite communications. While data delivery is straightforward, there are challenges to making

Fig. 2 Some near real-time passive acoustic monitoring platforms, including (**a**) moored buoy, (**b**) profiling float, (**c**) Slocum ocean glider, and (**d**) wave glider. Photographs (**a**, **b**, **d**) copyright Woods Hole Oceanographic Institution, (**c**) copyright The Nature Conservancy

acoustic measurements very close to the surface because of surface noise (unless the goal is to directly measure surface noise, such as waves or precipitation). Depending on the mooring design, these cables may or may not carry the strain of the buoy (i.e., mechanically connect the buoy to the anchor). For instruments that are placed deeper in the water column or at the seafloor, armored EM cables can be used to both connect the surface buoy to the anchor and transfer data from the depth to the surface. Inductive modems can also be used to transfer data from anywhere along a continuous mooring cable to the surface buoy. A signal can be induced in the mooring cable by an instrument through the coupler that attaches the instrument to the mooring cable, and that signal is received through a similar mechanism at the surface buoy. This system has been used in some moored ocean observatory systems (e.g., [109]). Data can also be transmitted from depth to the surface buoy via an acoustic modem [27, 28], although transfer speeds are low (up to 625 bytes per second).

As one example of meeting the challenges of quieting a mooring and delivering data from an underwater passive acoustic instrument to the surface buoy, the Woods Hole Oceanographic Institution (WHOI) developed a mooring design that uses a surface buoy, stretch hoses, subsurface flotation, EM cable, urethane-jacketed chain, and a bottom structure on which the acoustic instrument is mounted (Fig. 3). The patented hose (EOM Offshore) can stretch to nearly twice its relaxed length; hence,

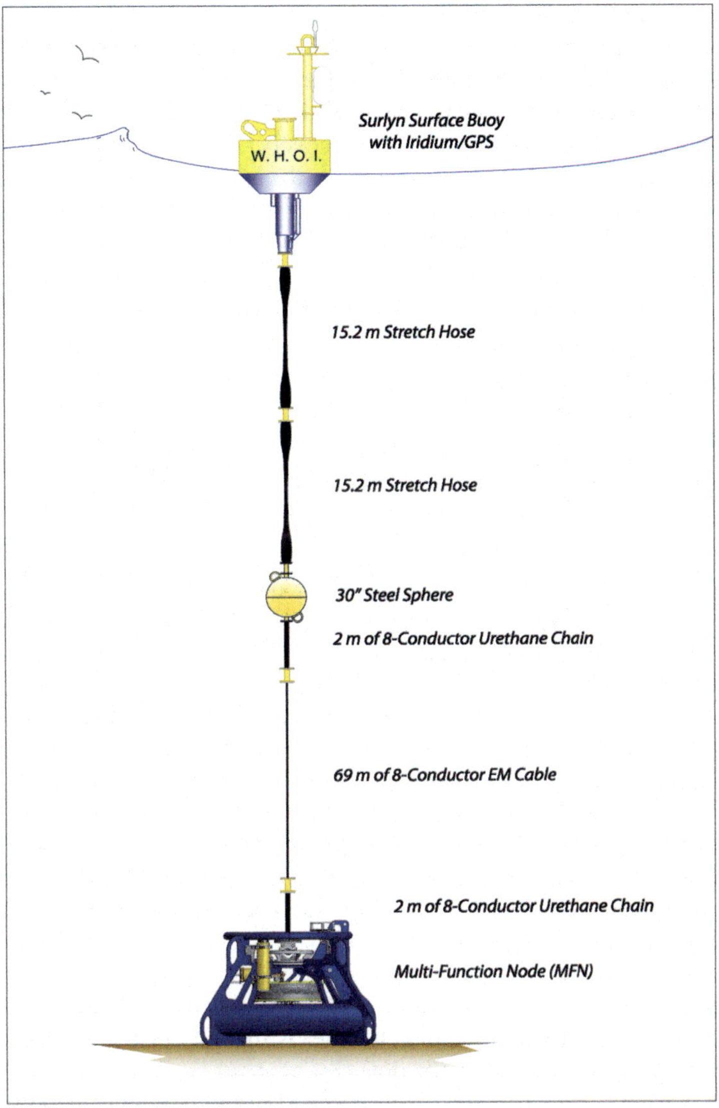

Fig. 3 Quiet moored buoy design for near real-time passive acoustic monitoring. The digital acoustic monitoring (DMON) instrument is mounted near the bottom on the multifunction node, and marine mammal detection data are transferred through the electromagnetic (EM) cable and stretch hoses to an electronics package in the surface buoy where the data are stored and transmitted to shore every 2 h via Iridium satellite communications

it is capable of decoupling the subsurface float from the motion of the surface buoy [78]. Helically wound conductors are embedded in the hose to allow EM signals to be passed through the hose despite its changeable length. Urethane-jacketed chains with integrated EM cables are used to further dampen both mooring motion due to the motion of the surface buoy as well as the transmission of cable strumming energy to the bottom structure. The bottom structure, termed the multifunction node (MFN), consists of an aluminum frame with integrated flotation that is weighted by an anchor that, in turn, is attached to an acoustic release and a spool of Spectra line that is used to retrieve the anchor upon recovery. A digital acoustic monitoring (DMON) instrument (see below) is mounted to the MFN at the seafloor, and it sends marine mammal detection and classification data through the EM cables and stretch hose to an electronics package in the surface buoy for transmission to shore every 2 h via Iridium satellite.

3.2 Mobile Nonnavigated Platforms

Floats have been used extensively for passive acoustic monitoring, including for real-time acoustic measurements (e.g., [61]). They can be used at the surface, at depth (with a tethered surface expression to allow data transfer), or in a profiling mode where they dive to depth and return occasionally to the surface to transfer data to shore. Most floats cannot be navigated, as they passively drift and are thus advected by local currents. Profiling floats (e.g., APEX, Teledyne Webb Research; SOLO/S3A, MRV Systems) can sometimes be crudely navigated by remaining at a depth of favorable currents or by parking (station-keeping) at the bottom with only occasional visits to the surface to transfer data (Fig. 2b). Surface floats are not commonly used for passive acoustic monitoring because of the noise produced by waves and precipitation at the surface; however, sonobuoys have been used extensively for monitoring underwater sound in real time. Matsumoto et al. [61] outfitted an APEX profiling float with a system to passively record audio, process the audio to detect beaked whale clicks, and to transmit those data to shore via Iridium in near real time. Despite the availability of near real-time beaked whale detections, the float was designed to be recovered after 2–3 weeks of operation to allow access to the archived audio. Profiling floats are typically used in physical oceanography applications for much longer periods (years; Fig. 1) and are intended to be expendable [84]. For such missions, near real-time data transfer is the only practical way to access acoustic information collected by the float.

Sonobuoys have been used since World War II in military applications to detect and track underwater sounds [39], as well as to study marine geological structure (e.g., [42]), monitor seismic activity (e.g., [59]), and measure meteorological conditions (e.g., [75]). Considerable research has also been done with sonobuoys to detect and track marine mammals in real time (e.g., [52, 54, 98, 111]). Sonobuoys are ship or air deployable, and separate upon entering the water into a hydrophone section and a surface float section that are connected with a conducting cable. The surface float has an integrated antenna that permits radio transmission of audio in real time to a nearby receiver. Reception range is on the order of a few tens of kilometers for a nearby ship, depending on the height of the receiving antenna. In addition to omnidirectional versions, sonobuoys can also be equipped with directional frequency

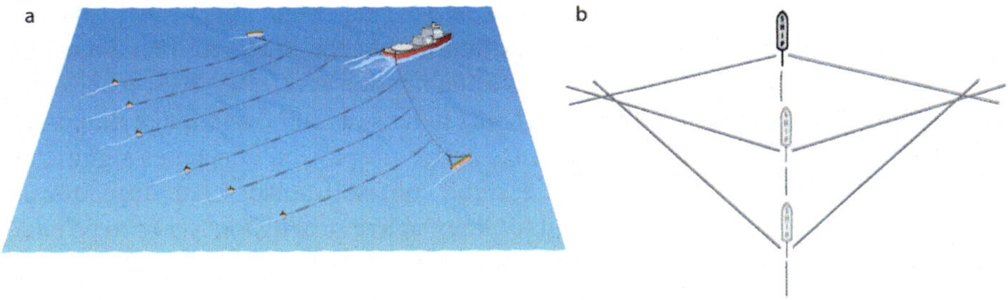

Fig. 4 (**a**) Ship towing multiline towed hydrophone array and air gun for seismic exploration. (**b**) Depiction of localization of a presumed stationary sound source from a single-line towed hydrophone array where bearings from the ship are estimated via beamforming, and multiple bearings from the moving ship allow localization from bearing crossings. Note left-right ambiguity in the localization using a single-line towed hydrophone array

analysis and recording (DIFAR) capability to provide bearings to low-frequency sounds [62]. The endurance of sonobuoys is limited to hours or tens of hours, depending on the configuration, and while battery life limits this duration, the need to attend the sonobuoy with a nearby receiving system can also limit the operational endurance of the system (i.e., a ship or aircraft needs to be in radio range to receive the audio).

3.3 *Mobile Navigated Platforms*

Mobile platforms that are capable of real-time passive acoustic monitoring consist of human-occupied platforms (ships), autonomous underwater vehicles (AUVs), and autonomous surface vehicles (ASVs). Despite significant self-noise, ships towing hydrophone arrays are used routinely for seismic and marine mammal surveys. Hydrophone arrays can be single-line or multiline arrays; the former are used for marine mammal detection and localization, and the latter for seismic surveys (Fig. 4a). The array configuration allows detection of sounds with directivity via beamforming, and so both detection and bearing estimation are possible. For marine mammal surveys, multiple bearing estimates for an assumed stationary source from the moving ship allow estimation of source location using bearing crossings (Fig. 4b).

Autonomous underwater and surface vehicles all operate on similar principles of navigation. Each vehicle has some way of ascertaining its position, either by a global positioning system (GPS) receiver if the vehicle visits the surface, or by acoustic localization of the vehicle itself if it remains constantly submerged. A set of onboard waypoints is used to traverse a course, and a pilot can often update this list of waypoints in near real time; however, it is the vehicle's responsibility to determine how to navigate between the waypoints. There are three primary means of propulsion: active, buoyancy-driven, and wave-driven. Many AUVs have motors and propellers to actively provide forward motion; hence, they can move quickly, and only the strongest currents impact their navigation (e.g., REMUS, Kongsberg Hydroid). They have limited endurance (hours to tens of hours) and range (hundreds of meters to kilometers), however, because active propulsion consumes significant power. Moreover, these AUVs produce significant self-noise. Like ships, towed hydrophone arrays are being developed for these active propulsion AUVs to provide real-time detection and bearings to acoustic sources. For many applications, a nearby ship must attend the AUV to receive real-time data and to facilitate recovery after a relatively short deployment.

Buoyancy-driven ocean gliders (e.g., Slocum glider, Teledyne Webb Research; Seaglider, Kongsberg; Spray glider, Scripps Institute of Oceanography) make small changes in their volume to alternately become more or less dense than the surrounding seawater, allowing them to sink or float, respectively (Fig. 2c). They also have short wings to provide lift, which generates forward movement during both the dive and climb. To move forward, ocean gliders must continuously profile through the water column in a sawtooth pattern. Typical travel speeds in quiescent waters are slow, roughly 0.35–0.40 knots, and their navigation can be severely affected by strong currents. Self-noise is very low for these platforms when diving and climbing, consisting primarily of mechanical noises associated with occasional rudder adjustments. The most significant advantage of ocean gliders is their long endurance. Since adjusting volume does not consume much power, gliders can remain at sea for weeks to many months, depending on the sensor payload and battery configuration. Passive acoustic recording has been conducted with ocean gliders since the mid 2000s [67, 8], including some applications to detect, classify, and report marine mammal sounds in near real time [11, 12, 44]. Because of their mobility, low self-noise, and native radio and satellite communication systems, ocean gliders are ideal for passive acoustic monitoring and the transfer of derived acoustic measurements from sea to shore.

The wave glider (Liquid Robotics, Inc.) is a unique ASV that uses the energy of waves for propulsion and the energy of the sun for recharging batteries; so, its endurance is theoretically unlimited (in practice, biofouling and mechanical wear limits the endurance of the wave glider to several months). It consists of a surface float shaped like a surfboard with solar panels, navigation and communications packages, and GPS and Iridium satellite antennas (Fig. 2d). A 7 m tether with embedded conductors connects the surface float to the sub, a relatively simple structure with vanes and a rudder. As waves lift the float, the vanes on the sub pivot down so that the sub moves up and forward; as waves drop the float, the vanes on the sub pivot up so that the sub moves down and forward. The sub is always moving forward while dragging the surface float along with it, and the rudder on the sub allows the navigation computer to direct the glider's movements. The wave glider moves faster than the buoyancy-driven gliders, allowing it to cover more ground with similar endurance. Self-noise is a challenging problem for doing passive acoustic monitoring on wave gliders, as there is considerable motion associated with the propulsion mechanism, including the vanes, a spring mechanism meant to act as a shock absorber, and the tether mount. Moreover, the relatively fast speed of the wave glider imparts flow noise to a mounted or towed hydrophone. Vehicle noises can be quieted with mechanical techniques, and flow noise can be minimized with appropriate hydrophone fairings. Real-time audio or derived acoustic measurements can be transferred through the tether's conductors to a payload bay in the surface float, and a custom electronics package is needed to store and transfer these data to ship or shore by radio, cellular, or satellite communication systems.

4 Measurements

Ambient noise, the soundscape of the ocean, is the result of both biotic and abiotic sources. Biotic sources include soniferous fish, crustaceans, and marine mammals. Abiotic sounds can be further divided into natural and anthropogenic (human-caused) sources. The former includes mid-frequency sounds from wind and rain, and low-frequency sounds from underwater earthquakes and volcanoes and icebergs. The latter includes primarily low-frequency sounds from ships, oil and gas exploration, pile driving, and even nuclear explosions. In this section, we focus on how sounds are used to provide information about sources of both biotic and abiotic signals in the ocean.

4.1 Biotic

Perhaps the best-known producers of sound in the ocean are marine mammals, ranging in size from the 1.5-m harbor porpoise (*Phocoena phocoena*), which produces ultrasonic clicks (120–140 kHz; [7]), to the 30-m blue whale (*Balaenoptera musculus*), which produces infrasonic moans (10–30 Hz; [97, 103]). Sound is used for two purposes by marine mammals: communication and environmental sensing [104, 105]. Marine mammals are social animals, often organized in groups. Because the ocean is optically opaque during daytime and dark at nighttime, individual members of groups cannot rely on visual contact to maintain cohesion. Sound travels much farther in water than light; so, marine mammals have evolved the ability to communicate with one another acoustically. The toothed whales and dolphins (odontocetes) communicate over a few kilometers using mid-frequency whistles (2–35 kHz; [55, 76]), while baleen whales (mysticetes) communicate over tens and possibly even hundreds of kilometers using low-frequency moans (10–2000 Hz; [79]). Odontocetes also use echolocation, the repeated production of short-duration broadband clicks and the reception and characterization of the click echoes, to detect and capture prey [6], measure their altitude above the seafloor [33, 41], and classify other objects ([69]; including other marine mammals, [32]).

Mysticetes do not use echolocation (although see [99]), but they are capable of sensing their environment with sound by monitoring the echoes of their vocalizations [104]; for example, Arctic-adapted bowhead whales likely use the echoes from their calls to detect thin sea ice through which they can break to breathe air [22, 30].

Many fish and invertebrates can also produce sound for a variety of purposes. Fish sounds are produced at frequencies between 20 and 4000 Hz by two mechanisms: drumming on a swim bladder or stridulation [1, 87]. The swim bladder is used primarily for buoyancy, but the sonic muscles flanking the swim bladder in many species are used to expand and contract the swim bladder to make a drumming sound [93]. Stridulation is the process by which two body parts are rubbed together to make a sound, such as with the pectoral fin and pectoral girdle in marine catfishes [49]. Fish use these sounds to attract a mate, repel a competitor, or as a fright response. Marine invertebrates such as lobsters, crabs, urchins, mussels, scallops, and shrimps can also produce sound, either incidentally during movement or feeding, or purposefully via stridulation or other mechanisms. For example, lobsters produce sound with fundamental frequencies around 5 kHz via a stick-and-slip method similar to a bowed violin [51, 77]. A snapping shrimp generates a high-speed water jet and cavitation bubble by the rapid closing of its large snapper claw, and the collapse of this cavitation bubble creates a loud broadband sound [108]. The water jet is used both as a weapon to stun prey or attack predators, and for intraspecific communication [37]. In some regions, the sounds of invertebrates such as snapping shrimp can dominate the soundscape.

4.2 Abiotic Sources – Natural and Anthropogenic

Natural abiotic sources of sound in the ocean can emanate from (1) the atmosphere in the form of weather (wind and rain), (2) the bottom of the ocean as seismicity from earthquakes and underwater volcanoes, or (3) sea ice via glacier calving, iceberg grounding, or wind-driven and current-driven shear of sea ice. The acoustic energy produced by each of these sources is significant, and can be readily quantified and measured using acoustic observations.

Wind and rain increase ambient noise levels between 200 Hz and 30 kHz for wind and up to 50 kHz for rain [112], allowing the direct measurement of the duration and amplitude of these atmospheric phenomena via passive acoustic monitoring [4, 47, 56, 58, 71, 74, 81, 106, 112]. Acoustic measurements of these sources allow for near-field estimates of wind speed or rainfall amounts in remote or difficult-to-measure regions over smaller areas and shorter timescales than is possible via satellite measurements. The acoustic signatures of both wind and rain are distinctive and can be detected well below the surface of the ocean [24]. Wind injects noise into the ocean through the creation and breaking of waves at the surface. Likewise, the contribution of rainfall to the underwater sound field is the result of raindrops impacting the water's surface, with different sized drops having distinct acoustic signatures [13, 64]. Ma and Nystuen [56, 57] developed an algorithm that uses received sound pressure level and bandwidth to determine rainfall rate.

Acoustic monitoring can also be used to study underwater seismicity and other geophysical processes, including underwater earthquakes, landslides, seafloor spreading events, and volcanic eruptions [14, 15, 25, 26]. These signals are of very low frequency, with fundamentals well below 100 Hz that can last many tens of seconds. Although seismicity occurs within the Earth's crust, acoustic energy from geophysical processes propagates into the water column and can therefore be detected by hydrophones. Acoustic data have been used to establish the duration and amplitude of seismic events and map the patterns of seismicity at geologically active sites over long time and space scales [26, 34]. In addition to monitoring persistent seismicity in remote regions throughout the world's oceans, hydroacoustic data can be used to identify the characteristics of underwater events that may result in tsunamis, therefore allowing for better prediction or earlier detection of such events [17, 36, 100, 113].

A relatively recent development in the use of passive acoustics to understand abiotic sources of sound is the study of sounds produced by drifting and grounded Antarctic icebergs [16, 101]. Widely spaced hydrophone arrays can be used to track the paths of drifting icebergs [21, 68]. The acoustic signature of an iceberg grounding can be distinguished from a free-floating iceberg that is breaking up [21]. Ashokan et al. [5] have recently studied ice calving and ice bobbing noise in the Arctic Kongsfjord. Low-frequency tremor signals from grounded, drifting, and disintegrating icebergs have been shown to increase ambient noise levels in the Southern Ocean seasonally, and these signals can be detected as far north as the equator in some oceans [60].

In addition to the natural sources of noise in the ocean, passive acoustic monitoring can listen for anthropogenic, or human-caused, sources. These sound sources are relatively novel in the oceans, having only emerged in the nineteenth century as artifacts of industrial development. Since this time, anthropogenic sources have increased ambient noise levels in all oceans and over long timescales. Anthropogenic sources that have been monitored with passive acoustic data include shipping, atomic explosions, and oil and gas exploration and extraction. These sources all tend to be of relatively low frequency (<1000 Hz) and are therefore detectable over distances from tens to thousands of kilometers.

Shipping, particularly commercial shipping, injects low-frequency sound into the water from cavitation of air bubbles created when the propeller spins [85]. This produces blade lines that are quasi-tonal low-frequency bands with harmonics, the fundamental frequency of which can be used to determine the propeller blade rate of a ship. Commercial shipping is the chief contributor to underwater ambient noise levels from ~5 to 1000 Hz in most of the world's oceans and has increased with each decade since the 1960s [3, 63, 86].

Passive acoustic monitoring has also been used to monitor illegal nuclear explosions. The Comprehensive Nuclear Test Ban Treaty (CTBT) organization has, in addition to land-based seismometers, underwater hydroacoustic stations as part of their International Monitoring System (IMS). This system is designed to detect and localize the source and magnitude of nuclear explosions to inform the global community of illegal testing [20]. Although designed to detect the infrasonic signals of such explosions, the data accumulated by the IMS have been used to detect baleen whales [29, 88, 95, 96] and study global noise levels [65, 66].

Another dominant source of low-frequency ambient noise in most oceans is seismic air gun explosions used during oil and gas prospecting and, to a lesser extent, scientific research. Air guns, usually deployed in arrays, produce very loud sounds by releasing compressed air that creates an impulsive, low-frequency (~1–188 Hz) signal designed to penetrate the ocean floor. The reflected sound of these pulses is received by a towed hydrophone array and used to estimate the bottom and sub-bottom composition of the ocean floor in the quest for oil and gas deposits [83]. Scientists also use the data to determine the structure and dynamics of the Earth's crust. These signals can be detected at distances of well over 1000 km [73, 102]. In the Atlantic, air gun signals are recorded year-round from oil and gas surveys that have been acoustically localized in both the northern and southern hemispheres and the eastern and western Atlantic [72].

Interestingly, undersea substrate properties can also be studied using new techniques that include both direct measurements and inversion schemes using passive acoustic data and an understanding of the physics of signal propagation through complex media (i.e. [70, 89]). Seabed properties, such as critical angle and reflection loss, can be estimated directly from passive array measurements of ambient noise produced by wind and ships [31, 80]. Sanjana et al. [89] used a vertical line array of 12 hydrophones in the northern Indian Ocean to obtain ambient noise measurements in conjunction with wind speed and rainfall information, and they found that the critical angle derived from the noise measurements matched well with sediment samples acquired at the same time as the experiment. Sanjana et al. [90] then conducted a geoacoustic inversion experiment to further estimate surface and subsurface seabed acoustic parameters, including sound speed, density, attenuation, and layering, and again found that the derived sediment characteristics matched well with the core sample data collected during the experiment.

5 Experience

The capability to detect the many different sound sources that comprise ocean ambient noise, be they biotic or abiotic, has expanded significantly over the past 40 years. For the most part, the great majority of studies have used passive acoustic recordings that were analyzed after being retrieved from underwater instrumentation. Near real-time studies have been fewer but are becoming more common as the required technology matures. In this section, we provide examples of the use of real-time technology that has permitted rapid response to an event or events that could only have been detected through the use of passive acoustic monitoring.

5.1 *Marine Mammal Monitoring in Real Time*

A number of systems have been built in recent years to detect and report marine mammal occurrence in near real time from autonomous platforms [11, 18, 44, 61, 92, 94, 107]. These systems have focused on baleen and beaked whales, as conservation needs are greatest for these taxa. The North Atlantic right whale (*Eubalaena glacialis*) is critically endangered and is prone to ship strikes and fishing gear entanglements in its calving and feeding habitats along the east coast of USA and Canada [45, 46, 48]. A network of buoys developed by Cornell University has been used near Boston, MA, USA, to detect the presence of right whales and relay that information to mariners via a government-sponsored advisory system [18, 94, 107]. The system runs an in-situ detector to identify a particular call that right whales produce (the upcall), and sends audio snippets to shore of putative right whale calls via satellite communications. On shore, a human analyst listens to the snippet and assesses whether the sound was truly a right whale call.

The Cornell University Bioacoustics Research Program pioneered near real-time detection and reporting from a moored buoy by focusing on a single call type produced by a single species in a single location. Automated detection and reporting from mobile autonomous platforms show great promise for assessing the spatial distribution of vocalizing animals [44, 61], but both moored and mobile systems need to be expanded to simultaneously detect the calls of a wide variety of species. Baumgartner et al. [11] report the use of a near real-time system that is capable of detecting, classifying, and reporting the presence of fin (*Balaenoptera physalus*), humpback (*Megaptera novaeangliae*), sei (*Balaenoptera borealis*), and North Atlantic right whales in near real time from Slocum electric gliders. The system uses a generalized detection and classification system [9] that generates pitch tracks for detected sounds, classifies each pitch track via discriminant function analysis, and relays both pitch tracks and classification information to shore via Iridium satellite. On shore, an analyst can review both the classification information as well as the context in which calls are detected to determine the occurrence of each of the four baleen whale species. Baumgartner et al. [12] describe an extension of this system for detecting bowhead and beluga whales as well as bearded seals from Slocum gliders in the Arctic.

5.2 *Seismic Activity Monitoring*

Monitoring underwater seismicity from earthquakes and volcanoes has provided important information about hydrothermal venting, the chemistry of the global ocean, seafloor spreading, and the risk of tsumanigenic earthquakes based on subduction zone characteristics. Real-time passive acoustic monitoring has been used in numerous applications to detect volcanic activity and seafloor spreading events and to direct shipboard responses to document, characterize, and study these

Fig. 5 (**a**) NIOT vertical line array surface buoy with radio frequency (RF) transmitting antenna, (**b**) land-based receiving antenna, and (**c**) real-time display of power spectral density estimates

events [25, 26]. Interdisciplinary research of such events in near real time has provided novel insights into the linkages among geological perturbations, local chemistry, and biological processes in remote difficult-to-study regions. This includes discoveries of active hydrothermal plumes and associated microbial communities that, while ephemeral in time and space, can stimulate broad changes in the local environment [23, 38, 40, 43].

5.3 *Real-Time Ambient Noise Monitoring*

To obtain long-term near real-time ambient noise data, an automated system (Fig. 5) has been developed and deployed by the National Institute of Ocean Technology (NIOT), Ministry of Earth Sciences, Government of India [50]. The system has been deployed in shallow waters off both the east and west coasts of India. The system (for which a patent has been filed) is comprised of a vertical line array of omnidirectional hydrophones in an oil-filled polyurethane tube with tilt sensors. Initially, a 12-element array (design frequency 5 kHz) cabled to a surface buoy was used, which was later enhanced to 21 elements (design frequency 10 kHz). The data acquisition modules and battery pack are housed in the surface buoy along with data communication modules. The mooring line is comprised of subsea floats, marker floats, and sinker weight. An acoustic pinger designed for offshore use is mounted with the array to aid in locating the system. The hydrophone array is calibrated at the Acoustic Test Facility of NIOT.

Since each element of the array samples simultaneously, the amount of raw data acquired is very large and poses problems for real-time transmission. To meet this challenge, a program was developed for the digital acquisition system to compute power spectral density (PSD) estimates, which are small enough to be transmitted in real time. Different communication modes have been tested, including Wi-Fi, radio frequency (RF), general packet radio service (GPRS), and Indian National Satellite System (INSAT) (patent to be filed). The real-time PSD estimates enable the identification and classification of noise sources such as rain, shipping, and wind. With a proper database of ship noise, the type of ship being monitored could be determined in real time. Further work in this area will enable the system to be used as an intruder detection system. Wi-Fi communication is feasible only for short range, with a maximum distance from sea to shore of <2 km. Data are transmitted from the surface buoy to shore and then relayed via high-speed Internet to NIOT. Radio frequency signals can be used for long-range transmission of data, with a maximum distance of up to 15 km given clear line of sight. Data are transmitted from an RF modem in the surface buoy to an RF receiving system on shore where, as with Wi-Fi, data are relayed to NIOT via high-speed Internet. PSD estimates can also be transmitted directly from the surface buoy to NIOT using GPRS over very long ranges and do not require a shore station intermediary. Finally, although there are data transmission size limitations that preclude the sending of PSD estimates, INSAT can be used to transmit data on the status of the real-time system directly to NIOT.

One of the challenges of autonomous systems that have a surface expression in shallow waters is protecting the hardware from human interference. At present, manned watchkeeping boats are used to monitor the installation to prevent damage or theft of the equipment. A GPRS system is placed on the watchkeeping boat to monitor its movement and ensure uninterrupted functioning of the system. Because the system is attended, the raw acoustic data can be downloaded from the installation at regular intervals. The voluminous amount of data acquired is then subjected to initial processing and further detailed analysis to understand site-specific characteristics of ambient noise, including wind speed/rainfall estimation as well as bioacoustic, geoacoustic, and anthropogenic signals [81, 82, 91].

5.4 *The Future of Real-Time Passive Acoustic Monitoring*

Real-time passive acoustic monitoring provides a very different capability than archival recorders; it enables action. Whether detecting an earthquake, nuclear explosion, ship, or whale, a human is alerted to that detection within seconds to hours, and there is an opportunity to do something with that information. Real-time systems will not take the place of archival recorders; there are many instances when real-time information is not necessary, because no response is planned or required.

But as detection algorithms expand and improve in accuracy, and instrumentation capable of running those algorithms in-situ become more available, real-time passive acoustic monitoring will only grow along with the myriad of applications that can take advantage of it. Industrial activities such as oil and gas exploration and extraction, dredging, shipping, and wind farm construction introduce loud sounds into the ocean that have an impact on marine animals [83]. Real-time passive acoustic monitoring offers the opportunity to monitor those sounds and reduce or halt activities when sounds reach a particular instantaneous or cumulative threshold. Similarly, the intensity of industrial activities can be adjusted based on the presence of marine mammals detected in near real time. There are only a few examples of using such real-time systems for mitigation of anthropogenic activities today, but it seems inevitable that real-time passive acoustic technology will be mandated in future regulatory frameworks.

Our changing climate will force changes in the distribution of marine organisms. For highly mobile predators like marine mammals, changes in distribution may be dramatic, occurring quickly over large spatial scales. No observing system exists today that is capable of documenting and studying these changes in marine mammal distribution. However, there is an outstanding example of a global autonomous observing system that has been built to monitor ocean variability driven by climate change: the ARGO profiling float program [84]. Through technical innovation and international cooperation, the ARGO program now consists of a few thousand expendable profiling floats distributed from tropical to subarctic waters monitoring ocean heat content in near real time. Imagine a similar array of hundreds of long-endurance autonomous platforms reporting in-situ detections of marine mammals. Such a listening array would be able to monitor changes in marine mammal distribution over time, as well as the temporal and spatial distribution of ocean noise throughout nearly the entire world's ocean. Because these platforms have long endurance and are expendable, archived audio could never be retrieved; hence, near real-time detection, classification, and reporting are central to this grand vision of a global marine mammal observing system.

References

1. Amorim MCP (2006) Diversity of sound production in fish. In: Ladich F, Collin SP, Moller P, Kapoor BG (eds) Communication in fishes, vol 1. Science Publishers, Enfield, pp 71–105
2. André M, van der Schaar M, Zaugg S, Houegnigan L, Sánchez AM, Castell JV (2011) Listening to the deep: live monitoring of ocean noise and cetacean acoustic signals. Mar Pollut Bull 63:18–26. https://doi.org/10.1016/j.marpolbul.2011.04.038
3. Andrew RK, Howe BM, Mercer JA, Dzieciuch MA (2002) Ocean ambient sound: comparing the 1960s with the 1990s for a receiver off the California coast. ARLO 3:65. https://doi.org/10.1121/1.1461915
4. Ashokan M, Latha G, Ramesh R (2014) Analysis of shallow water ambient noise due to rain and derivation of rain parameters. Appl Acoust 88:114–122

5. Ashokan M, Latha G, Thirunavukkarasu A, Raguraman G, Venkatesan R (2016) Ice berg cracking events as identified from underwater ambient noise measurements in the shallow waters of Ny-Alesund, Arctic. Pol Sci 10:140–146
6. Au WWL (1993) The sonar of dolphins. Springer-Verlag, New York
7. Au WWL, Kastelein RA, Rippe T, Schooneman NM (1999) Transmission beam pattern and echolocation signals of a harbor porpoise (*Phocoena phocoena*). J Acoust Soc Am 106:3699–3705
8. Baumgartner MF, Fratantoni DM (2008) Diel periodicity in both sei whale vocalization rates and the vertical migration of their copepod prey observed from ocean gliders. Limnology Oceanography 53:2197–2209
9. Baumgartner MF, Mussoline SE (2011) A generalized baleen whale call detection and classification system. J Acoust Soc Am 129:2889–2902
10. Baumgartner MF, Freitag L, Partan J, Ball K, Prada K (2008) Tracking large marine predators in three dimensions: the real-time acoustic tracking system. IEEE J Ocean Eng 33:146–157
11. Baumgartner MF, Fratantoni DM, Hurst TP, Brown MW, Cole TVN, Van Parijs SM, Johnson M (2013) Real-time reporting of baleen whale passive acoustic detections from ocean gliders. J Acoust Soc Am 134:1814–1823
12. Baumgartner MF, Stafford KM, Winsor P, Statscewich H, Fratantoni DM (2014) Glider-based passive acoustic monitoring in the Arctic. Mar Technol Soc J 40(5):40–51
13. Black PG, Proni JR, Wilkerson JC, Samsury CE (1997) Oceanic rainfall detection and classification in tropical and subtropical mesoscale convective systems using underwater acoustic methods. Mon Weather Rev 125:2014–2042
14. Caplan-Auerbach J, Dziak RP, Bohnenstiehl DR, Chadwick WW, Lau TK (2014) Hydroacoustic investigation of submarine landslides at West Mata volcano, Lau Basin. Geophys Res Lett 41:5927–5934. https://doi.org/10.1002/2014GL060964
15. Chadwick WW Jr, Cashman KV, Embley RW, Matsumoto H, Dziak RP, de Ronde CEJ, Lau TK, Deardorff ND, Merle SG (2008) Direct video and hydrophone observations of submarine explosive eruptions at NW Rota-1 volcano, Mariana arc. J Geophys Res 113:1–23. https://doi.org/10.1029/2007JB005215
16. Chapp E, Bohnenstiehl DR, Tolstoy M (2005) Sound-channel observations of ice-generated tremor in the Indian Ocean. Geochem Geophys Geosyst 6:1–14. https://doi.org/10.1029/2004GC000889
17. Chierici F, Pignagnoli L, Embriaco D (2010) Modeling of the hydroacoustic signal and tsunami wave generated by seafloor motion including a porous seabed. J Geophys Res Oceans 115. https://doi.org/10.1029/2009JC005522
18. Clark CW, Calupca T, Gillespie D, Von der Heydt K, Kemp J (2005) A near-real-time acoustic detection and reporting system for endangered species in critical habitats. J Acoust Soc Am 117:2525
19. Clay CS, Medwin H (1977) Acoustical oceanography: principles and applications. Wiley, New York. Chapter A9
20. Clery D (2009) Test ban monitoring: no place to hide. Science 325:382–385. https://doi.org/10.2307/20536670
21. Dziak R, Fowler M, Matsumoto H, Bohnenstiehl D, Park M, Warren K, Lee WS (2013) Life and death sounds of iceberg A53a. Oceanography 26. https://doi.org/10.5670/oceanog.2013.20
22. Ellison WT, Clark CW, Bishop GC (1987) Potential use of surface reverberation by bowhead whales, *Balaena mysticetus*, in under-ice navigation: preliminary considerations. Rep Int Whaling Comm 37:329–332
23. Embley R, Chadwick WW, Jonasson IR, Butterfield DA, Baker ET (1995) Initial results of the rapid response to the 1993 CoAxial event: relationships between hydrothermal and volcanic processes. Geophys Res Lett 22:143–146
24. Evans DL, Watts DR, Halpern D, Bourassa S (1984) Ocean winds measured from the seafloor. J Geophys Res 89:3457–3461. https://doi.org/10.1002/(ISSN)2156-2202

25. Fox CG, Dziak RP (1998) Hydroacoustic detection of volcanic activity on the gorda ridge, february–march 1996. Deep-Sea Res II Top Stud Oceanogr 45:2513–2530. https://doi.org/10.1016/S0967-0645(98)00081-2
26. Fox CG, Radford WE, Dziak RP, Lau TK (1995) Acoustic detection of a seafloor spreading episode on the Juan de Fuca Ridge using military hydrophone arrays. Geophys Res Lett 22:131–134
27. Freitag L, Grund M, Partan J, Singh S, Koski P, Ball K (2005) The WHOI Micro-Modem: an acoustic communications and navigation system for multiple platforms. OCEANS 2005. Proceedings of MTS/IEEE, Washington, DC, 17–23 Sept 2005, pp 1086–1092
28. Gallimore E, Partan J, Vaughn I, Singh S, Shusta J, Freitag L (2010) The WHOI Micromodem-2: a scalable system for acoustic communications and networking. OCEANS 2010. Seattle, 20–23 Sept 2010, pp 1–7
29. Gavrilov AN, McCauley R (2013) Acoustic detection and long-term monitoring of pygmy blue whales over the continental slope in southwest Australia. J Acoust Soc Am 134:2505–2513. https://doi.org/10.1121/1.4816576
30. George JC, Clark C, Carroll GM, Ellison WT (1989) Observations on the ice-breaking and ice navigation behavior of migrating bowhead whales (*Balaena mysticetus*) near Point Barrow, Alaska, Spring 1985. Arctic 42:24–30
31. Gervaise C, Kinda BG, Bonnel J, Stéphan Y, Vallez S (2012) Passive geoacoustic inversion with a single hydrophone using broadband ship noise. J Acoust Soc Am 131:1999. https://doi.org/10.1121/1.3672688
32. Gol'din P (2014) 'Antlers inside': are the skull structures of beaked whales (Cetacea: Ziphiidae) used for echoic imaging and visual display? Biol J Linn Soc 113:510–515
33. Goold JC, Jones SE (1995) Time and frequency domain characteristics of sperm whale clicks. J Acoust Soc Am 98:1279–1291
34. Goslin J, Lourenco N, Dziak RP, Bohnenstiehl DR, Haxel J, Luis J (2005) Long-term seismicity of the Reykjanes Ridge (North Atlantic) recorded by a regional hydrophone array. Geophys J Int 162:516–524. https://doi.org/10.1111/j.1365-246X.2005.02678.x
35. Greene C, McLennan M, Mcdonald T, Jakubczak R, Richardson W (2004) Directional frequency and recording (DIFAR) sensors in seafloor recorders to locate calling bowhead whales during their fall migration. J Acoust Soc Am 116:799–813
36. Hanson JA, Bowman JR (2005) Dispersive and reflected tsunami signals from the 2004 Indian Ocean tsunami observed on hydrophones and seismic stations. Geophys Res Lett 32. https://doi.org/10.1029/2005GL023783
37. Herberholz J, Schmitz B (1999) Flow visualisation and high speed video analysis of water jets in the snapping shrimp (*Alpheus heterochaelis*). J Comp Physiol A 185:41–49
38. Holden JF, Summit M, Baross JA (1998) Thermophilic and hyperthermophilic microorganisms in 3-30° C. FEMS Microbiol Ecol 25:33–41
39. Holler R, Horbach A, McEachern J (2006) Not ready for retirement: the sonobuoy approaches age 65: aircraft-launched antisubmarine warfare sensors redefine their roles and continue to monitor the ocean. Sea Technol 47(11):10–14
40. Huber JA, Butterfield DA, Baross JA (2003) Bacterial diversity in a subseafloor habitat following a deep-sea volcanic eruption. FEMS Microbiol Ecol 43:393–409
41. Jaquet N, Dawson S, Douglas L (2001) Vocal behavior of male sperm whales: why do they click? J Acoust Soc Am 109:2254–2259
42. Jones EJ, McMechan GA, Zeng X (2015) Seismic evidence for crustal underplating beneath a large igneous province; the Sierra Leone Rise, equatorial Atlantic. Mar Geol 365:52–60
43. Kelley DS, Baross JA, Delaney JR (2002) Volcanoes, fluids, and life at mid-ocean ridge spreading centers. Annu Rev Earth Planet Sci 30:385–491. https://doi.org/10.1146/annurev.earth.30.091201.141331
44. Klinck H, Mellinger DK, Klinck K, Bogue NM, Luby JC, Jump WA, Shilling GB, Litchendorf T, Wood AS, Schorr GS, Baird RW (2012) Near-real-time acoustic monitoring of beaked whales and other cetaceans using a Seaglider. PLoS One 7:e36128. https://doi.org/10.1371/journal.pone.0036128

45. Knowlton AR, Kraus SD (2001) Mortality and serious injury of northern right whales (*Eubalaena glacialis*) in the western North Atlantic Ocean. J Cetacean Res Manag 2(Special Issue):193–208
46. Knowlton AR, Hamilton PK, Marx MK, Pettis HM, Kraus SD (2012) Monitoring North Atlantic right whale *Eubalaena glacialis* entanglement rates: a 30 yr retrospective. Mar Ecol Prog Ser 466:293–302
47. Knudsen VO, Alford RS, Emling JW (1948) Underwater ambient noise. J Mar Res 7:410–429
48. Kraus SD, Brown MW, Caswell H, Clark CW et al (2005) North Atlantic right whales in crisis. Science 309:561–562
49. Ladich F, Bass AH (1998) Sonic/vocal motor pathways in catfishes: comparisons with other teleosts. Brain Behav Evol 51:315–330
50. Latha G (2010) Autonomous ambient noise measurement system for shallow water applications. INDO-US Workshop on Shallow Water Acoustics, National Institute of Oceanography, Dona Paula, Goa, India. 3–4 Feb 2010
51. Latha G, Senthilvadivu S, Venkatesan R, Rajendran V (2005) Sound of shallow and deep-water lobsters: measurements, analysis, and characterization (L). J Acoust Soc Am 117:2720–2723
52. Laurinolli MH, Hay AE, Desharnais F, Taggart CT (2003) Localization of North Atlantic right whale sounds in the Bay of Fundy using a sonobuoy array. Mar Mamm Sci 19:708–723
53. Lin Y-T, Newhall AE, Lynch JF (2012) Low-frequency broad- band sound source localization using an adaptive normal mode back- propagation approach in a shallow-water ocean. J Acoust Soc Am 131:1798–1813
54. Ljungblad D, Thompson P, Moore SE (1982) Underwater sounds recorded from migrating bowhead whales, Balaena mysticetus, in 1979. J Acoust Soc Am 71:477–482. https://doi.org/10.1121/1.387419
55. Lammers MO, Au WWL, Herzing DL (2003) The broadband social acoustic signaling behavior of spinner and spotted dolphins. J Acoust Soc Am 114:1629–1639
56. Ma B, Nystuen JA (2005) Passive acoustic detection and measurement of rainfall at sea. J Atmos Ocean Tech 22:1225–1248
57. Ma B, Nystuen JA (2007) Detection of rainfall events using underwater passive aquatic sensors and air-sea temperature changes in the tropical pacific ocean. Mon Weather Rev 135:3599–3612. https://doi.org/10.1175/MWR3487.1
58. Ma BB, Nystuen JA, Lien R-C (2005) Prediction of underwater sound levels from rain and wind. J Acoust Soc Am 117:3555–3565. https://doi.org/10.1121/1.1910283
59. Macdonald KC, Mudie JD (1974) Microearthquakes on the Galapagos Spreading Centre and the seismicity of fast-spreading ridges. Geophys J R Astron Soc 36:245–257
60. Matsumoto H, Bohnenstiehl DR, Tournadre J, Dziak RP, Haxel JH, Lau TKA, Fowler M, Salo SA (2014) Antarctic icebergs: a significant natural ocean sound source in the Southern Hemisphere. Geochem Geophys Geosyst 15:3448–3458. https://doi.org/10.1002/2014GC005454
61. Matsumoto H, Jones C, Klinck H, Mellinger DK, Dziak RP, Meinig C (2013) Tracking beaked whales with a passive acoustic profiler float. J Acoust Soc Am 133:731–740
62. McDonald MA (2004) DIFAR hydrophone usage in whale research. Can Acoust 32:155–160
63. McDonald MA, Hildebrand JA, Wiggins SM (2006) Increases in deep ocean anibient noise in the northeast pacific west of San Nicolas Island. California J Acoust Soc Am 120:711–718. https://doi.org/10.1121/1.2216565
64. Medwin H, Nystuen JA, Jacobus PW, Ostwald LH, Snyder DE (1992) The anatomy of underwater rain noise. J Acoust Soc Am 92:1613–1623
65. Miksis-Olds JL, Bradley DL, Maggie Niu X (2013) Decadal trends in Indian Ocean ambient sound. J Acoust Soc Am 134:3464–3475. https://doi.org/10.1121/1.4821537
66. Miksis-Olds JL, Nichols SM (2016) Is low frequency ocean sound increasing globally? J Acoust Soc Am 139:501–511. https://doi.org/10.1121/1.4938237
67. Moore SE, Howe BM, Stafford KM, Boyd ML (2007) Including whale call detection in standard ocean measurements: application of acoustic Seagliders. Mar Technol Soc J 41:53–57

68. Müller C, Schlindwein V, Eckstaller A, Miller H (2005) Singing Icebergs. Science 310:1299. https://doi.org/10.1126/science.1117145
69. Nachtigall PE (1980) Odontocete echolocation performance on object size, shape and material. In: Busnel R-G, Fish JF (eds) Animal sonar systems. Springer US, Boston, pp 71–95
70. Najeem S, Noufal KK, Latha G (2016) Estimation of seabed properties using ambient noise from shallow water sites of the Indian Continental Shelf. Mar Geod 39:21–31
71. Najeem S, Sanjana MC, Latha G, Edwards P (2015) Wind induced ambient noise modelling and comparison with field measurements in Arabian Sea. Appl Acoust 89:101–106
72. Nieukirk SL, Mellinger DK, Moore SE, Klinck K, Dziak RP, Goslin J (2012) Sounds from airguns and fin whales recorded in the mid-Atlantic Ocean, 1999–2009. J Acoust Soc Am 131:1102–1112. https://doi.org/10.1121/1.3672648
73. Nieukirk SL, Stafford KM, Mellinger DK, Dziak RP, Fox CG (2004) Low-frequency whale and seismic airgun sounds recorded in the mid-Atlantic Ocean. J Acoust Soc Am 115:1832–1843. https://doi.org/10.1121/1.1675816
74. Nystuen JA (1996) Acoustical rainfall analysis: rainfall drop size distribution using the underwater sound field. J Atmos Ocean Tech 13:74–84
75. Nystuen JA, Selsor HD (1997) Weather classification using passive acoustic drifters. J Atmos Ocean Technol 14:656–666
76. Oswald JN, Rankin S, Barlow J (2008) To whistle or not to whistle? Geographic variation in the whistling behavior of small odontocetes. Aquat Mamm 34:288–302
77. Patek SN (2001) Spiny lobsters stick and slip to make sound. Nature 411:153–154
78. Paul W, Bocconcelli A (1994) Drifting buoy systems using rubber stretch hose. Sea Technol 35(11):37–44
79. Payne R, Webb D (1971) Orientation by means of long range acoustic signaling in baleen whales. Ann N Y Acad Sci 188:110–141
80. Quijano JE, Dosso SE, Dettmer J, Zurk LM, Siderius M, Harrison CH (2012) Bayesian geoacoustic inversion using wind-driven ambient noise. J Acoust Soc Am 131:2658–2667. https://doi.org/10.1121/1.3688482
81. Ramji S, Latha G, Rajendran V, Ramakrishnan S, S. (2008) Wind dependence of ambient noise in shallow water of Bay of Bengal. Appl Acoust 69:1294–1298
82. Ramji S, Latha G, Ramakrishnan S (2009) Estimation and interpolation of underwater low frequency ambient noise spectrum using artificial neural networks. Appl Acoust 70:1111–1115
83. Richardson WJ, Greene CR Jr, Malme CI, Thomson DH (1995) Marine mammals and noise. Academic Press, San Diego
84. Riser SC, Freeland HJ, Roemmich D, Wijffels S, Troisi A, Belbéoch M, Gilbert D, Xu J, Pouliquen S, Thresher A, Le Traon P-Y, Maze G, Klein B, Ravichandran M, Grant F, Poulain P-M, Suga T, Lim B, Sterl A, Sutton P, Mork K-A, Joaquín Vélez-Belchí P, Ansorge I, King B, Turton J, Baringer M, Jayne SR (2016) Fifteen years of ocean observations with the global Argo array. Nat Clim Chang 6:145–153
85. Ross DG (1976) Mechanics of underwater noise. Pergamon Press, New York
86. Ross DG (1993) On ocean underwater ambient noise. Acoust Bull 18:5–8
87. Rountree RA, Gillmore RG, Goudey CA, Hawkins AD, Luczkovich JJ, Mann DA (2006) Listeing to fish. Fish Oceanogr 31:433–446
88. Samaran F, Stafford KM, Branch TA, Gedamke J, Royer J-Y, Dziak RP, Guinet C (2013) Seasonal and geographic variation of southern blue whale subspecies in the Indian Ocean. PLoS One 8:e71561. https://doi.org/10.1371/journal.pone.0071561.t001
89. Sanjana MC, Latha G, Mahanty MM (2013) Seabed characteristics from ambient noise at three shallow water sites in Northern Indian Ocean. J Acoust Soc Am 134:EL366–EL372. https://doi.org/10.1121/1.4818936
90. Sanjana MC, Latha G, Potty GR (2016) Geoacoustic inversion in shallow tropical waters for a silty and sandy seabed. J Acoust Soc Am 139:EL38–EL44. https://doi.org/10.1121/1.4942545

91. Sanjana MC, Latha G, Thirunavukkarasu A (2010) Fluctuation and variability of shallow water ambient noise from time series measurements. Fluctuation Noise Lett 9:193–202
92. Simard Y, Bahoura M, Park CW, Rouat J, Sirois M, Mouy X, Seebaruth D, Roy N, Lepage R (2006) Development and experimentation of a satellite buoy network for real-time acoustic localization of whales in the St. Lawrence. Proceedings of Oceans 2006 MTS/IEEE, Boston, 18–21 Sept 2006, 6 pp
93. Skoglund CR (1961) Functional analysis of swim-bladder muscles engaged in sound production of the toadfish. J Biophys Biochem Cytol 10:187–200
94. Spaulding E, Robbins M, Calupca T, Clark C, Tremblay C, Waack A, Warde A, Kemp J, Newhall K (2009) An autonomous, near-real-time buoy system for automatic detection of North Atlantic right whale calls. J Acoust Soc Am 125:2615
95. Stafford KM, Bohnenstiehl DR, Tolstoy M, Chapp E, Mellinger DK, Moore SE (2004) Antarctic-type blue whale calls recorded at low latitudes in the Indian and eastern Pacific Oceans. Deep-Sea Res I Oceanogr Res Pap 51:1337–1346. https://doi.org/10.1016/j.dsr.2004.05.007
96. Stafford KM, Chapp E, Bohnenstiel DR, Tolstoy M (2011) Seasonal detection of three types of "pygmy" blue whale calls in the Indian Ocean. Mar Mamm Sci 27:828–840. https://doi.org/10.1111/j.1748-7692.2010.00437.x
97. Stafford KM, Fox CG, Clark DS (1998) Long-range acoustic detection and localization of blue whale calls in the northeast Pacific Ocean. J Acoust Soc Am 104:3616–3625. https://doi.org/10.1121/1.423944
98. Stafford KM, Moore SE, Laidre KL, Heide-Joergensen M-P (2008) Bowhead whale springtime song off West Greenland. J Acoust Soc Am 124:3315–3323
99. Stimpert AK, Wiley DN, Au WWL, Johnson MP, Arsenault R (2007) 'Megapclicks': acoustic click trains and buzzes produced during night-time foraging of humpback whales (Megaptera novaeangliae). Biol Lett 3:467–470
100. Synolakis CE, Bardet JP, Borrero JC, Davies HL, OKAL EA, Silver EA, Sweet S, Tappin DR (2002) The slump origin of the 1998 Papua New Guinea tsunami. Proc R Soc A Math Phys Eng Sci 458:763–789. https://doi.org/10.1098/rspa.2001.0915
101. Talandier J, Hyvernaud O, Okal EA (2002) Long-range detection of hydroacoustic signals from large icebergs in the Ross Sea, Antarctica. Earth Planet Sci Lett 203:16–16. https://doi.org/10.1016/S0012-821X(02)00867-1
102. Thode A, Kim KH, Greene CRJ, Roth E (2010) Long range transmission loss of broadband seismic pulses in the Arctic under ice-free conditions. J Acoust Soc Am 128:E181–E187. https://doi.org/10.1121/1.3479686
103. Thompson PO, Findley LT, Vidal O, Cummings WC (1996) Underwater sounds of blue whales, *Balaenoptera musculus*, in the Gulf of California, Mexico. Mar Mamm Sci 12:288–293
104. Tyack PL (1997) Studying how cetaceans use sound to explore their environment. In: Owings DH, Beecher MD, Thompson NS (eds) Communication. Springer US, Boston, pp 251–297
105. Tyack PL (1999) Communication and cognition. In: Reynolds JE III, Rommel SA (eds) Biology of marine mammals. Smithsonian Institution Press, Washington, DC, pp 287–323
106. Vagle S, Large WG, Farmer DM (1990) An evaluation of the wotan technique of inferring oceanic winds from underwater ambient sound. J Atmos Ocean Tech 7:576–595
107. Van Parijs SM, Clark CW, Sousa-Lima RS, Parks SE, Rankin S, Risch D, Van Opzeeland IC (2009) Management and research applications of real-time and archival passive acoustic sensors over varying temporal and spatial scales. Mar Ecol Prog Ser 395:21–36
108. Versluis M, Schmitz B, von der Heydt A, Lohse D (2000) How snapping shrimp snap: through cavitating bubbles. Science 289:2114–2117
109. Wallinga JP, Pettigrew NR, Irish JD (2003) The GoMOOS moored buoy design. OCEANS 2003. Proceedings, San Diego, 22–26 Sept 2003, pp 2596–2599

110. Watkins WA, Schevill WE (1971) Four hydrophone array for acoustic three-dimensional location. Woods Hole Oceanographic Institution, Woods Hole, MA, Technical Report 71-60, p 33
111. Watkins WA, Moore KE, Tyack P (1985) Sperm whale acoustic behaviors in the Southeast Caribbean. Cetology 49:1–15
112. Wenz GM (1962) Acoustic ambient noise in the ocean: spectra and sources. J Acoust Soc Am 34:1936–1956
113. Yuan X, Kind R, Pedersen HA (2005) Seismic monitoring of the Indian Ocean tsunami. Geophys Res Lett 32:L15308. https://doi.org/10.1029/2005GL023464

Data Return Aspects of CODAR and WERA High-Frequency Radars in Mapping Currents

Yonggang Liu, Clifford R. Merz, Robert H. Weisberg, Benjamin K. O'Loughlin, and Vembu Subramanian

Abstract Two types of high-frequency (HF) radar systems, long-range CODAR SeaSonde and medium-range WERA, are concurrently operated on the West Florida Coast for the purpose of observing coastal ocean currents and waves. In this chapter, we examine the data return aspect of HF radar performance, using radial currents measured with the CODAR SeaSonde and WERA systems at the same site origin – Venice, Florida. Based on the data collected during February 2 – 5 March, 2014, our analysis revealed that the two HF radar systems exhibited complicated data return variations in both the spatial and temporal domains. Even though data return was generally higher near the site origin rather than in the outer band of the offshore radar footprint, it was unevenly distributed across the bearing angles. The long-range CODAR tended to have more data return in the northern half of its footprint, while the medium-range WERA's data return was more evenly distributed across the bearing angles. Both radar systems exhibited diurnal and synoptic variations in data return; however, the peak performance hours differed. The 4.90 MHz CODAR system tended to have a higher data return during the daytime hours, while the 12.58 MHz WERA system tended to return more data during nighttime hours. The CODAR system exhibited increased data return performance during the conditions of high sea state, while the WERA system's performance did not exhibit an obvious sea state relationship with waves measured using an offshore Waverider buoy.

Y. Liu (✉) • C.R. Merz • R.H. Weisberg
College of Marine Science, University of South Florida,
140 7th Avenue South, St. Petersburg, FL, USA
e-mail: yliu@mail.usf.edu

B.K. O'Loughlin
College of Marine Science, University of South Florida,
140 7th Avenue South, St. Petersburg, FL, USA

US Coast Guard Academy, New London, CT, USA

V. Subramanian (deceased)
Southeast Coastal Ocean Observing Regional Association (SECOORA),
Charleston, SC, USA

R. Venkatesan et al. (eds.), *Observing the Oceans in Real Time*, Springer Oceanography, DOI 10.1007/978-3-319-66493-4_11

1 Introduction

As shore-based remote sensing systems, high-frequency (HF) radars can map real-time surface currents, waves, and winds in a relatively larger spatial area than moored stations in coastal oceans (e.g., [28]), and play an increasingly important role in coastal ocean observing systems (e.g., [15]). Their widespread applications are found around the world's coastal oceans (e.g., [13]).

The Ocean Circulation Group of the University of South Florida began coastal ocean observation activities on the West Florida Shelf (WFS) with a nonreal-time single Acoustic Doppler Current Profiler (ADCP) mooring in 1993, which developed into an ADCP mooring array consisting of surface and subsurface buoys with 13 moorings at its peak time (e.g., [17, 18, 32, 33, 35]). These gradually evolved into a real-time Coastal Ocean Monitoring and Prediction System (COMPS) program in 1998 (e.g., [23]) for the purpose of improving the understanding and predictive capability of coastal ocean circulation and dynamics (e.g., [34, 35]), as well as marine environmental applications (e.g., [16, 37, 38]). The COMPS asset consists of offshore buoys, coastal stations, HF radars, and models. The COMPS program maintains and operates two types of HF radars on the West Florida Coast, the Coastal Ocean Dynamics Application Radar (CODAR; [3]) and the Wellen Radar (WERA; [11]), all looking over the ADCP mooring array. Evolution of the WFS HF radar systems can be found in the work of Merz et al. [24].

Three long-range CODAR SeaSonde HF radar systems are located in the cities of Redington Shores, Venice, and Naples, respectively, operating at a nominal frequency of 4.90 MHz and a fixed measurement bandwidth of 25.734 kHz. Two medium-range WERA HF radars are located within Pinellas County's Fort De Soto Park and at the US Coast Guard Auxiliary Station 86 in Venice, respectively. These WERA radars operate between frequencies of 12.275 and 13.20 MHz with varying bandwidths using the "Listen Before Talk" adaptive noise procedure [25]. The locations and footprints of these HF radars are provided in Fig. 1. Detailed configuration settings of these HF radars are documented by Liu et al. [19]. Various experiments, assessments, and applications of the HF radars on the WFS can be seen in a series of publications by the Ocean Circulation Group at the University of South Florida [4, 6, 7, 9, 19–22, 25, 29].

A performance metric of HF radar in measuring surface currents is the accuracy of the HF radar current velocity. This is usually evaluated in terms of the root-mean-square differences (rmsd) between the HF radar velocity data and other measurements. The accuracies of the two HF radar systems on the WFS have been assessed using the concurrent moored ADCP current velocity data in terms of the rmsd between the HF radar and the ADCP radial currents, bearing offsets and radial velocity uncertainties [19, 20]. Despite differences in a variety of aspects between the direction-finding CODAR SeaSonde (long-range, effective-sensed "centroid" depth of Bragg scatter of 2.4 m, integration time of 4 h, and idealized antenna patterns) and the beamforming WERA (median-range, effective-sensed "centroid" depth of Bragg scatter of 0.9 m and integration time of 1 h), both HF radar systems demonstrated good surface current mapping capability when intercompared [19].

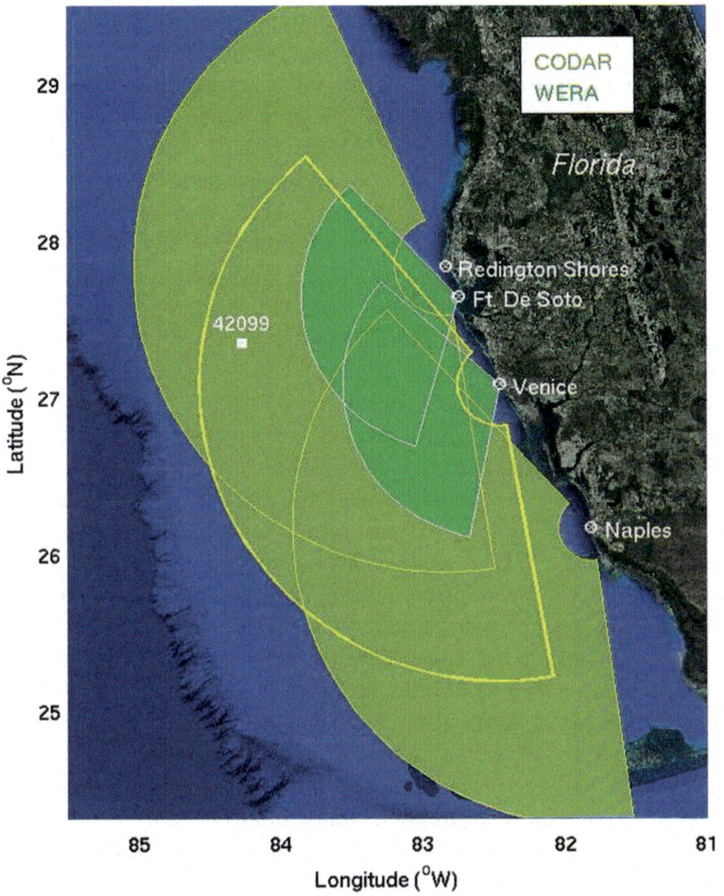

Fig. 1 High-frequency radar radial footprints on the West Florida Shelf maintained by the COMPS program of University of South Florida. CODAR SeaSonde sites: Redington Shores, Venice, and Naples; WERA sites: Ft. DeSoto and Venice. Also shown is the location of a Waverider buoy (42099) maintained by the CDIP of Scripps Institution of Oceanography

They generally agreed with the ADCPs to within rmsd for hourly radial velocity components of 5.1–9.2 and 3.8–6.5 cm/s for SeaSonde and WERA, respectively, and within rmsd for 36-h low-pass filtered radial velocity components of 2.8–6.0 and 2.2–4.3 cm/s for SeaSonde and WERA, respectively. The differences between the velocities measured with the HF radar and the ADCP are sufficiently small on this low-energy shelf that much of these rmsd values may be accounted for by the expected measurement differences due to the horizontal, vertical, and temporal sampling differences of the ocean current observing systems used [19]. Based on the findings of these studies, the HF radar current velocity data are further used to evaluate satellite altimetry-derived velocity product on the WFS [22].

Another performance metric of HF radar current mapping is data return. The idealized radial current data return would be full of the HF radar footprint; however, only partial coverage of the footprint can actually be returned. Liu et al. [20] examined the data return of the CODAR SeaSonde systems on the WFS and discussed possible environmental factors affecting the HF radar performance. Both the low-energy sea state (currents and waves) and the unfavorable surface wave directions are the main limiting factors for the 4.90 MHz CODAR SeaSonde HF radar observations of currents on the WFS. By examining the HF radar radial velocities at low wave energy, it is found that the data returns decrease rapidly for significant wave heights smaller than 1 m, and that the rms differences between the HF radar and ADCP radials are degraded when the significant wave height is smaller than 0.3 m [20]. As a result, the long-range CODAR SeaSonde systems have relatively better data return in winter months than summer months on the WFS. This is particularly evident during the summer months when the overall sea state condition is calm and a reduced amount of current data is collected. To improve data returns in this low-energy environment (e.g., [20, 36]), special data processing procedure is implemented: Five successive raw cross spectra (with 2048 points FFT, output every 34 min) are averaged over a total of 4 h, with the averaging interval advanced each hour. These 4-h averaged radials are then archived as hourly time series.

Data returns from the WERA systems also change with time and vary from site to site. Merz et al. [25] noticed different filling rates of data on the storage disks for the two WERA sites, and ascribed these to site-to-site differences in external electromagnetic (EM) background noise levels (ionospheric and/or Radio Frequency Interference [RFI]) [10, 12]. Through a series of experiments using different transmitting frequencies (fixed or variable) and bandwidths, they reported that WERA's data coverage and signal-to-noise ratio increase if the frequency-adaptive "Listen before Talk Mode" software [12] is used to dynamically adapt the center transmit frequency (Tx) and widest measurement bandwidth to react to local changes in the EM propagation characteristics present at the time of measurement [25].

This chapter is a follow-on study of Liu et al. [20], Merz et al. [25] on the performance of the HF radars in collecting valid current data. We focus on the spatial and temporal variations of the data returns, and try to examine their relationship with various marine environmental variables, and compare the data return characteristics of the two HF radar systems, the long-range CODAR SeaSonde and the medium-range WERA systems. The rest of the chapter is arranged as follows: observational data information is provided in Sect. 2, data return analysis results are reported in Sect. 3, and Sect. 4 contains a brief summary.

2 Data

2.1 HF Radar Radial Current Data

USF's Venice site HF radars are used in this study, because both a long-range CODAR SeaSonde and a medium-range WERA are colocated at the same location, and have been since 2010. The direction-finding CODAR SeaSonde HF radar system utilizes separate transmit (Tx) and receive (Rx) antennas, operates at nominal frequency of 4.90 MHz, and observes radial velocity components (radials) at an effective average Bragg depth of 2.4 m (e.g., [27, 31]), with nominal range and bearing resolutions of 5.8 km and 5°, respectively. Different from CODAR SeaSonde, the phased-array WERA system has four antennas arranged in a rectangular transmit array configuration and 12 antennas configured alongshore in linear receive line array. The WERA's transmit frequency and bandwidth are determined by the "Listen before Talk" procedure, that is, it first scans the entire Federal Communications Commission (FCC)-licensed 1-MHz operational band prior to each acquisition to determine the region of lowest noise level and corresponding widest measurement bandwidth. For this performance analysis, we chose the time period of February 2–March 5, 2014, corresponding to one of the experiments when the operating frequency was set to a fixed frequency of 12.58 MHz and a 100 kHz bandwidth [25]. Since the CODAR operating frequency is fixed, this will allow better comparability between the two HF radar systems. The WERA system also uses 2048 spectral Doppler bins in the FFT. The radials and vector totals are processed using standard WERA-supplied software with a nominal range and bearing of 1.5 km and 8°, respectively, at broadside (90°), changing to 1.5 km and 20° at ±60°, respectively. The WERA data are archived every 20 min. An example of radial currents measured with the two HF radar systems is shown in Fig. 2.

Data processing of the HF radar radial currents follows those in previous studies [19, 22, 25]. The hourly CODAR SeaSonde radials are interpolated onto uniform radial grids (bearing and range intervals of 5° and 5.8 km, respectively) using a

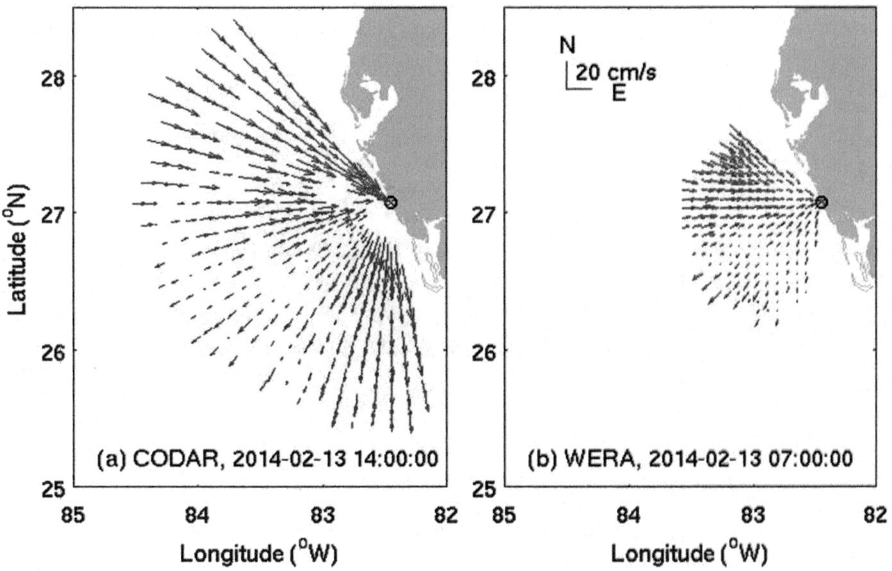

Fig. 2 Snapshots of HF radar radial currents measured with (**a**) CODAR SeaSonde and (**b**) WERA systems at Venice in the case of good data return. Time stamps in UTC

community toolbox HFR_Progs (https://cencalarchive.org/~cocmpmb/COCMP-wiki/index.php/HFR_Progs_download_page), which is the implementation of the least squares method, as described by Lipa and Barrick [14]. Outliers (unusually large speeds >50 cm s^{-1}) are removed prior to the interpolation. The 20-min WERA radial currents are processed by the WERA software on a rectangular grid. Data quality control and outlier removal are implemented using the "accuracy" threshold following Liu et al. [19].

2.2 *Ocean Surface Wave Data*

A Waverider buoy is deployed offshore from St. Petersburg, Florida, at a location (27°20′29″ N 84°16′20″ W) that is outside of the WERA footprint but within the CODAR footprint. The buoy is about 200 km from the site origin at Venice, Florida (Fig. 1). The ocean wave data are furnished by the Coastal Data Information Program (CDIP), Integrative Oceanography Division, operated by the Scripps Institution of Oceanography (http://cdip.ucsd.edu/), under the sponsorship of the US Army Corps of Engineers and the California Department of Parks and Recreation. The data are available through National Oceanic and Atmospheric Administration's (NOAA) National Data Buoy Center (NDBC) website (http://www.ndbc.noaa.gov/), assigned as buoy number 42099. The hourly wave data include significant wave height (calculated as the average of the highest one-third of all of the wave heights during the 20-min sampling period), dominant wave period, average wave period of all waves during the 20-min period, and dominant wave direction. More information can be seen in the "Wave Measurements" section at http://www.ndbc.noaa.gov/wave.shtml.

2.3 *Wind Data*

Wind data are obtained from the NOAA NDBC Coastal-Marine Automated Network (C-MAN) station VENF1 at Venice, Florida (online at http://www.ndbc.noaa.gov/). The anemometer height is 11.6 m above mean sea level. The hourly wind data are extracted from the standard meteorological data and converted to oceanographic convention for further analysis.

3 HF Radar Data Return

One indicator of HF radar performance in data return is the spatial coverage of radial currents (called "radials" in the HF radar community) achieved over time. HF radar coverage is defined as the number of sectors returning valid data each time,

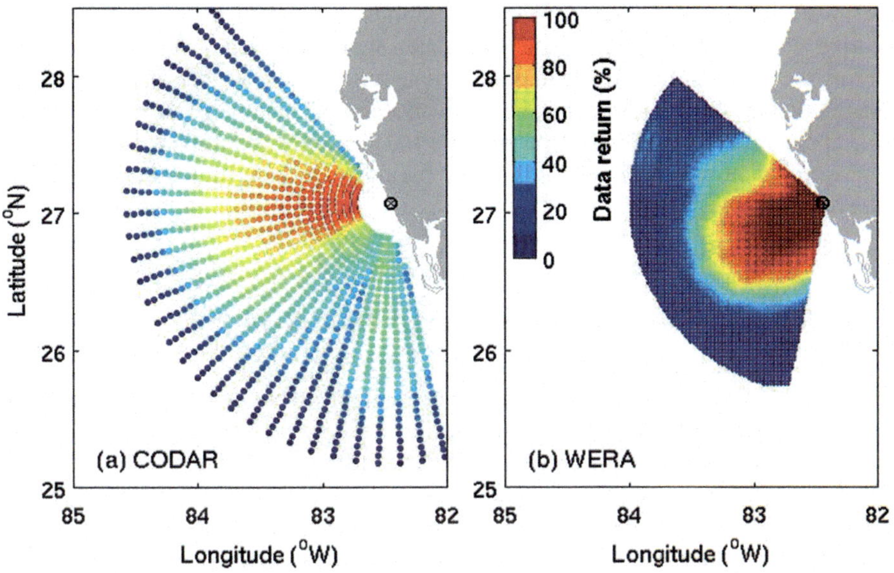

Fig. 3 Percent coverage maps of HF radar radial currents measured with (**a**) CODAR SeaSonde and (**b**) WERA systems at Venice during February 2–March 5, 2014

following previous studies (e.g., [8, 19, 20, 30]). Here, the coverage is used to quantify the radial current data returns. We analyze the spatial patterns and temporal variations of the coverage of the two HF radar systems, and examine their relationship to the environmental variables.

3.1 *Spatial Patterns of HF Radar Data Return*

For each radar radial sector, the total number of valid radial current data divided by the record length during the selected time period (February 3– March 3, 2014) is shown as a percentage number in Fig. 3. The percent coverage maps show the spatial distribution of the overall data return in the HF radar footprints.

Both radars generally have higher data returns near the site origin than in the outer band of the radar footprint, which is expected for all the HF radar ocean current observations. Radial current data returns are found to be unevenly distributed across the bearing angles. The long-range CODAR tends to have more data return in the northern half of its footprint than the other area of the footprint. This is due to distortions in the antenna response pattern caused by the near-field antenna environment. It is confirmed in our previous study [20] that the uneven distribution of the coverage across the bearing angles corresponds to the distortion of the measured patterns from the ideal antenna patterns (e.g., [2]). Higher coverage is often seen over certain sectors with protruding (measured) antenna patterns [20]. In contrast, the medium-range WERA's data return exhibits a more even distribution across bearing angles, with its high data return area more centered within its footprint. During this analysis period, the maximum data return rate was 99% and 90% for the WERA and the CODAR systems, respectively. We note that the CODAR system did not report valid data between the time period February 3–6, 2014, because of a receive antenna cable failure. This 3-day period is excluded from computing the data return in Fig. 3.

3.2 Temporal Variation of HF Radar Data Return

Time series of wind, wave, and HF radar data return are shown in Fig. 4. There were three synoptic weather front events during this analysis period, around 5–6, 13–14, and 27–28 of February 2014, respectively. The strong winds associated with these weather fronts incurred high waves on the shelf. We examine the relationship between the environmental variables and the data return from the HF radars.

Using the return signal scattered off the ocean surface, HF radars provide a means for mapping fields of near-surface ocean current velocities toward or away from the receive antennas. HF radars operate in the 3–30 MHz frequency band, providing over-the-horizon capabilities by virtue of ground wave or sky wave propagation. When a HF radio wave signal is transmitted from a radar and reaches the ocean surface, a portion of the incident energy is scattered back toward the source by ocean waves (mostly wind-driven gravity waves) whose wavelength is exactly one-half of the transmitted electromagnetic wavelength. This backscattering produces an energy spectrum at the receiver, which is used to infer radial currents according to Bragg's Scattering Law and the Doppler-induced frequency-shift theory (e.g., [1, 5]). The Bragg scattering peak for the 4.90 MHz radar corresponds to a deep-water surface gravity wave of approximately 0.23 Hz (a 4.4 s wave period), while that for the 12.58 MHz radar is about 0.36 Hz (2.7 s wave period). Our purpose is to examine the relationship between the presence of the relevant ocean surface gravity waves and the HF radar's data return. Thus, the two wave periods (4.4 s and 2.7 s) are shown as dashed and dashed-and-dot lines, respectively, in Fig. 4d.

Time series of the radial coverage show considerable temporal variation, modulated at both high and low frequencies. Data return (coverage) from the long-range CODAR system exhibits both diurnal and synoptic weather and ionospheric band variations (Fig. 4e). The wind at Venice shows some diurnal changes (sea breeze) at Venice; however, the significant wave height at the buoy 42099 does not have any obvious diurnal variations. CODAR coverage synoptic variation corresponds well with both the wind and the significant wave height. During the days with stronger winds and higher sea states, the CODAR tends to return more data. These CODAR data return findings are consistent with our previous study [20]. To examine the relationship between CODAR data return with sea state, we compute conditional average of the coverage as binned by the wave height measured with the Waverider buoy 42099. The average CODAR radial coverage grows with wave height (Fig. 5).

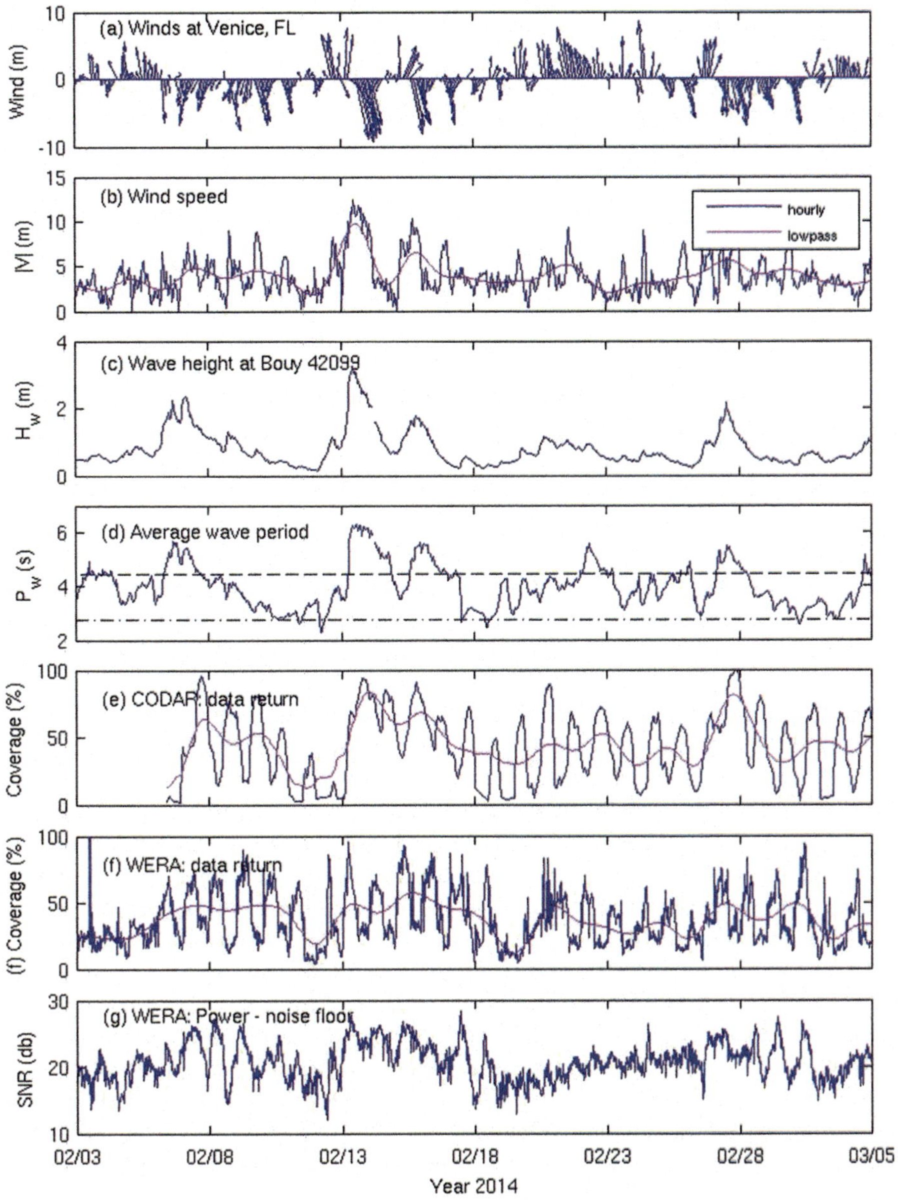

Fig. 4 Time series of wind, wave and HF radar data return during 2 February–5 March 2014. (**a**) Two-hourly winds measured at Venice, FL, (**b**) hourly and 48-h lowpass filtered wind speed at Venice, (**c**) hourly wave height measured by the Waverider buoy 42099, (**d**) hourly averaged wave period at buoy 42099 (wave periods of 4.4 s and 2.7 s are also indicated as dashed and dash-dotted lines, respectively), (**e**) hourly and 48-lowpass filtered data return (percent coverage) of CODAR SeaSonde system at Venice, (**f**) 20-min and 48-h lowpass filtered data return of WERA system at Venice, and (**g**) 20-min spatially averaged SNR of the WERA system, defined as the difference between power and noise floor

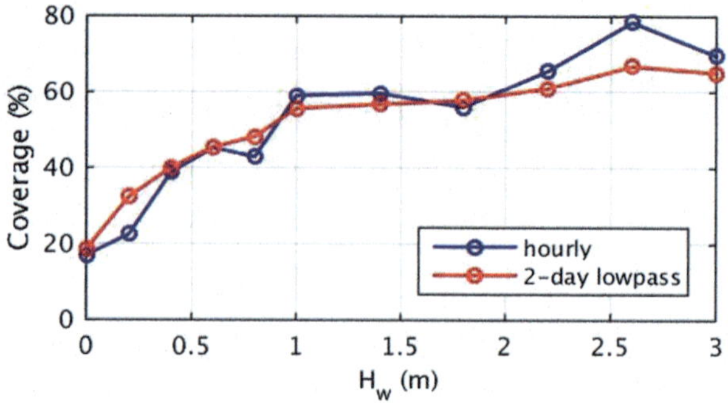

Fig. 5 Conditional averages of hourly and 2-day lowpass filtered CODAR coverage as binned by the wave height (H_w) measured with the Waverider buoy 42099

The coverage decreased rapidly when wave height becomes less than 1 m. This is consistent with previous findings on the WFS [20].

Data return from the medium-range WERA system also exhibits diurnal and low-frequency variations (Fig. 4f). Different from the coverage of the CODAR system, the low-frequency variation of the WERA coverage does not obviously correspond with the synoptic weather events. For example, during February 13–14 front event, the CODAR's data return was generally elevated to a higher level, while the WERA's data return seemed to stay on its own way – exhibiting mostly diurnal changes. Also, the data returns do not have an obvious relationship with the average wave periods. So, the average wave period may not be a good indicator of the wave energy presence at particular frequencies. Note that the original 20-min time series is used for WERA in Fig. 4f; so, the line is not as smooth as that for CODAR.

There are some time lags between the data returns from the two HF radar systems, and the coverages of the two systems are almost antiphased in many occasions if the diurnal peaks are focused on (Fig. 4e and f). To further examine this time lag, we compute average hourly data return for all 24 h, as shown in Fig. 6, which indicates the general diurnal variations of the HF radar data returns. The long-range CODAR system has higher data return during daytime (during 14–23 h UTC, i.e., 9–18 h local time), which is consistent with the previous findings for the CODAR total currents [26]. In contrast, the medium-range WERA system tends to return more data during nighttime (3–12 h UTC). There is a time lag of about 11–12 h.

Note that the CODAR data were processed to return more data based on an average of the cross-spectral data during a time window of 4 h. To be more comparable, the 20-min WERA radial velocity time series are further smoothed using a 4-h running mean filter (averaging among the adjacent 13 data points) to produce hourly time series. The updated figure is shown in Fig. 7. The WERA system's data return is generally increased for all the 24 h (Fig. 7). The lowest average data returns of the two HF radars are about the same (~30%), but at different hours (7–9 versus 19–20 h UTC, respectively, for the CODAR and WERA systems). The CODAR system exhibits higher data returns during its peak hours than observed for the WERA system, but the good data return time window (coverage >50%) is narrower (10 versus 14 h).

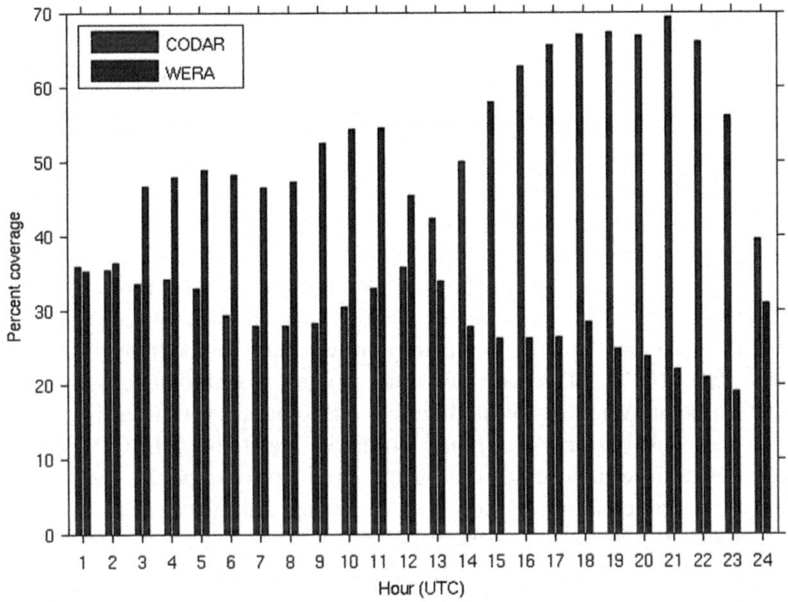

Fig. 6 Average hourly HF radar data return (percent coverage) for CODAR SeaSonde and WERA systems at Venice during 2 February–5 March 2014. The statistics are based on the hourly CODAR data and the 20-min WERA data, respectively

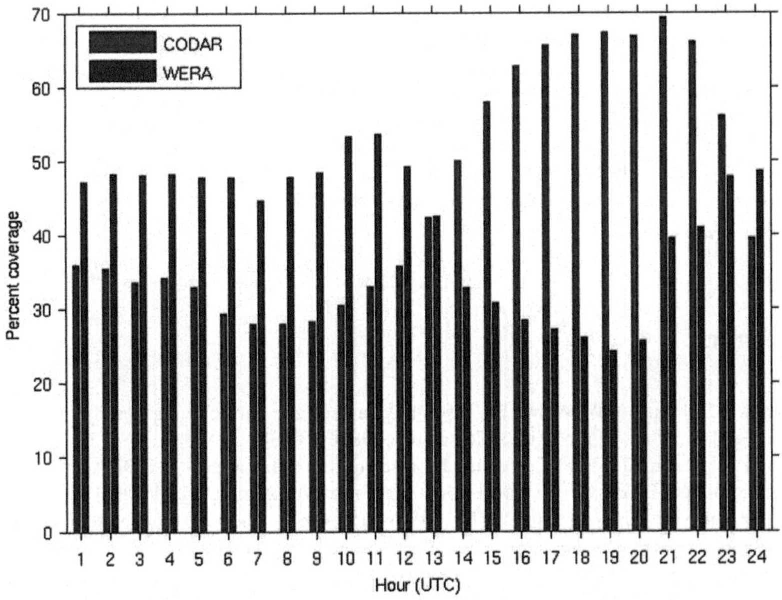

Fig. 7 Same as Fig. 6 except that the 20-min WERA data are time-averaged using a 13-point running-mean filter and subsampled hourly. The CODAR and WERA data are both hourly and comparable

4 Summary

Data return aspects of HF radar performance were examined using radial currents measured with the 4.90 MHz long-range CODAR SeaSonde and 12.58 MHz medium-range WERA systems at the same site origin – Venice, Florida. Based on the data collected during February 2–March 5, 2014, on the West Florida Shelf, we found out that the two HF radar systems exhibit complicated data return variations in both spatial and temporal domains. Even though the data return was generally higher near the site origin than in the outer band of the radar footprint, it was found to be unevenly distributed across the bearing angles. The long-range CODAR tended to have more data return in the northern half of its footprint, while the medium-range WERA's data return was more evenly distributed across the bearing angles. Both radar systems exhibited diurnal and synoptic variations in data return; however, the peak performance hours differed. The 4.90 MHz CODAR system tended to have more data return during the daytime hours, while the 12.58 MHz WERA system tended to return more data during the nighttime hours. The CODAR system exhibited increased data return performance during conditions of high sea state, while the WERA system's performance did not exhibit an obvious sea state relationship with waves measured using an offshore Waverider buoy. Investigations will continue into the examination of site differences in external EM background noise levels as well as variations in diurnal and synoptic data returns.

Acknowledgments The University of South Florida (USF) COMPS operates along the Gulf of Mexico's West Florida Coast and was implemented in 1977 as a State of Florida legislative initiative. The COMPS program receives partial support from the Southeast Coastal Ocean Observing Regional Association (SECOORA) through the US IOOS Office operated by NOAA (Award No. NA11NOS0120033). WERA HF Radar equipment was originally acquired with USF internal R&D funds. Partial support was also provided by NASA Ocean Surface Topography Science Team (OSTST) (# NNX13AE18G). Ocean wave data were furnished by the Coastal Data Information Program (CDIP), Integrative Oceanography Division, operated by the Scripps Institution of Oceanography, under the sponsorship of the US Army Corps of Engineers and the California Department of Parks and Recreation. Wind data were downloaded from NOAA NDBC. This is CPR Contribution 48.

References

1. Barrick DE, Evans MW, Weber BL (1977) Ocean surface currents mapped by radar. Science 198:138–144. https://doi.org/10.1126/science.198.4313.138
2. Barrick DE, Lipa BJ (1986) Correcting for distorted antenna patterns in CODAR ocean surface measurements. IEEE J Ocean Eng 11:304–309. https://doi.org/10.1109/JOE.1986.1145158
3. Barrick DE, Lipa BJ, Crissman RD (1985) Mapping surface currents with CODAR. Sea Technol 26:43–48
4. Barth A, Alvera-Azcárate A, Weisberg RH (2008) Assimilation of high-frequency radar currents in a nested model of the West Florida shelf. J Geophys Res 113:C08033. https://doi.org/10.1029/2007JC004585

5. Crombie DD (1955) Doppler spectrum of sea echo at 13.56Mc./s. Nature 175:681–682
6. Dzvonkovskaya D, Merz CR, Liu Y, Weisberg RH, Helzel T, Petersen L (2014) Initial surface current measurements on the West Florida Shelf using WERA HF ocean radar with multiple input multiple output (MIMO) synthetic aperture. Proceedings of MTS/IEEE OCEANS'14, Newfoundland. https://doi.org/10.1109/OCEANS.2014.7003235
7. Dzvonkovskaya D, Helzel T, Petersen L, Merz CR, Liu Y, Weisberg RH (2014) Initial results of ship detection and tracking using WERA HF ocean radar with MIMO configuration. Radar symposium. https://doi.org/10.1109/IRS.2014.6869265
8. Emery BM, Washburn L, Harlan JA (2004) Evaluating radial current measurements from CODAR high-frequency radars with moored current meters. J Atmos Ocean Technol 21:1259–1271. https://doi.org/10.1175/1520-0426(2004)021<1259:ERCMFC>2.0.CO;2
9. Gomez G, Helzel T, Petersen L, Kniephoff M, Merz CR, Liu Y, Weisberg RH (2014) Real-time quality control of current velocity data on individual grid cells in WERA HF radar. MTS/IEEE OCEANS'14, Taipei. https://doi.org/10.1109/OCEANS-TAIPEI.2014.6964502
10. Gurgel K-W, Barbin Y, Schlick T (2007) Radio frequency interference techniques in FMCW modulated HF radars. Proceedings of MTS/IEEE Ocean'07, Aberdeen
11. Gurgel K-W, Antonischki G, Essen H-H, Schlick T (1999) Wellen radar (WERA): a new ground-wave HF radar for ocean remote sensing. Coast Eng 37:219–234. https://doi.org/10.1016/S0378-3839(99)00027-7
12. Helzel T (2007) FMCW radar "WERA." Helzel Messtechnik GmbH Internal Doc. 19 Oct 2007, 6 pp.
13. Harlan J, Terrill E, Hazard L, Keen C, Barrick D, Whelan C, Howden S, Kohut J (2010) The integrated ocean observing system high-frequency radar network: status and local, regional, and national applications. Mar Technol Soc J 44:122–132. https://doi.org/10.4031/MTSJ.44.6.6
14. Lipa BJ, Barrick DE (1983) Least-squares methods for the extraction of surface currents from CODAR crossed-loop data: application at ARSLOE. IEEE J Ocean Eng 8:226–253. https://doi.org/10.1109/JOE.1983.1145578
15. Liu Y, Kerkering H, Weisberg RH (eds) (2015) Coastal ocean observing systems. Elsevier (Academic Press), London, UK, 461 pp, isbn:978-0-12-802022-7. http://www.sciencedirect.com/science/book/9780128020227
16. Liu Y, MacFadyen A, Ji ZG, Weisberg RH (eds) (2011) Monitoring and modeling the deepwater horizon oil spill: a record-breaking enterprise, Geophysical monograph, vol 195. American Geophysical Union, Washington, DC, 271 pp, isbn: 978-0-87590-485-6. https://doi.org/10.1029/GM195
17. Liu Y, Weisberg RH (2005) Patterns of ocean current variability on the West Florida shelf using the self-organizing map. J Geophys Res 110:C06003. https://doi.org/10.1029/2004JC002786
18. Liu Y, Weisberg RH (2012) Seasonal variability on the West Florida shelf. Prog Oceanogr 104:80–98. https://doi.org/10.1016/j.pocean.2012.06.001
19. Liu Y, Weisberg RH, Merz CR (2014) Assessment of CODAR and WERA HF radars in mapping currents on the West Florida shelf. J Atmos Ocean Technol 31:1363–1382. https://doi.org/10.1175/JTECH-D-13-00107.1
20. Liu Y, Weisberg RH, Merz CR, Lichtenwalner S, Kirkpatrick GJ (2010) HF radar performance in a low-energy environment: CODAR SeaSonde experience on the West Florida shelf. J Atmos Ocean Technol 27:1689–1710. https://doi.org/10.1175/2010JTECHO720.1
21. Liu Y, Weisberg RH, Shay LK (2007) Current patterns on the West Florida shelf from joint self-organizing map analyses of HF radar and ADCP data. J Atmos Ocean Technol 24:702–712. https://doi.org/10.1175/JTECH1999.1
22. Liu Y, Weisberg RH, Vignudelli S, Roblou L, Merz C (2012) Comparison of the X-TRACK altimetry estimated currents with moored ADCP and HF radar observations on the West Florida shelf. Adv Space Res 50:1085–1098. https://doi.org/10.1016/j.asr.2011.09.012

23. Merz CR (2001) An overview of the Coastal Ocean Monitoring and Prediction System (COMPS). In: MTS/IEEE oceans 2001: an ocean odyssey, vol. 2. MTS and IEEE, pp 1183–1187. https://doi.org/10.1109/OCEANS.2001.968281
24. Merz CR, Weisberg RH, Liu Y (2012) Evolution of the USF/CMS CODAR and WERA HF radar network. In: Oceans 2012, IEEE, pp 1–5. https://doi.org/10.1109/OCEANS.2012.6404947
25. Merz CR, Liu Y, Gurgel K-W, Peterson L, Weisberg RH (2015) Effect of radio frequency interference (RFI) noise energy on WERA performance using the "listen before talk" adaptive noise procedure. In: Coastal ocean observing systems. Elsevier (Academic Press), London, UK, pp 229–247. https://doi.org/10.1016/B978-0-12-802022-7.00013-4
26. O'Loughlin BK (2016) Evaluation of Search and Rescue Planning Tools on the West Florida Shelf, M.Sc. thesis, University of South Florida
27. Paduan JD, Graber HC (1997) Introduction to high-frequency radar: reality and myth. Oceanography 10:36–39. https://doi.org/10.5670/oceanog.1997.18
28. Paduan JD, Kosro PM, Glenn SM (2004) A national coastal ocean surface current mapping system for the United States. Mar Technol Soc J 38:102–108. https://doi.org/10.4031/002533204787522839
29. Pan C, Zheng L, Weisberg RH, Liu Y, Lembke C (2014) Comparisons of different ensemble schemes for glider data assimilation on West Florida shelf. Ocean Modell 81:13–24. https://doi.org/10.1016/j.ocemod.2014.06.005
30. Shay LK, Graber HC, Ross DB, Chapman RD (1995) Mesoscale surface current structure detected by HF radar. J Atmos Ocean Technol 12:881–900
31. Stewart RH, Joy JW (1974) HF radio measurements of surface currents. Deep-Sea Res 21:1039–1049. https://doi.org/10.1016/0011-7471(74)90066-7
32. Weisberg RH, Black BD, Yang H (1996) Seasonal modulation of the west Florida continental shelf circulation. Geophys Res Lett 23:2247–2250
33. Weisberg RH, Black BD, Li Z (2000) An upwelling case study on Florida's west coast. J Geophys Res 105:11459–11469
34. Weisberg RH, He R, Liu Y, Virmani J (2005) West Florida shelf circulation on synoptic, seasonal, and inter-annual time scales. In: Circulation in the Gulf of Mexico: observations and models, Geophysical monograph American Geophysical Union, Washington, DC, 161:325–347. https://doi.org/10.1029/161GM23
35. Weisberg RH, Liu Y, Mayer DA (2009) West Florida shelf mean circulation observed with long-term moorings. Geophys Res Lett 36:L19610. https://doi.org/10.1029/2009GL040028
36. Weisberg RH, Liu Y, Merz CR, Virmani JI, Zheng L (2012) A critique of alternative power generation for Florida by mechanical and solar means. MTS J 46(5):12–23. https://doi.org/10.4031/MTSJ.46.5.1
37. Weisberg RH, Zheng L, Liu Y, Lembke C, Lenes JM, Walsh JJ (2014) Why no red tide was observed on the west Florida continental shelf in 2010. Harmful Algae 38:119–136. https://doi.org/10.1016/j.hal.2014.04.010
38. Weisberg RH, Zheng L, Liu Y, Murawski S, Hu C, Paul J (2016) Did deepwater horizon hydrocarbons transit to the west Florida continental shelf? Deep Sea Res II 129:259–272. https://doi.org/10.1016/j.dsr2.2014.02.002

Part V
Data

Sensor Performance and Data Quality Control

Sébastien P. Bigorre and Nancy R. Galbraith

Abstract This chapter discusses the optimization of meteorological and oceanographic measurements from ocean surface moorings. First, we give guidelines and procedures for the selection of appropriate instruments and their preparation (calibration, configuration, integration and testing). Then, we present critical steps to evaluate and improve the data quality. We also discuss the limitations and benefits of real-time data, including various data quality control steps. Finally, we briefly compare two commonly used telemetry systems – ARGOS and Iridium.

In this chapter, we describe the ways to assess and optimize meteorological and oceanographic measurements from oceanic surface moorings. Surface moorings have been providing measurements of air-sea interaction for a long time, allowing the calibration of satellite measurements and numerical climate models. These measurements have been well characterized [3, 6] and shown to be very accurate [11]. Although focused on surface moorings, our discussion can be extended to other types of platforms.

1 Optimizing Observations

1.1 Instrument Selection

To collect a good dataset, a sensor must first preserve its physical integrity and measuring capability. To mitigate the risks associated with the harsh marine environment, a range of safeguards are available: titanium housings protect against corrosion and high pressure exposure; load-cages (Fig. 1) shield sensors on the mooring line from fouling with fishing gear or fish bites; copper hardware or antifouling chemical devices hinder biofouling; desiccant and new O-rings placed inside instrument

S.P. Bigorre (✉) • N.R. Galbraith
Woods Hole Oceanographic Institution, Woods Hole, MA 02543, USA
e-mail: sbigorre@whoi.edu

R. Venkatesan et al. (eds.), *Observing the Oceans in Real Time*, Springer Oceanography, DOI 10.1007/978-3-319-66493-4_12

Fig. 1 From top to bottom, left to right: (**a**) surface buoy with meteorological instruments inside crash bar; note the missing propellers on wind anemometers, probably from wave impacts; (**b**) crash bar damaged by collision and personnel onboard the buoy to evaluate replacement sensors; (**c**) bird droppings on radiation sensors and bird-wire on precipitation gauge; (**d**) subsurface instrument at recovery, tagged for identification, showing heavy biofouling and damaged protecting load bar; (**e**) ADCP at recovery showing some trace of biofouling, fishing gear fouling on its broken load bar; (**f**) conductivity sensor at recovery with barnacles on instrument but rather clean conductivity cell thanks to antifouling plug

housings control moisture and corrosion. Instruments can also be improved for specific conditions, thanks to simple adaptations: filter around a humidity sensor to protect from condensation and spray; bird-wire on precipitation gauges or wind and radiation sensors; ventilation shield on air temperature sensors to reduce the diurnal heating, and so on.

In some instances, different types of sensors are more adapted to specific conditions. For example, propeller-vane anemometers are the most commonly used wind sensors because of their lower cost and power demand. But sonic wind sensors have a smaller surface area and no moving parts, making them less susceptible to damage from wave impacts or corrosion and more resistant in high sea states (Fig. 1). Similarly, current measurements in the deep ocean with little backscatter material may be better observed using mechanical sensors such as Vector Measuring Current Meters compared to Doppler current meters [7].

Finally, sensor accuracy is a key factor when selecting an instrument. Manufacturer's specifications are often optimistic, being based on tests in laboratory conditions, and some degradation of the sensor's performance is to be expected once deployed in the field. Other factors that will guide the selection of an instrument include cost, integrability, and manufacturer's customer service.

1.2 Instrument Preparation

Once an instrument is acquired, and before it is deployed, the user can optimize its future performance through careful calibration, configuration, integration, and testing.

1.2.1 Calibration

Calibrating instruments prior to deployment is recommended in most situations but may not always be possible. The decision depends in part on what level of accuracy is needed for the measurements. For example, signal variability in the ocean mixed layer is much higher than that below the thermocline, and environmental noise may therefore dominate instrument error in the upper ocean. Some sensors also tend to be more stable than others and may not require the same frequency of calibration and maintenance. Typically, temperature sensors are quite stable, whereas humidity or conductivity sensors tend to drift more rapidly over time. Conductivity drift is typically due to environmental factors that change the geometry of the borosilicate cell (e.g., biofouling, scouring of the glass from current-driven sediment flow [2]). It is often useful to understand how calibrations are made and in particular their range of validity. For example, relative humidity sensors are typically calibrated using polynomial fits in laboratory conditions at ambient room temperature. However, field conditions may differ greatly and induce calibration errors. If the calibration is made in-house, there may be more opportunity to calibrate the

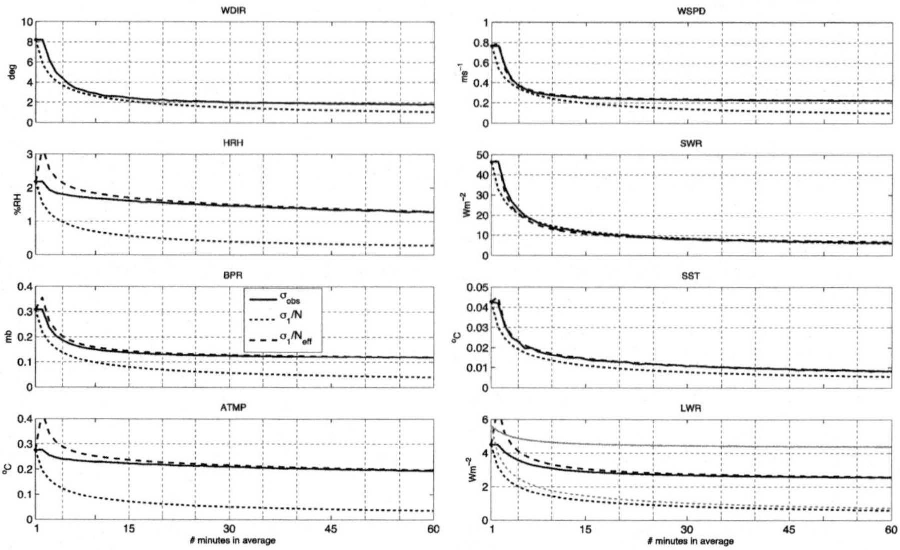

Fig. 2 Standard deviation of the difference between two ASIMET measurements, for each bulk variable, as a function of a number of samples (N) averaged together: Data (*solid lines*), $\sigma_1/\sqrt{N}$ (*dotted lines*), and $\sigma_1/\sqrt{N_{eff}}$ (*dashed lines*); σ_1 is standard deviation of 1-minute time series and N_{eff} is effective number of independent variables ([3] ©American Meteorological Society. Used with permission)

complete instrument, sensor, and integrated components (e.g., electronic board, housing). Checking the sensors used as standards of reference for calibration is also a good practice to identify and prevent any long-term drift (e.g., black paint aging on pyrgeometers).

1.2.2 Configuration

Once calibration is done, the user can select the sampling configuration. The choice is often a trade-off between battery limitations and measurement accuracy. Factors that affect battery duration (e.g., environmental temperature, vertical orientation) should therefore be taken into account.

To decrease the error associated with environmental or sensor noise, it is good practice to average multiple data samples. This averaging can be done in the instrument, in postprocessing, or both. In meteorology, it is common to average measurements over several minutes, a timescale associated with turbulent motion in the lower atmosphere. This is illustrated in Fig. 2, using meteorological data collected on a surface mooring deployed in the Gulf Stream for 1 year. The error was estimated using the standard deviation of the difference between measurements from the duplicate sensors on the buoy. Air-Sea Interaction METeorology (ASIMET, [12]) sensors were used which provided data with 1-minute resolution. For each meteorological variable, 1-minute samples were averaged together to produce a new estimate. As the averaging ensemble increased, the high-frequency signal was smoothed out and decreased the noise level. The plot shows that for most variables, using a 10-min average gets rid of a large portion of the random errors. Averaging further does little to improve the accuracy and could indicate the presence of biases, which are insensitive to averaging. The decrease in error is less than predicted by the central limit theorem for independent random errors. In fact, errors are correlated due to processes such as flow distortion or platform motion. For ASIMET[1] sensors, the 1-min data is itself an average from multiple readings. Performance of Acoustic Doppler Current Meters and Profilers (ADCM and ADCP) relies heavily on the configuration setup used by the user. The number of pings should resolve the wave-induced motion and minimize aliasing of longer swells. A rule of thumb is to resolve at least 10 swell cycles and average them out.

1.3 Instrument Integration

1.3.1 Mooring Design

To collect a good dataset, one must first ensure the mooring itself will maintain its integrity and location. The mooring design should take into account static and dynamic loads created by currents and wave surge at the location of the planned deployment. Adding fairing on the mooring wire [15] reduces the drag on the line induced by vortex shedding from water flowing past the mooring wire. When integrating metallic components to a mooring line (and surface mooring), special care must be taken to minimize corrosion by adding insulating components such as plastic bushings and sacrificial anodes. For meteorological sensors on a surface buoy, having a crash bar on the outside perimeter of the meteorological tower (Fig. 1) protects the instruments from collision by boats and unintended impacts during deployment and recovery.

1.3.2 Burn-in and Telemetry Testing

Functionality of instruments and data collection system should be tested prior to deployment. For meteorological data, one can simply install the surface buoy and its instruments outdoor and let the data collection system (instruments, loggers, telemetry) operate in the deployment configuration. This burn-in test can be done in some open space with minimal disturbances (e.g., good wind exposure, no shadows, little magnetic field). During burn-in, any faulty element (sensor, logger, cable, memory card) should be replaced with a new one and a set of spare sensors prepared. Control procedures can also be put in place, such as filling and draining rain gauges with

[1] http://www.whoi.edu/instruments/viewInstrument.do?id=12827

known quantities of water or checking the horizontal field-of-view of radiation sensors using a high precision level. Compasses (typically associated with wind sensors) can be validated with a buoy spin procedure, during which the buoy is oriented in several known directions in sequence, while readings from the compasses are monitored [4]. The buoy spin should be done in an area with minimum magnetic disturbance (e.g., docks in ports are typically difficult due to the dense metallic lattice in the reinforced concrete). Once the buoy spin is completed, compasses should not be moved until deployment so that their orientation relative to the buoy remains the same (this information provides a proxy for the buoy orientation, which may become relevant if flow distortion needs to be evaluated).

Often, instruments and other components must be shipped in containers where heat and shock are highly probable. These conditions can sometimes lead to sensors being damaged or fall out of calibration. It is therefore recommended to do another short burn-in prior to the deployment cruise. Telemetry at this stage is very much valuable since it is convenient and nonintrusive. Once the burn-in is completed, the clocks of the instruments should be checked and memory cards initialized. Then, final preparations of the sensors can be done prior to deployment.

1.3.3 Deployment Preparations

Sensors that will be deployed in the photic zone should be coated with antifoulants that will hinder the growth of algae and barnacles. Barnacles can grow inside any sharpness or gap, such as joints, which should be covered with tape and treated with antifoulants. Desitin®, a cream used for skin rash, is an efficient and benign antifoulant that can be applied on the transducers of ADCPs. For sensors with components made of metals with different electrical potentials, risks of corrosion should be minimized. One should inspect the sensor for signs of chipped paint that could compromise electrical insulation. Sacrificial anodes can be installed to mitigate the risk of corrosion. However, once in the water, these anodes create a local galvanic current that may change the local levels of oxygen and conductivity and distort readings of nearby sensors.

2 Data Quality Assurance

2.1 Data Quality Evaluation

2.1.1 Telemetry Monitoring

After deployment and before recovery, instruments can be monitored through telemetry to evaluate their performance. Although telemetered data often have degraded quality due to bandwidth limitations (e.g., hourly averaged or decimated), they can be used to identify problems such as biases or incomplete measurements.

If there is indication that some instrument is not performing as expected (e.g., incomplete or inaccurate data) following its deployment, the problem should be documented, and several actions can be taken. If weather permits, a buoy ride (Fig. 1) can be organized for personnel to board the buoy and evaluate more closely the cause of the problem; in many instances, replacing a faulty sensor with a spare (recently calibrated and evaluated during the burn-in) is the best solution. Sometimes, two-way telemetry can be used to communicate with the measuring system and correct the problem remotely. If recovery is near, an intercomparison with measurements from a different platform should be undertaken. Once recovered, a faulty instrument can be further evaluated by monitoring its performance next to another sensor with known calibration.

2.1.2 Intercomparison

A simple way to evaluate the measurements from a mooring is to compare them with similar measurements from nearby platforms (e.g., another mooring or a ship). Analyzing the differences between these measurements will help identify their causes and possible corrections. Some of these differences are inherent to the environment itself (e.g., natural heterogeneity combined with the spatial separation of the measurements) and have no relation to sensor accuracy. Differences due to sensor performance (e.g., drift) or measurement bias (e.g., flow distortion) are often correctable. We will discuss here a couple of examples, showing how intercomparison was key in identifying and correcting errors in the observations. The buoys used in the following observations were all Modular Ocean Buoys (MOB, see [15]) with 2.7 m diameter. Each MOB supports a meteorological tower on which instruments can be mounted about 3 m above sea level.

Case 1: Humidity Drift

In this example, a surface mooring was about to be recovered after a yearlong deployment. Since this was a mooring turn-around, a new mooring (let us call it S12) was first deployed, and the older mooring (S11) and its instrumentation, identical to S12, were recovered a couple of days later. During this overlap time, the research ship was stationed ½ nm downwind of each buoy for 24 h. Instrumentation available on the ship consisted of the ship's own sensors and freshly calibrated sensors that the mooring research group had installed temporarily on the ship in prevision for the intercomparison. The comparison of relative humidity (HRH) measurements between the different platforms is shown in Fig. 3, which plots simple statistics of nighttime data. Ship measurements were first adjusted to the buoy height, using the COARE algorithm [5, 9, 10]. HRH measurements from three identical ASIMET sensors on the S11 buoy were low relative to values from a different sensor (Lascar) on the same buoy, similar sensors on the new S12 buoy, and the ship's sensor. After recovery of S11, its buoy was secured on the ship's fantail, and a freshly calibrated ASIMET sensor (HRH SN231; this was a spare sensor for S12 and had a high bias (2–3%RH), as identified during burn-in) was added next to the

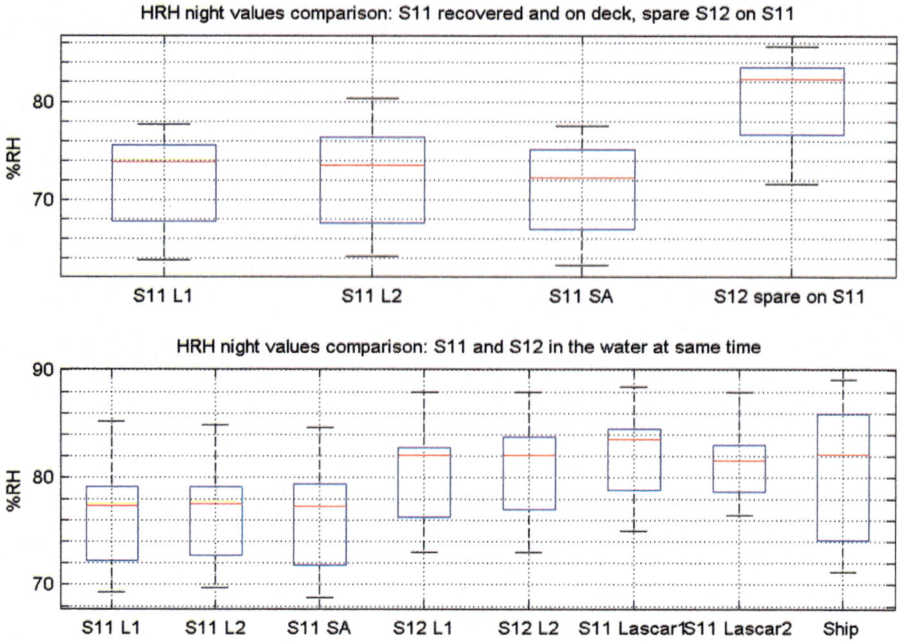

Fig. 3 Statistics of nighttime relative humidity (HRH) values. Median (*red*), 25th and 75th percentiles (edges, *blue boxes*), extreme values not considered outliers (*black whiskers*). *Upper*: period when S11 was on aft deck of R/V Melville. *Lower*: period when both the old and the new buoys (S11 and S12, respectively) were in the water. The main sensors (L1, L2, and *SA*) on the older buoy (S11) were low. All other measurements available were higher: including one sensor on S11 (Lascar), all sensors on new buoy (S12), and the ship's sensor

suspicious instruments on the buoy. After a few hours, the data collected confirmed the low bias on the ASIMET instruments from the recovered S11. Upper panel shows S11 ASIMET sensors were about 8%RH lower than SN231. The low bias observed on the old ASIMET was therefore about 5%RH, much beyond the nominal ASIMET accuracy (2%RH). Note that the two Lascar sensors deployed on S11 compared well with S12 and the ship values.

A postrecovery calibration of the three ASIMET HRH deployed on S11 was done 2 months after recovery and confirmed the low biases (3–4%RH) of all three sensors. A similar calibration was also done for air temperature, which indicated very little drift. On the other hand, the Lascar humidity sensors did not drift and agreed well with freshly calibrated sensors at recovery. Figure 4 shows the HRH difference between the ASIMET and Lascar sensors had a linear trend toward more negative values. This indicates that the ASIMET HRH sensors drifted toward lower values most likely linearly with time. Since the ASIMET HRH sensors have better precision and resolution than the Lascar sensors, it was desirable to correct the ASIMET drift. A linear correction was therefore issued, using zero offsets at deploy-

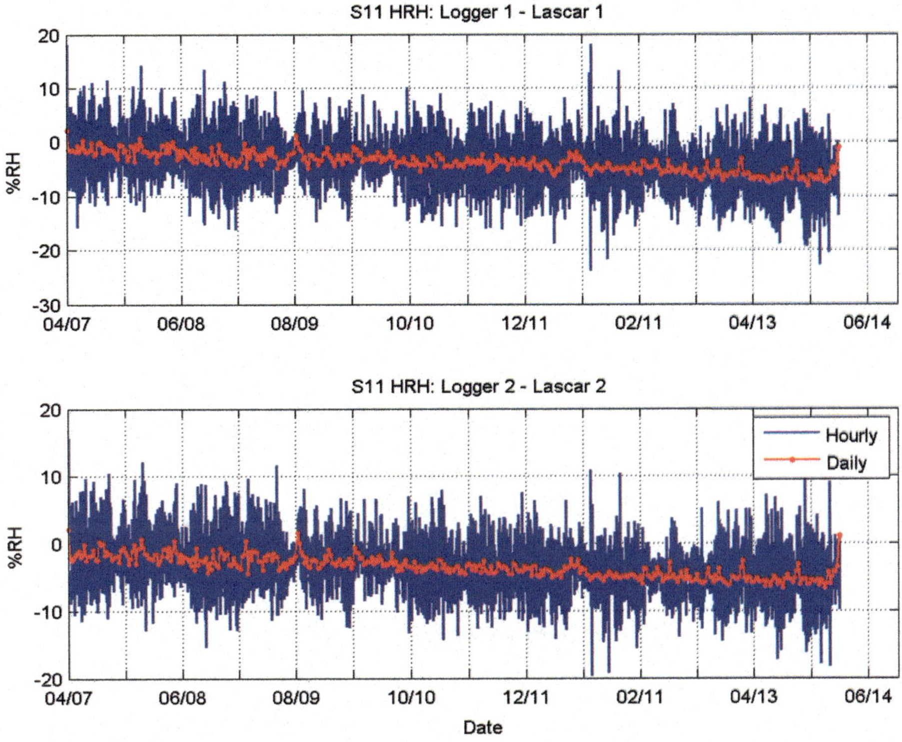

Fig. 4 Difference between ASIMET and Lascar relative humidity (HRH) sensors during S11 deployment, from April 2011 until May 2012. *Blue* hourly data. *Red* daily data

ment (based on burn-in and intercomparison during deployment cruise) and offsets given by postrecovery calibration near the end of the dataset.

Case 2: Wind Flow Distortion

In this case, we were able to detect and quantify flow distortion created on a surface mooring and issued a correction for the bias it introduced in the wind speed measurement. The first clue that flow distortion was occurring came from the observation that two identical wind anemometers, deployed both on the upstream face of the buoy but on opposite sides, measured a wind direction that was slightly but consistently offset with each other. The vanes of the wind sensors were also consistently offset by about 15°, which was also seen during visual inspection of the deployed buoy. The offset indicated that the horizontal airflow was slightly diverging between the two sensors. Figure 5 schematizes the flow distortion for a case where the wind impinges on the forward port quarter of the buoy (the buoy vane is designed to orient the buoy face into the wind, but other factors can interfere), and streamlines get deflected around sensors and the meteorological tower above the buoy. According to Bernoulli's principle, we should also expect a slight increase in pressure and lower wind speed near the upstream sensor where the flow diverges.

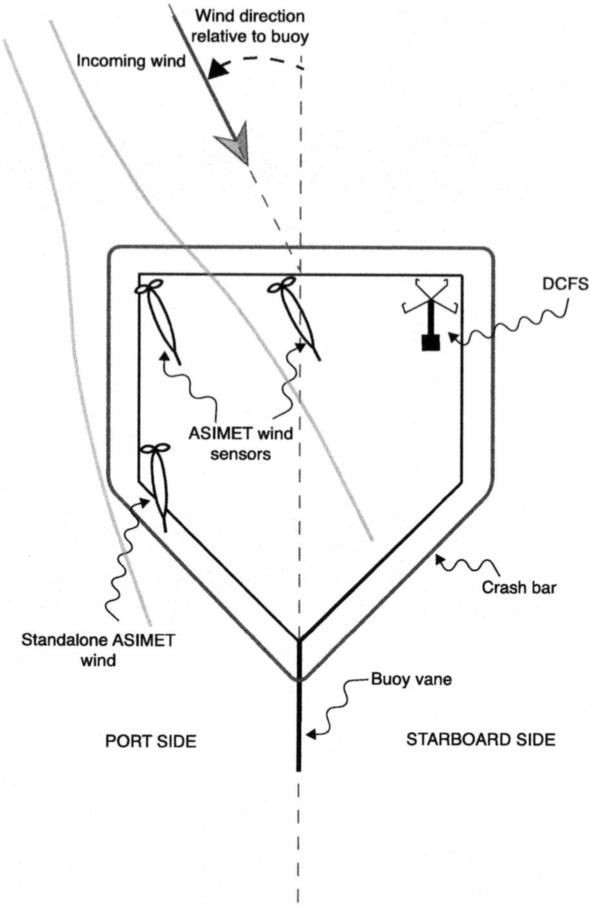

Fig. 5 Schematic of flow distortion around meteorological tower mounted on a surface buoy when wind impinges on the buoy with an oblique angle. This schematic (Reproduced from Bigorre et al. [3];©American Meteorological Society. Used with permission) represents the configuration used in CLIMODE experiment, which included a Direct Covariance Flux System (DCFS)

We quantified this flow distortion by plotting the wind speed measured at two different locations on the buoy as a function of the wind direction relative to the buoy (Fig. 6). Wind sensors were identical, both deployed on the forward face of the buoy but opposite sides. As the wind direction varied from dead ahead (0° angle) to several points on the portside bow (positive) or starboard bow (negative), the relative difference in wind speed between the two sensors varied linearly with respect to the wind angle, changed sign, and reached a maximum of about 5%.

Next clue came from the CLIMODE experiment with a similar MOB platform and wind sensors, but that included in addition a 3D wind sensor and motion package (Direct Covariance Flux System, or DCFS), used for turbulence measurements. The vertical component of the wind measured by the DCFS increased with wind speed, consistent with vertical flow distortion. The DCFS also measured attitude angles and angular rates, enabling the rotation of the wind vector from the pitched buoy frame of reference to a horizontal one. We also observed that the buoy did not always face perfectly the incoming wind, but rather had a small oblique angle (20–

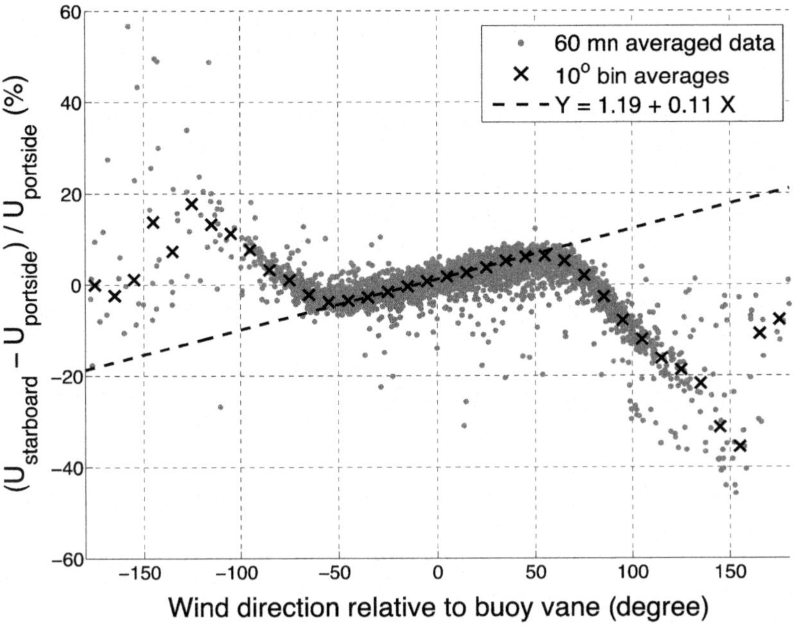

Fig. 6 Typical lateral flow distortion observed between two R. M Young propeller-vane wind sensors placed on the upstream face but on opposite sides of the meteorological tower on a Modular Ocean Buoy. The plot shows the relative difference in wind speed between each sensor as a function of the wind angle relative to the face of the buoy (Bigorre et al. [3] ©American Meteorological Society. Used with permission)

30°) away from the wind. Computational Fluid Dynamics (CFD) simulations [8] of the air flow distortion around MOB platforms show that pressure forcing on the buoy well is not aligned with the buoy vane and is responsible for the oblique angle between the buoy face and incoming wind. The CFD simulations also confirm the lateral and vertical flow distortion.

Finally, we conducted a short-term deployment using a similar buoy configuration next to a fixed air-sea interaction tower (ASIT). Wind measurements from the ASIT had minimal flow distortion and gave a good estimate of true wind that was compared with the buoy data. This comparison confirmed that the buoy measurements were quite accurate, all within 5% of the ASIT values, and that the upstream sensor on the buoy measured slightly lower wind speed, consistent with Bernoulli's principle.

Wind flow distortion can also affect the ventilation of temperature sensors during low wind periods and create different diurnal heating depending on the location of these sensors on a buoy. Once these effects are detected, there may be ways to correct for them. When multiple sensors are present, one may simply use the sensor with the least diurnal warming. Another possibility is to characterize and correct for the heat island effect using aspirated sensors [1].

The two examples described here emphasize the importance of having duplicate sensors on the buoy (or any other observing platform). Not only does it provide a backup dataset if a sensor fails or performs poorly, but it also gives the opportunity to identify the measurement biases. Of course, these biases should be confirmed using other independent measurements, such as alternate platforms, postrecovery calibrations, or numerical simulations. We will now describe in more detail some of the steps that form the intercomparison between ship and buoy measurements.

During intercomparison, the ship should keep station ½ nm and downwind of the buoy for 24 h to cover a full diurnal cycle [5]. On sunny days, this allows for the detection of level offsets on the radiation sensors as the sun moves from one side of their field-of-view to another. It is also long enough that several CTD casts can be undertaken to compare with the nearby mooring while resolving the tidal signal. It is best to have meteorological sensors mounted high up on the bow mast of the ship where flow distortion and heat island effects are lower. One should also monitor and document the conditions around these sensors during the day (e.g., shade from surrounding structures), their known biases (e.g., seawater temperature from thermosalinographs is typically biased warm due to heat transfer inside the ship's piping system), their calibration dates, heights, and sampling schemes. Finally, measurements of air temperature and humidity, wind speed, barometric pressure, and sea surface temperature are valid only at the height of the air or water parcel that is sampled; these measurements must be adjusted to the height of the corresponding sensors on the mooring before a direct comparison can be done. This height adjustment can be made using bulk algorithms [9, 10] that describe the surface boundary layers. Figure 7 shows how the height adjustment is used to construct a wind speed vertical profile to which wind measurements made at different heights can be compared [4, 5]. Bulk algorithms often use potential temperature, and it is therefore necessary to convert back to in-situ temperature to compare with buoy measurements. Sometimes, the ship data processing system converts the barometric pressure to the mean sea level by default. Once the heights are adjusted, some difference may remain, due to the physical horizontal separation between the ship and the buoy, flow distortion (especially on the ship), heat island effect, and other sources of error (shading, calibration, accuracy, etc.). Using nighttime data can minimize some of these factors.

For oceanographic sensors, one or several CTD casts can be done near but at a safe distance from the mooring. Another strategy is to deploy the instruments from the mooring on the CTD rosette and do a calibration cast with bottled samples [13]. At each stop, the CTD should be stopped for several minutes until the rosette has stabilized and all sensors have equilibrated to the surrounding temperature. If sensors with longer response time are calibrated, the soaking time should be extended (e.g., oxygen sensors may require up to 20 min). This strategy has the advantage of calibrating sensors in batch with the same seawater parcel at each stop, thus removing environmental heterogeneity as an error factor. Current measurements from the mooring can also be compared with the ones from the ship ADCP. However, errors from ship motion increase while on station, and such a comparison may serve only as a validation check.

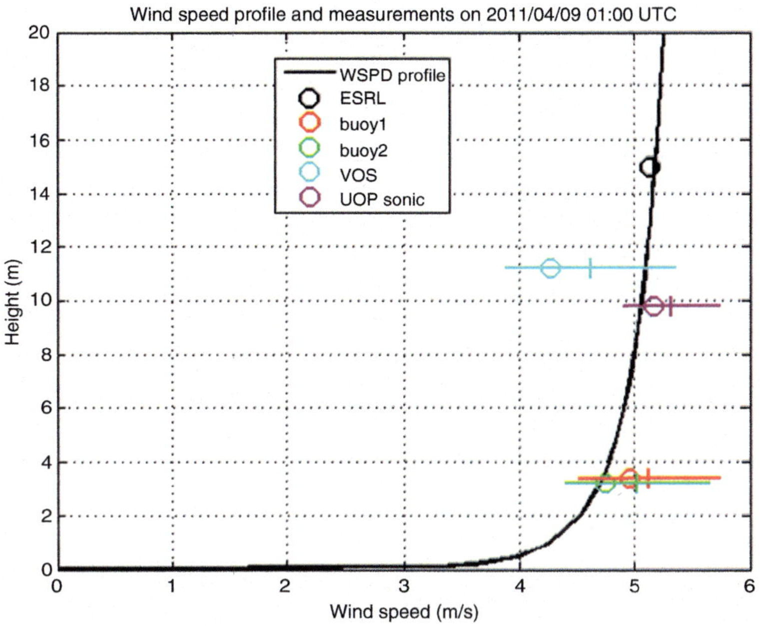

Fig. 7 Typical wind speed measured on ship and buoy. Vertical profile from COARE 3.0 (*black line*), using hourly averaged value for time 01:00 UTC (nighttime). Concomitant measurements from all wind sensors available are denoted by open circles. *Small vertical bars* indicate the average distance between the profile and measurements over the 24 h of the intercomparison; the horizontal lines denote the standard deviation

2.1.3 Postrecovery Procedures

When the ship is close to the buoys, pictures should be taken to record the physical integrity of sensors after deployment and before recovery, the attitude of the buoy (orientation into the wind, tilt), the presence of birds, traces of vandalism, or any other disturbance that may influence measurements. During recovery, pictures and notes must be taken to record any anomaly about the instruments, check the time of recovery for each sensor, and confirm their serial numbers and position on the mooring line (Fig. 1). This information may be useful during data processing if a sensor's clock is off or its actual depth is different from the planned one. After recovery, the instruments that have a temperature sensor can be plunged in a bath with cold water at a known precise time. The temperature spike can be used to synchronize the instrument clock, in case it drifted and no other time reference was available (instruments with no battery left at recovery usually do not report their time). Some of these steps may not be relevant in some situations (e.g., instruments synchronized to a master atomic clock) but exemplify the need for backup procedures when some element of the observation system fails.

2.2 Data Processing

Once instruments are recovered, documented, evaluated visually, cleaned, and passed through the recovery procedures, all their available data need to be downloaded. Then, the data processing can start and the first step is to evaluate how long and useful the dataset is. If the data content is not as expected, the user can decide to troubleshoot the problem or send the instrument back to the manufacturer for a diagnosis and another attempt at extracting the data. Once all the available data have been retrieved, the instrument can be postcalibrated and repaired if needed. The postcalibration results should be transmitted back to the person in charge of data processing, so that corrections to measurements biases can be implemented.

3 Telemetry and Real-Time Data

3.1 Limitations and Benefits of Real-Time Data

The primary limitation of most telemetry systems is that they are unable to provide sensor data at the same resolution as internally recorded data, due to bandwidth limits and cost. These barriers are easing somewhat over time, but continue to be the limiting factor in most telemetry systems that rely on satellite systems.

Telemetered data provides the buoy operator with information on the status of the platform and of the deployed sensors, and provides environmental data that can be used for real-time applications such as weather models and other scientific purposes. Perhaps equally important, it provides a copy of the data that can be used in case of lost or damaged instruments unable to communicate upon recovery.

Platform status information includes buoy position and transmission or power problems. It may include GPS position or lower resolution positions provided by the satellite systems themselves, such as Argos and Iridium. Real-time position information allows the buoy operator to detect breaks in the mooring line and to quickly begin planning for recovery. For buoys also transmitting environmental data, a smaller change in location, such as anchor drag, is an important piece of metadata for proper use of the environmental data.

Loss of communication may indicate anything from a catastrophic failure on the buoy to a broken antenna wire; understanding failure modes and symptoms may allow the operator to determine how serious the problem is, and whether a replacement needs to be considered.

Sensor status can range from sensor noise or drift to outright failure or loss of communication with the telemetry system. Again, understanding failure modes can help diagnose the problem and determine if intervention is needed. For situations where a particular sensor is critical to the mission, having sensor status in real time facilitates planning for sensor replacement. Alternatively, knowing that a sensor has simply drifted can enable a correction to the real-time data before publication.

Where multiple identical sensors are reporting in real time, this correction can be done with confidence.

Real-time *environmental data* provide information on local conditions. These data can be used in many ways, by mariners and weather forecast modelers, for those planning mooring operations in the area, and for those who need the data for other science applications.

3.2 Telemetry Systems Overview

Several options for satellite transmission of data in real time are currently available; a brief description of two commonly used systems is presented here, but there are others in use at many research institutions. Cabled observatories provide all the benefits of satellite telemetry, with fewer limitations (principally bandwidth and power), but are not practical in open ocean settings, and are not discussed here.

The Argos System[2] was created in 1978, under an agreement between the French Space Agency (CNES), the US National Oceanic and Atmospheric Administration (NOAA), and the US National Aeronautics and Space Administration (NASA). Argos is managed by the CLS Group, a CNES subsidiary. Several other international space agencies now actively participate in the Argos system, including the European Organization of the Exploitation of Meteorological Satellites (EUMETSAT) and the Indian Space Research Organization (ISRO).

Argos Platform Terminal Transmitters, or PTTs, continuously transmit short-duration messages (of less than 1 s) to the satellites; a series of ground stations collect the messages and route them to data centers and on to the owners, using an ID that is part of the message. Data can be passed through unchanged, or preprocessed by Service Argos; it can be forwarded to the buoy operator as an email message, collected via FTP from one of the Service Argos servers, or accessed via a web interface.

Two of the best features of the Argos system are the low power consumption and small size of the PTTs; these features have been pushed by the wildlife tracking community to the great benefit of buoy operators. Argos is, however, characterized by relatively short message length, and it consists primarily of one-way communication, from the buoy to shore. Messages may contain 256 bits of sensor data or less, and are transmitted more or less continuously without reply from the satellite. Plans have been underway for adding two-way communication, but the number of satellites capable of this is (or will be) limited, and there have been delays in implementing this feature.

A single PTT can accommodate multiple IDs; so, rolling buffers can be used to increase the payload. On the other hand, because of the small number of satellites, which are polar-orbiting, buoys near the equator may have gaps in the data unless multiple buffers are used to retransmit older data.

---

[2] Argos System: http://www.argos-system.org/

Due to the communications frequency, nominally 401.65 MHz, there are sometimes sensor interference problems; choosing the antenna and PTT carefully will help avoid this problem. Interference is sometimes seen during setup and testing, possibly due to reflections on buildings and other surrounding structures, and then disappears once a buoy is deployed.

Argos provides position information based on Doppler shift in the transmissions received from the PTT. The frequency measured by the satellite component of the system shifts upward as the satellite approaches, and downward as it moves further away. Along with the PTT's position, a quality flag is generated, based on the number of messages received and the elevation of the satellite.

Iridium[3] has a longer message capacity as well as two-way communication and offers several choices for data routing. Binary data can be accommodated by the use of Short Burst Data (SBD) messages, which are essentially delivered as attachments to email messages. Options for higher bandwidth users include Router-Based Unrestricted Digital Internetworking Connectivity Solutions (RUDICS). RUDICS uses an optimized circuit switched data channel and circuit switching technology, enabling a host application to originate and terminate numerous connections simultaneously. Different Iridium modem models have different SBD message lengths; the smaller 960X models do not use satellite beam-switching, and this impacts their ability to maintain a connection with a satellite, as compared to the larger 952X modems. On the other hand, the 960X modems have lower power consumption and smaller footprints, making them useful on open-ocean buoys.

Other telemetry components on the mooring may include inductive or acoustic subsystems that bring data from depth to the buoy where it can be forwarded to the satellite system. Off-the-shelf products from Sea-Bird Electronics and Teledyne-RDI, among others, are available for these tasks.

3.3 *Monitoring Data Output and Quality*

3.3.1 Monitoring Techniques

The importance of having the ability to rapidly assess the buoy cannot be overstated. The main purpose of the monitoring system should be confirmation of buoy position, followed by sensor health and data quality. For those monitoring multiple buoys, it is far preferable to automate several checks: these include a check for any buoy that is outside its watch circle, by calculating the distance of the most recent position from the anchor position, and a check for data that is not updating at the expected rate. Either of these problems should lead to an alarm being raised, usually by an email being sent to the appropriate team member indicating the details of the problem with the buoy.

[3] Iridium: https://www.iridium.com/

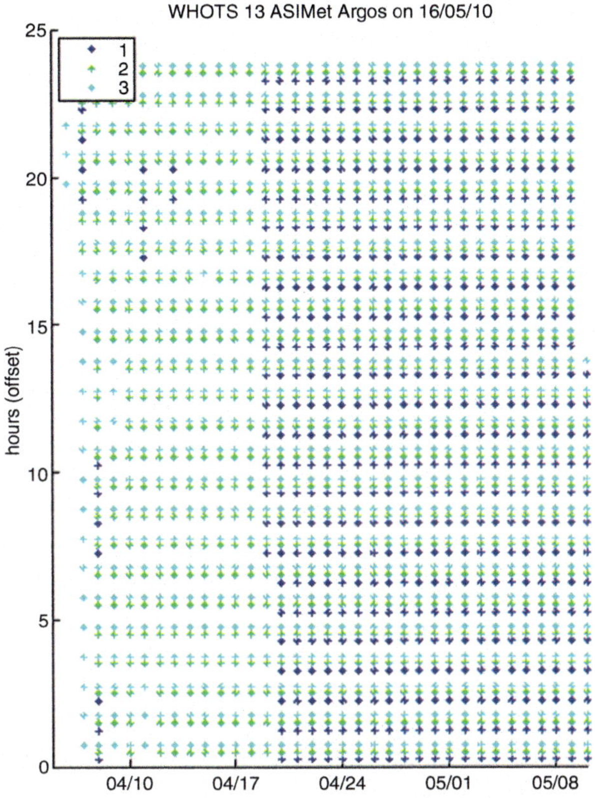

Fig. 8 Throughput plot showing for three different sensors the status of received record for each hourly transmission of the day

To be efficient, the monitoring system should quickly produce easily interpreted outputs, raise alerts about serious problems, and be very easy to view from any computer. A web interface is a great tool for this and can increase the "number of eyes" on the telemetry data. Regardless of the telemetry system or the delivery protocol used to access the data messages, an automated system that creates a log helps insure continuous processing. So, if data are arriving via email, a script can be run that checks for incoming data on the mail server, and downloads them to a processing directory; Python has several useful modules that help with handling email. If data are accessed via ftp, a utility like Expect can be incorporated into a script to look for new data on a regular basis. Web-based data are ideally handled by a program such as Lynx, which can be incorporated into a script and run without user intervention. Matlab also has a web interface that can copy text to a file using HTTP protocol.

The automated real-time processing should check the date of the most recent record for each instrument to detect and differentiate transmission gaps from failures on the instrument side. A throughput test can be implemented that plots the time of received records for every instrument at each hour of the day, as in Fig. 8.

3.3.2 Quality of Real-Time Data

There are many reasons for lower data quality in real-time data: communication errors between components on the buoy, transmission errors in the satellite system, sensor problems, and the innate lower resolution of telemetered data. Much work has been done to standardize and document quality control procedures for real-time data, because these data are an important part of the Global Ocean Observing System. QARTOD,[4] a multi-agency effort to address the Quality Assurance and Quality Control issues of the Integrated Ocean Observing System (IOOS) and broader international community, publishes manuals recommending procedures for real-time processing. NDBC [14] has also published useful manuals on this topic. A series of quality control steps might be as follows:

Gross error check: detects communication and outright sensor errors – these might use record length, or a checksum field in the record, and can look for values outside the limits of what a given sensor can measure.

Range check: compares values against pre-established limits that are set for the location and season.

Spike check: also called a time continuity check, which looks at change over time of a variable. Some instruments also use a "flag value" indicating no data; this can be removed during spike check.

Stuck value check: only needed for systems where a sensor or communication failure leads to an unchanging value.

Duplicate sensor consistency check: used when multiple identical sensors are deployed; this test chooses between the sensors if they disagree by more than a defined amount.

Most of these checks should have the capacity to be overridden when storms or other unusual events are anticipated that would generate out of the ordinary, but valid, measurements. The second category of automated checks identifies data that may not be grossly in error, but for some reason, suspect. Transmission errors account for a high percentage of errors in some types of data transfers. In the Argos system, where redundant copies of messages are often received, quality can be improved by comparing multiple records with the same time stamps.

Before real-time data are published, they can be quality-controlled to assess their usefulness. However, it is important that the buoy operator also has access to the raw "as-received" telemetered data, to detect problems in the instruments and, maybe more importantly, errors in the quality control processing. Sharing real-time sensor data via the Global Telecommunications System (GTS), operated by the World Meteorological Organization (WMO), is fairly easy to do and ensures that the data are available to, and discoverable by, all who need them.

[4] Quality Assurance of Real Time Ocean Data: https://ioos.noaa.gov/project/qartod/

References

1. Anderson SP, Baumgartner MF (1998) Radiative heating errors in naturally ventilated air temperature measurements made from buoys. J Atmos Ocean Technol 15:157–173
2. Ando K, Matsumoto T, Nagahama T, Ueki I, Takatsuki Y, Kuroda Y (2005) Drift characteristics of a moored conductivity–temperature–depth sensor and correction of salinity data. J Atmos Ocean Technol 22:282–291
3. Bigorre SP, Weller RA, Edson JB, Ware JD (2013) A surface mooring for Air–Sea interaction research in the Gulf stream. Part II: analysis of the observations and their accuracies. J Atmos Ocean Technol 30(3):450–469
4. Bigorre SP, Weller RA, Lord J, Galbraith N, Whelan S, Coleman J, Contreras MP, Aguilera C (2014) Stratus 13: thirteenth setting of the Stratus Ocean Reference Station. Cruise On Board RV Ron Brown, February 25 – March 15, 2014, Valparaiso, Chile – Arica, Chile. WHOI Technical Report 2014–06
5. Bradley EF, Fairall CW (2006) A guide to making climate quality meteorological and flux measurements at sea. NOAA technical memorandum OAR PSD-311. Earth System Research Laboratory, Physical Sciences Division, Boulder
6. Colbo K, Weller RA (2009) The accuracy of the IMET sensor package in the subtropics. J Atmos Ocean Technol 26(9):1867–1890
7. Drozdowski A, Greenan BJW (2013) An intercomparison of acoustic current meter measurements in low to moderate flow regions. J Atmos Ocean Technol 30(8):1924–1939
8. Emond M, Vandemark D, Forsythe J, Plueddemann AJ, Farrar JT (2012) Flow distortion investigation of wind velocity perturbations for two ocean meteorological platforms. Woods Hole Oceanographic Institution, technical report 2012–02. doi: https://doi.org/10.1575/1912/5182
9. Fairall CW, Bradley EF, Rogers DP, Edson JB, Young GS (1996) Bulk parameterization of air–sea fluxes for Tropical Ocean-global atmosphere Coupled-Ocean atmosphere response experiment. J Geophys Res 101:3747–3764
10. Fairall CW, Bradley EF, Hare J, Grachev A, Edson J (2003) Bulk parameterization of air-sea fluxes: updates and verification for the COARE algorithm. J Clim 16:571–591
11. Flügge M, Edson JB, Reuder J (2012) Sensor movement correction for direct turbulence measurements in the marine atmospheric boundary layer. Energy Procedia 24:159–165. https://doi.org/10.1016/j.egypro.2012.06.097
12. Hosom DS, Weller RA, Payne RE, Prada KE (1995) The IMET (improved meteorology) ship and buoy systems. J Atmos Ocean Technol 12:527–540
13. Kanzow T, Send U, Zenk W, Chave AD, Rhein M (2006) Monitoring the integrated deep meridional flow in the tropical North Atlantic: long-term performance of a geostrophic array. Deep-Sea Res I 53(3):528–546. https://doi.org/10.1016/j.dsr.2005.12.007
14. NDBC (2009) NDBC technical document 09–02, Handbook of automated data quality control checks and procedures
15. Weller RA, Bigorre S, Lord J, Ware J, Edson J (2012) A surface mooring for air-sea interaction research in the Gulf stream. Part 1: mooring design and instrumentation. J Atmos Ocean Technol 29(9):1363–1376. https://doi.org/10.1175/JTECH-D-12-00060.1

Near Real-Time Data Recovery from Oceanographic Moorings

Richard P. Trask and J. Thomas Farrar

Abstract The chapter focuses on the basic principles and challenges of transmitting near real-time data from surface and subsurface moorings, and discusses designs and approaches used in the current generation of moorings.

1 Introduction

Oceanographic moorings are an essential tool for making sustained, repeated measurements at one location, and it is often necessary to transmit the data to shore in near real time. Moorings generally contain three basic elements, an anchor, flotation, and some type of tension member (e.g., wire rope) that connects the flotation to the anchor. Instrumentation can be attached to or placed in line with the tension member(s). The flotation is typically used to keep the tension members and instrumentation off the seafloor and situated in the water column, and the anchor is intended to keep the mooring in a specific location. Various types of hardware components are used to connect the basic elements.

If the mooring has a buoy floating on the ocean surface, it is referred to as a surface mooring. If there is no surface expression and the entire mooring is completely below the surface, it is referred to as a subsurface mooring. Some hybrid mooring designs contain elements of both surface and subsurface moorings.

The choice between use of a surface or subsurface mooring for a particular application has historically depended on a balance of engineering and scientific considerations. Surface moorings allow instruments to be mounted on the surface (e.g., anemometers on weather buoys) and in the upper ocean (e.g., measuring ocean surface currents), but mooring components on the ocean surface and in the upper 20 m of the ocean are exposed to the greatest stresses from waves, weather, currents, and vandalism. Subsurface moorings avoid a lot of difficulties encountered at the ocean surface, but, having no surface flotation, they do not allow measurements in

R.P. Trask (✉) • J.T. Farrar
Woods Hole Oceanographic Institution, Department of Physical Oceanography, Woods Hole, MA, USA
e-mail: rtrask@whoi.edu

R. Venkatesan et al. (eds.), *Observing the Oceans in Real Time*, Springer Oceanography, DOI 10.1007/978-3-319-66493-4_13

the ocean surface or atmosphere. Also, the lack of a surface buoy on subsurface moorings makes it less straightforward to transmit data back to shore.

Near real-time data are needed from oceanographic moorings for a growing variety of applications. Surface meteorological data and ocean surface data are used operationally for initialization and evaluation of weather and seasonal forecast models [4], for calibration and validation of satellite measurements [5], for operational purposes such as marine search and rescue, and for the shipping and fishing industries. There is also an increasing demand for near real-time data for use in scientific research – this demand is driven in part by the development of larger and more interconnected measurement campaigns and the expectation that data from several different types of platforms (e.g., moorings, satellites, floats, and drifters) should be available with comparable data latencies. An example of this kind of multiplatform experiment making heavy use of availability of near real-time data is the Salinity Processes Upper-Ocean Regional Study (SPURS) that took place in the subtropical Atlantic Ocean in 2012–2013 [1, 3].

The Ocean Observatories Initiative (OOI) is a project funded by the US National Science Foundation to maintain a network of observatories for measuring physical, chemical, geological, and biological properties in the ocean and surface atmosphere. It was designed to measure these properties across a range of scales and to transmit all data in near real time [2]. The observatory design relies on using a combination of surface and subsurface moorings as part of the observatory to collect the required measurements while resolving a range of spatial scales (Fig. 1).

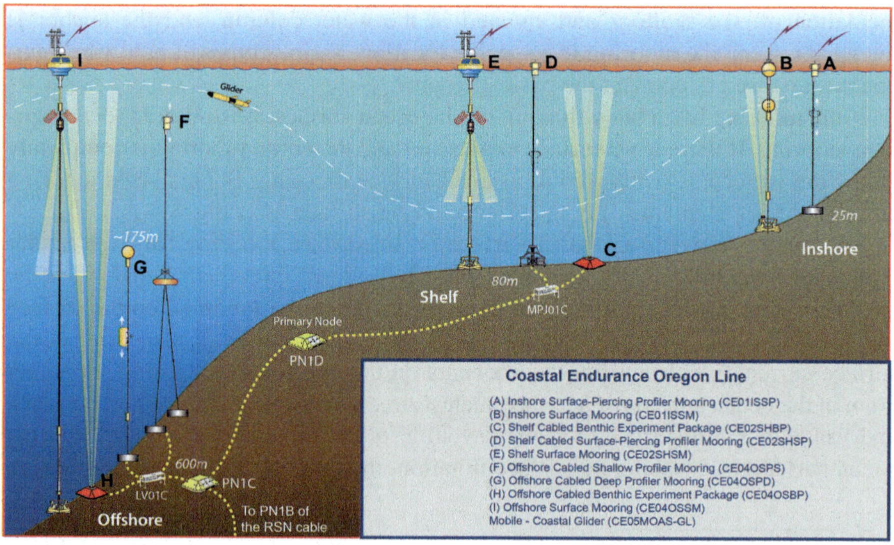

Fig. 1 An example of the way that the Ocean Observatories Initiative (OOI) uses a combination of surface and subsurface moorings to measure processes across a variety of spatial scales. A part of the so-called Endurance Array is shown, but different combinations of surface and subsurface moorings are used in other parts of OOI

It is a challenge to acquire near real-time data from both surface and subsurface moorings for the simple reason that it can be difficult to get the data to the surface where it can be transmitted wirelessly to satellite communication networks. In the case of subsurface moorings, they lack a surface expression from which data can be transmitted. Although the surface mooring has a platform from which data can be transmitted, the dynamic conditions at the ocean surface and the constant buoy motion make it difficult to maintain a reliable electrical connection through the air-sea interface. The OOI program has developed state-of-the-art approaches for addressing some of the challenges in transmitting near real-time data from the surface and subsurface moorings.

This chapter discusses the basic principles and challenges of transmitting near real-time data from surface and subsurface moorings, and includes a discussion of the design and approach used in the current generation of OOI moorings.

1.1 Deep Ocean Subsurface Moorings

The total length of a subsurface mooring is less than the water depth, so that the upper buoyancy is kept below the surface, away from the effects of ocean storms where there are high wind and wave conditions. It can therefore be constructed with components that are lighter, smaller, and less expensive than those that are needed on surface moorings.

Various types of flotation are used on subsurface moorings including glass balls in protective plastic covers, steel spheres, ceramic spheres, and syntactic foam bodies. Syntactic foam is made of small pressure-resistant glass microspheres embedded in an epoxy resin. It can be molded into custom shapes and can be engineered to meet specific buoyancy and service depth requirements. The OOI subsurface moorings typically use one or two large-diameter (1.32–1.62 m) syntactic foam spheres as the principal buoyancy on the Hybrid Profiler and Flanking moorings at the global sites (Fig. 2). The sphere at the top of the moorings has 2500 lb of positive buoyancy. If the mooring design includes a second sphere, it is located at a midwater depth and has a buoyancy between 800 and 1700 lb.

The mooring materials used on deep ocean subsurface moorings can include chain, plastic jacketed wire rope, synthetic rope, and connecting hardware. If the water is sufficiently deep and the design includes both wire rope and synthetic line, the wire rope is usually extended to a depth of at least 1500 m and frequently to as deep as 2000 m for fish bite protection. Wire rope is less susceptible to serious damage from fish bite than a synthetic material and is often used throughout the entire mooring. Galvanized 3 × 19 wire rope (3 twisted strands, each with 19 individual wires) with an extruded plastic jacket is widely used for oceanographic applications. The wire rope is designed to have minimal rotation when under load, so that it is less likely to develop loops should the tension drop quickly. If a loop were to form and the tension increased rapidly, there would be potential for a kink to form, which would weaken the wire.

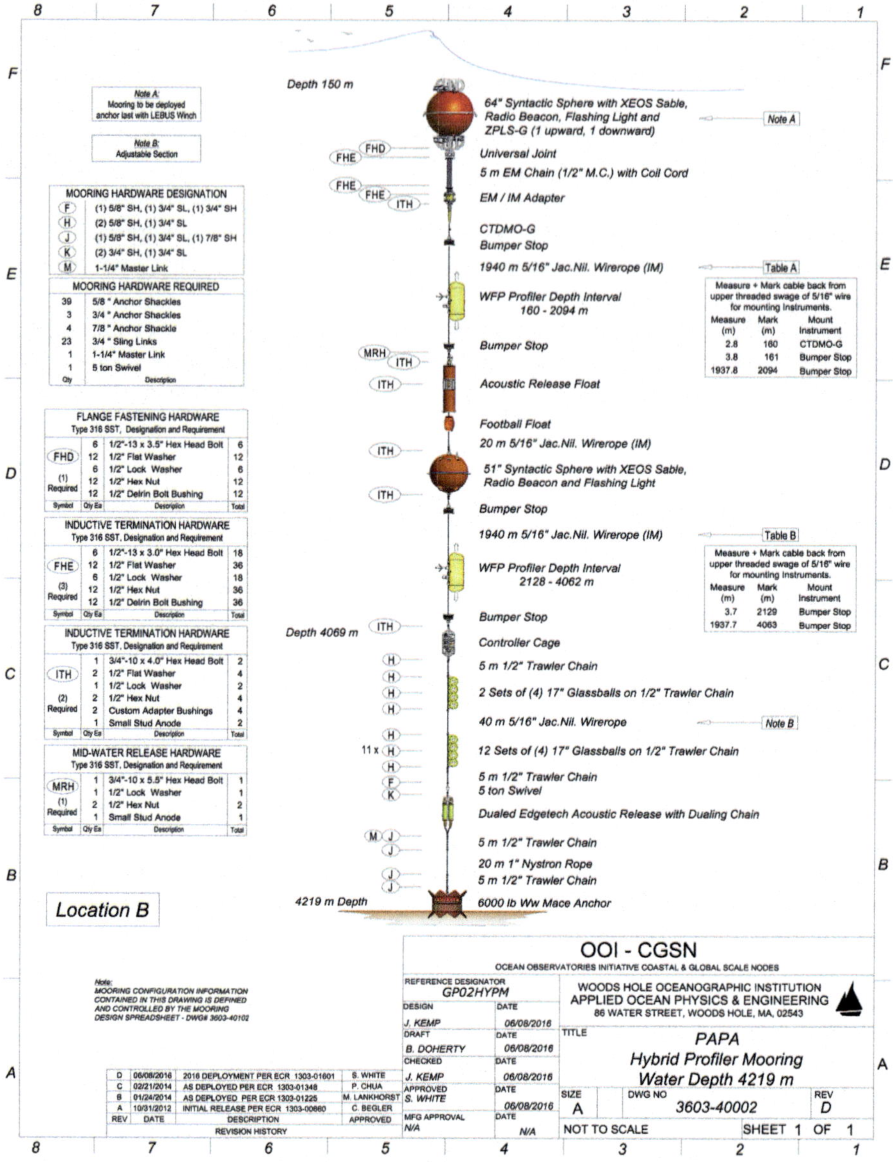

Fig. 2 A mooring schematic of an OOI Hybrid Profiler mooring. The mooring was designed by J. Kemp (WHOI) for the OOI program. Note the two-sphere design and the two wire-following profilers (denoted WFP in the drawing)

The wire is terminated with fittings that are slid onto the bare end of the wire and pressed or swaged using a hydraulic press. The swaged sockets can have a closed eye to accept a shackle, or they can have a threaded end to bolt to an intermediate body. A strain-relief boot slides onto the swaged fitting and extends over the wire to reduce the bending that can occur at the end of the rigid fitting.

Subsurface moorings with instrumentation that record data internally typically use lengths of wire rope terminated with closed swage sockets and use shackles and links to mechanically connect the various components. Instruments are either clamped to the wire at discrete depths, shackled in series with the wire lengths, or in some cases use the wire rope as a guide, profiling the water column as it moves up and down the wire rope.

Typically, the data from instrumentation on subsurface moorings is stored inside each instrument and is only accessible after the mooring is recovered. For long deployments, this results in extended delays before the data are available for analysis. If an instrument fails while deployed, it goes undetected until the mooring is recovered. If the data quality could be monitored in near real time and a serious problem is detected, one could consider recovering the mooring, correcting the problem, and redeploying the instrument if it were feasible to do so. When the data only resides in the instrument, all could be lost if the mooring parts or the instrumentation are damaged. Relaying the data back to shore while the instruments are deployed assures that a portion of the data record is available even if the instrument is lost.

Various schemes have been attempted to extract the data without recovering the mooring. One subsurface mooring design utilized current meters equipped with a small low-power acoustic transmitter that transferred compressed data from the instrument to a receiver on the mooring at approximately 500 m depth. The acoustic receiver would forward these data to an array of expendable data capsules. In a typical scenario, a capsule would release every 6 months over a 5-year deployment interval. The capsule would float to the surface and transfer the data via satellite as it drifts on the surface.

The acoustic transmission of data is not without problems. The range of data transmission can be a limiting factor, and there are a number of error sources and failure modes. Inductive modem (IM) technology used in conjunction with jacketed 3×19 galvanized steel wire rope provides an alternative technique for transferring data. With this approach, the wire rope, which has been used reliably for decades as a mooring strength member, is also used as a data transmission line. The inductive modem system is more power-efficient than acoustic modems and offers reliable communication over greater distances. For typical seawater applications, the maximum transmission length is approximately 6000 m or nearly full ocean depth.

An inductive modem uses electrical current loops to transmit information; electric current flowing in a wire loop induces a current to flow in a loop that passes through it. In a single instrument scenario, there are three interconnected loops. The first loop is at the underwater instrument where a signal is sent from a modem to the internal winding of a cable coupler that encircles the mooring wire. This induces a signal in a secondary loop created by the mooring wire rope, which passes through the cable coupler. The conductive seawater closes the loop associated with the mooring wire. The signal is extracted at a controller by means of a third loop using a cable coupler and modem similar to that at the underwater instrument.

When multiple lengths of wire rope are used and the data are to be transferred inductively to a central controller, they must be connected mechanically *and* electrically. The termination used for this application consists of a 1″ threaded swage socket that is pressed onto the bare wire end of a length of jacketed wire rope. This creates a good mechanical and electrical connection with the wire rope. The threaded swage is then bolted to an IM termination body in a manner that isolates the threaded swage (and wire) from the termination body. There is no electrical connection between the two. At least one conductor of a multipin connector with pigtail is connected to the threaded socket using a tapped hole at the end of the socket, thereby creating an electrical connection with the wire rope. The remaining pins are connected to the stainless body for the seawater ground electrical connection. The IM termination body is then filled with a polyurethane compound to prevent the infiltration of seawater into the termination.

Custom hardware components are utilized to connect the stainless IM termination bodies to other components in-line in the section of the mooring where data are transferred inductively. The connector pigtails allow adjacent components to be connected, thereby maintaining the integrity of the electrical connections. With good electrical connectivity, the data from the subsurface instrumentation can be inductively transferred to a controller on the mooring.

In the OOI design, the critical link between the controller and the surface is made possible with gliders. The gliders maneuver up and down through the water column by changing their buoyancy. Their wings provide lift, moving them forward through the water. Due to their ability to move both vertically and horizontally, they move in a sawtooth path, moving diagonally up and down through the water column. When at the surface they acquire their position via GPS and can transmit data and receive commands via satellite.

The gliders are directed to the controller on the mooring. When in close proximity to the controller, the data are acoustically transferred from the controller to the glider, which shuttles the information to the surface and relays it to the shore via satellite in near real time.

1.2 Deep Ocean Surface Moorings

The surface mooring is unique in that it has a buoy floating on the surface which makes it possible to measure surface meteorology, telemeter data, and make very near surface upper ocean measurements. The buoy is connected to an anchor on the bottom, which is typically 5000 m deep using various tension components. The mooring materials used on a surface mooring resemble those of subsurface moorings; however, the component sizes are increased to withstand the larger forces and increased wear associated with ocean storms and high wind and wave conditions. Figure 3 shows a photograph to illustrate the differences between the hardware used on surface and subsurface moorings at WHOI.

The materials include chain, plastic jacketed wire rope, synthetic line, and connecting hardware. As with the subsurface mooring, the wire rope may be extended to as deep as 2000 m for fish bite protection.

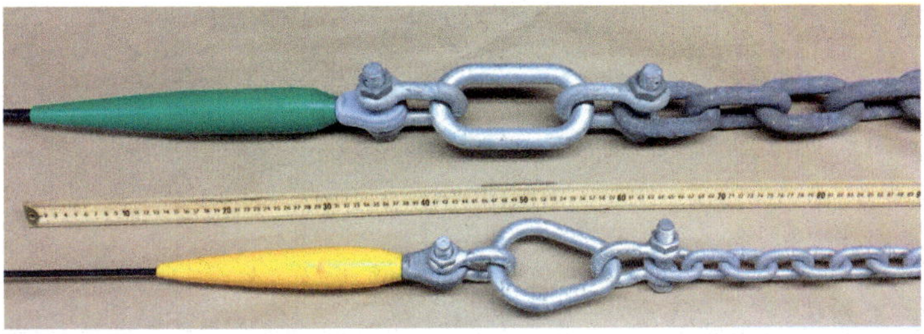

Fig. 3 The difference in hardware sizes is evident showing surface mooring hardware above and that typical of a subsurface mooring below

The buoys are fabricated in a variety of shapes and materials. Buoy shapes include the toroid or donut, the discus, and hemispherical. The oceanographic community has used the toroid-shaped buoy widely (e.g., for the surface buoys in the Tropical Atmosphere Ocean array in the Pacific; [4]). When greater buoyancy is required due to instrument load or environmental conditions, the discus-shaped buoy may be preferred. Buoy materials include aluminum, fiberglass, and various closed cell foams such as ionomer foam, which are extremely resistant to wear. The exterior surface of ionomer foam can be subjected to heat and pressure to produce a dense abrasion-resistant outer skin. The ionomer foam surface buoy used by OOI has approximately 10,000 lb of net buoyancy after deployment.

Data can be transferred along the jacketed wire rope inductively in a manner similar to that described for the subsurface mooring. In the case of the OOI surface moorings, data from instrumentation in the upper 1500 m are transferred inductively to a controller in an instrument frame approximately 10 m below the surface. The subsurface data can be transferred directly to the buoy and transmitted via satellite. It is however an engineering challenge, both mechanically and electrically, to make a reliable connection between the near-surface instrument frame and the surface buoy due to the constant motion of a surface-following buoy. The OOI surface moorings utilize an electromechanical (EM) chain assembly and a universal joint to make that connection. The chain assembly consists of a high-strength chain with a multiconductor cable wrapped around it in a spiral fashion. The chain with wrapped conductors is inserted in a marine exhaust hose, and urethane is injected into the hose creating a potted EM chain assembly. The EM chain connects the subsurface instrument frame to the universal joint at the base of the buoy (Figs. 4 and 5). A flange on each end of the EM chain bolts to a similar flange on the two connected components. The universal joint works to decouple the roll and pitch motion of the buoy and reduces the bending of the mooring components that are attached to the buoy. It has a central hole, which provides an unbending pathway from the mooring to the buoy through which conductors can pass. Pigtail connectors on each end of the EM chain complete the electrical connection between the instrument frame and the buoy.

With any surface mooring, there needs to be a degree of compliance built into the design. The design must be able to compensate for the vertical excursions of the

Fig. 4 Photograph showing the underside of a surface buoy to illustrate the mechanical/electrical connection between the subsurface components and the surface buoy. A special universal joint can be seen just under the buoy. The component bolted to the underside of the universal joint (*black hose*) is the potted EM chain assembly consisting of high-strength chain with a spiral wrapped cable all potted in a marine exhaust hose (Photo credit: Alex Dorsk, Woods Hole Oceanographic Institution)

buoy due to passing waves and swell along with changes due to the tides. The design must also be able to compensate for the lateral excursions of the buoy on the surface due to the drag forces associated with the ocean currents and not be pulled under when subjected to these forces. One means of providing compliance in deep ocean surface moorings is through the use of a synthetic material such as nylon. Nylon line has the elasticity to stretch when necessary, thereby maintaining the connection between the buoy on the surface and the anchor on the bottom.

Strong currents and surface waves impose forces on the surface mooring, which must be taken into consideration in the mooring design. There are many forces acting on a mooring causing tension in the mooring line. In general terms, the total tension is the result of a static contribution due to steady-state ocean currents plus a dynamic contribution due to wind and wave forcing.

Some of the most successful surface moorings utilize a design concept fashioned after the US National Data Buoy Center's "inverse catenary" mooring. The total length of the mooring materials in an inverse catenary mooring is 20–30% greater than the water depth. The ratio of the mooring length to water depth is called scope.

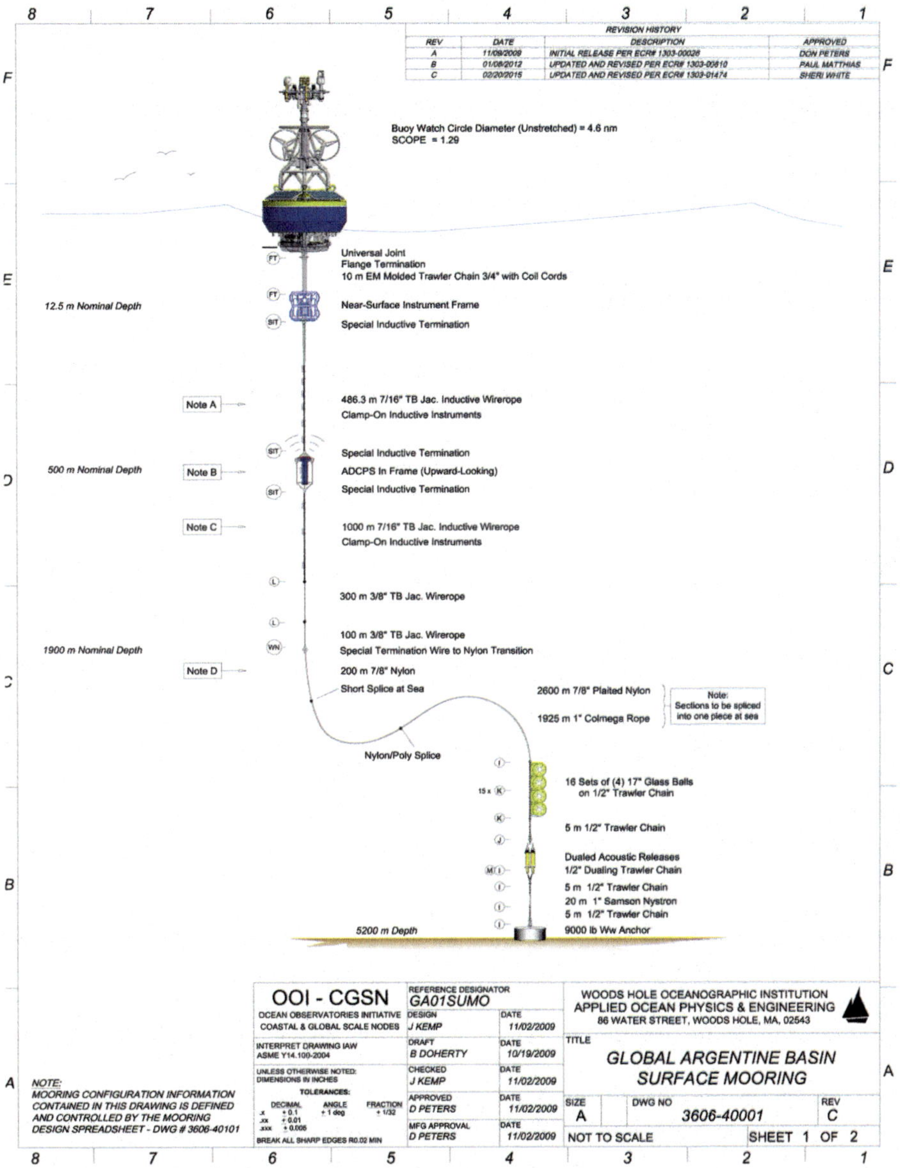

Fig. 5 Mooring schematic of an OOI deep ocean surface mooring. Features discussed in the text include the electromechanical (EM) chain and universal joint near the surface and the inverse catenary design utilizing nylon and Colmega rope in the lower portion of the mooring. Mooring design by J. Kemp (WHOI)

The scope of an inverse catenary design is typically 1.2 or greater. The typical inverse catenary design has wire in the upper part of the mooring connected to nylon line that is spliced to a buoyant synthetic line below. The buoyant synthetic line can be made of olefin copolymer or polyolefin resin fibers. When subjected to the same ocean current forcing, moorings with a scope typical of the inverse catenary design have lower static mooring tensions at both the buoy and the anchor than moorings with a lesser scope (scope = 1.1). During periods of low current, the positively buoyant line keeps the slightly negative nylon line from tangling with the mooring components below. The inverse catenary design lowers the static tensions, but the dynamic tension contribution to the total tension remains unchanged. Inverse catenary moorings can be used in a wide range of environmental conditions, as long as the water is deep enough that a "shallow water" design is not required (see next section).

In areas of the world's oceans where there are high wind and wave conditions, the dynamic loads may be so large that special consideration must be given to not only the strength of the mooring components but also to their fatigue properties. Tests on mooring hardware that had been used reliably in less dynamic environments showed that the hardware had a reduced service life due to fatigue when subjected to cyclic loads. The repeated application and partial relaxation of the load over sustained periods caused the components to either fail or show evidence of crack development. The load itself need not be excessive, but the repeated cyclic application of the load caused the part to fatigue. The tests also revealed that certain hardware shapes were more prone to failure than others. Components that are more fatigue-resistant in the range of expected tensions are now specified in the surface mooring design.

If a more fatigue-resistant hardware component is not available, the fatigue properties of the part are improved by shot peening. Shot peening is a process similar to sand blasting, but instead of using sharp angular sand the part is blasted with small spherical media called shot. The rounded shot dimples the surface that it strikes, causing the material below each dimple to be highly stressed in compression. The entire surface of the shot peened part has many overlapping dimples, thereby creating a surface layer with residual compressive stress. Parts tend to fail in fatigue due to the slow progression of cracks that develop when the part is pulled in tension. The compressive layer at the surface of the shot peened part impedes the development and propagation of surface cracks. Cracks do not tend to initiate or propagate in a compressive stress zone. Since cracks tend to start at the surface, a shot peened part will take longer to develop a crack, thereby increasing the fatigue life of the component.

Hardware such as shackles and links are not the only components susceptible to fatigue. Synthetic line that is cyclically loaded and flexed also shows signs of degradation. The sections of the mooring most affected are where the wire rope transitions to nylon line and in other areas where there tend to be abrupt changes in tension member stiffness.

The transition from wire rope to nylon line has over the years presented several design challenges. The wire rope is a stiff component, which in high sea state conditions can be forced past the nylon line causing the line to be flexed repeatedly. The extreme bending and flexing of the rope's synthetic fibers over the course of a 1-year deployment can lead to a 50% or more reduction in the strength of the rope. The amount of degradation depends on the mooring design, the deployment length, and the environmental conditions.

The wire-to-nylon transition is a fabricated component consisting of 100 m of wire rope shackled to a 200 m length of eight-strand nylon line. Several fabrication techniques have been employed to reduce the concentration of bending that occurs at the wire rope and nylon line junction. Some techniques that have worked well in more benign midlatitude environments developed problems in higher latitudes where more dynamic environments are encountered. The overall goal is to transition the stiffness of the mooring components from that typical of wire rope to the more flexible synthetics. A problem was first observed when the scope of the mooring design was increased. On early surface mooring designs, the junction between the wire and nylon was just the standard interconnecting hardware and nothing else. The nylon line on one mooring parted due to wear from coming in contact with the interconnecting hardware. The nylon line on four other moorings in the same array also displayed similar damage, but to a lesser degree. To minimize the observed damage, a hand-laid "boot" consisting of sheets of neoprene and tape was fashioned around the junction to keep the hardware from coming in contact with the nylon. This hand-made boot was approximately 1 m long and extended over both the nylon and the wire. Where it ended on the nylon created a point where the bending and flexing of the nylon became concentrated causing degradation in the nylon. In addition to the hand-made boot, the next innovation was a hand-applied urethane coating over approximately 10 m of the nylon to stiffen it sufficiently to transition the bending over a longer length, thereby not concentrating it in one area. The thickness of the coating is applied thicker near the boot and systematically reduced over the 10 m length. This has worked sufficiently well in relatively benign environments. In more dynamic environments where higher tensions are observed, a break in the urethane coating can occur due to the stretch of the nylon being greater than the elasticity of the urethane. When this occurs, the break in the coating creates a knuckle where the bending and flexing of the nylon is concentrated, and as a result there is degradation in the nylon line.

For more dynamic high-latitude environments, the upper part of the eight-strand nylon line has a 12-strand nylon overbraid, which starts at the top of the nylon and extends approximately 90 m down the nylon line. The overbraid is applied tight in the upper portion and transitions to a looser overbraid in the lower section. The entire overbraid section has a urethane coating applied. The overbraid stiffens the nylon, offering a transition from the stiff wire rope to the bare nylon. The boot over the junction between the wire rope and the nylon line, encapsulating the wire termination, the connecting shackle, and approximately 0.5 m of the overbraided nylon line is a molded urethane assembly. As part of the molding process, the parts are placed in an autoclave where elevated pressure and temperature aid in curing the urethane and produce a more standardized product than was possible when done by hand.

Another area of concern in the synthetic is where the nylon is spliced to the buoyant synthetic line. The synthetic fibers that make up the buoyant line are coarse, and when plaited together produce a rope that is stiffer than nylon. There is an abrupt change in the stiffness at the splice used to connect these two materials. The splice remains stiff, and any bending and flexing in the nylon immediately above the buoyant line occur close to the splice. After extended deployments this concentration of bending has the potential to damage the fibers and weaken the rope.

1.3 Shallow Water Surface Moorings

Shallow water surface moorings (water depth less than 600 m) are in some ways more challenging than deep ocean surface moorings since there is less depth in which to incorporate many of the same design features. An appropriate combination of compliance and fish bite resistant materials, which tend not to stretch, is still needed even though the water depth is less.

Compliance can be provided through the use of large-scope moorings in which the lower part of the mooring consists of a long length of chain lying on the bottom connecting the vertical part of the mooring to the anchor. The mooring responds to variations in water depth due to tides and high sea states by raising and lowering the chain. The constant vertical motion of the chain in and out of the bottom can cause considerable link-to-link wear, which tends to be concentrated in a relatively small area where the suspended chain contacts the bottom. The lateral motion of the buoy drags the chain over the bottom also causing wear to the chain due to abrasion with the bottom. One technique employed to reduce the chain wear is to encapsulate the chain in a heavy marine exhaust hose filled with polyurethane. This approach minimizes the link-to-link wear as well as the wear due to the chain being in direct contact with the bottom. Breaks in the hose and the polyurethane can still occur depending on the length of the deployment.

Another approach for incorporating compliance in shallow water moorings is through the use of "stretch hoses" to provide a compliant electromechanical connection. Fabricated from rubber and nylon, the stretch hose is capable of elongating to twice its original length under normal use and can have an ultimate breaking strength up to 10,000 lb. The stretch hose has conductors embedded in the hose, which provide pathways for both power and data.

The OOI program makes use of stretch hoses on moorings deployed in the coastal arrays where the water depths range from 25 to 600 m. For example, the OOI Pioneer Coastal Profiler moorings (Fig. 6) have characteristics of both subsurface and surface moorings. These moorings have a 64″ diameter syntactic foam sphere, which supports a wire-following profiler that moves along the mooring wire continuously sampling the water column. The link to the surface is by a submersible surface buoy that is connected to the top of the subsurface syntactic sphere using a stretch hose. The hose helps to decouple the buoy heave motion from the mooring below and minimizes snap loads. Data from the subsurface instruments are transferred to the surface buoy via the conductors in the stretch hose and then transmitted to shore via satellite. The wire-following profiler measures velocity, dissolved oxygen, conductivity, temperature, depth, fluorescence, and photosynthetically available radiation from 15 m below the surface to within 23 m off the bottom.

In addition to the profiler moorings, the OOI program also utilizes coastal surface moorings (Fig. 7) that consist of a series of stretch hoses between a multifunction node (MFN) on the seafloor, anchoring the mooring, and a near-surface instrument frame at 7 m depth below the surface. An electromechanical chain

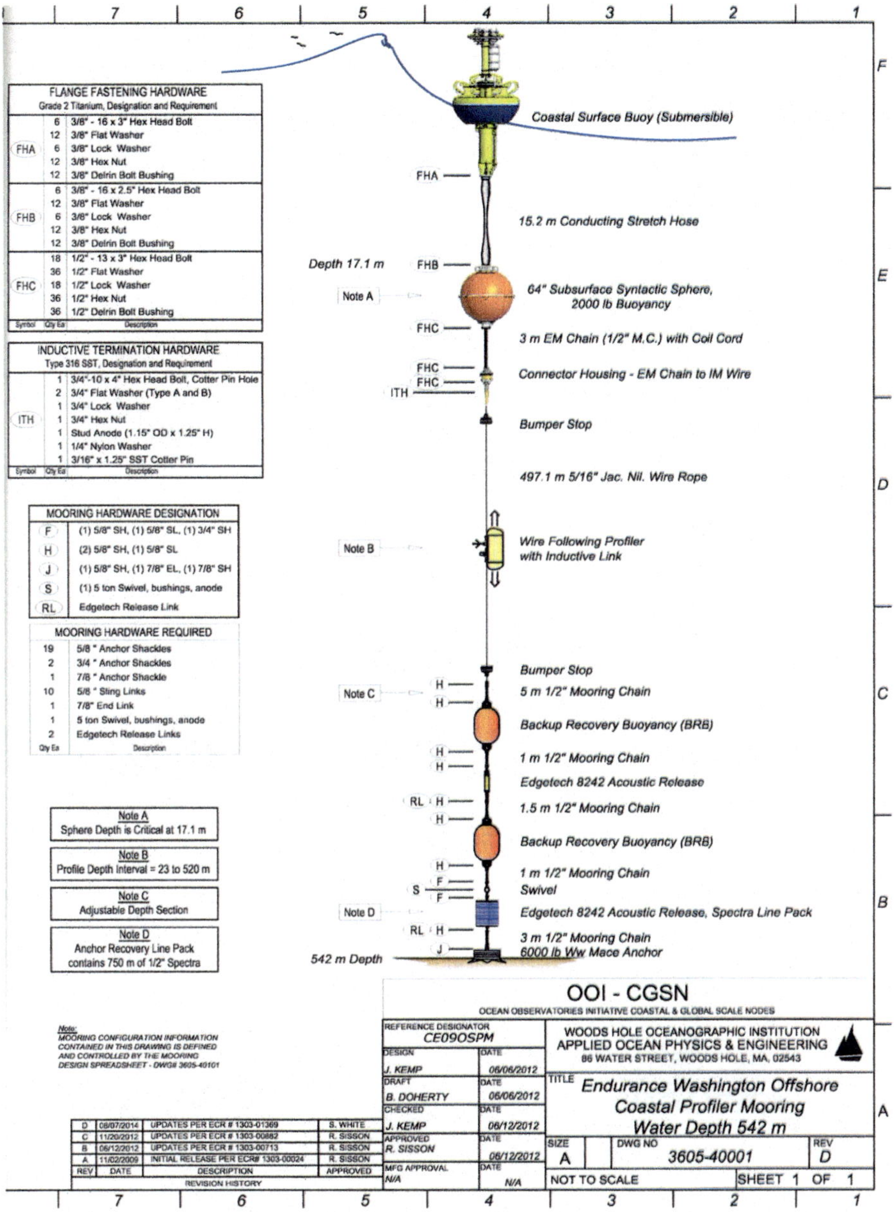

Fig. 6 Mooring schematic of an OOI Coastal Profiler Mooring. Key design features include: (1) the conducting stretch hose near the surface to provide a compliant, fish bite resistant, electromechanical connection to the buoy while isolating the lower portion of the mooring from the surface motion, and (2) the anchor recovery system near the bottom, consisting of two in-line acoustic releases and a line pack. The mooring was designed by J. Kemp

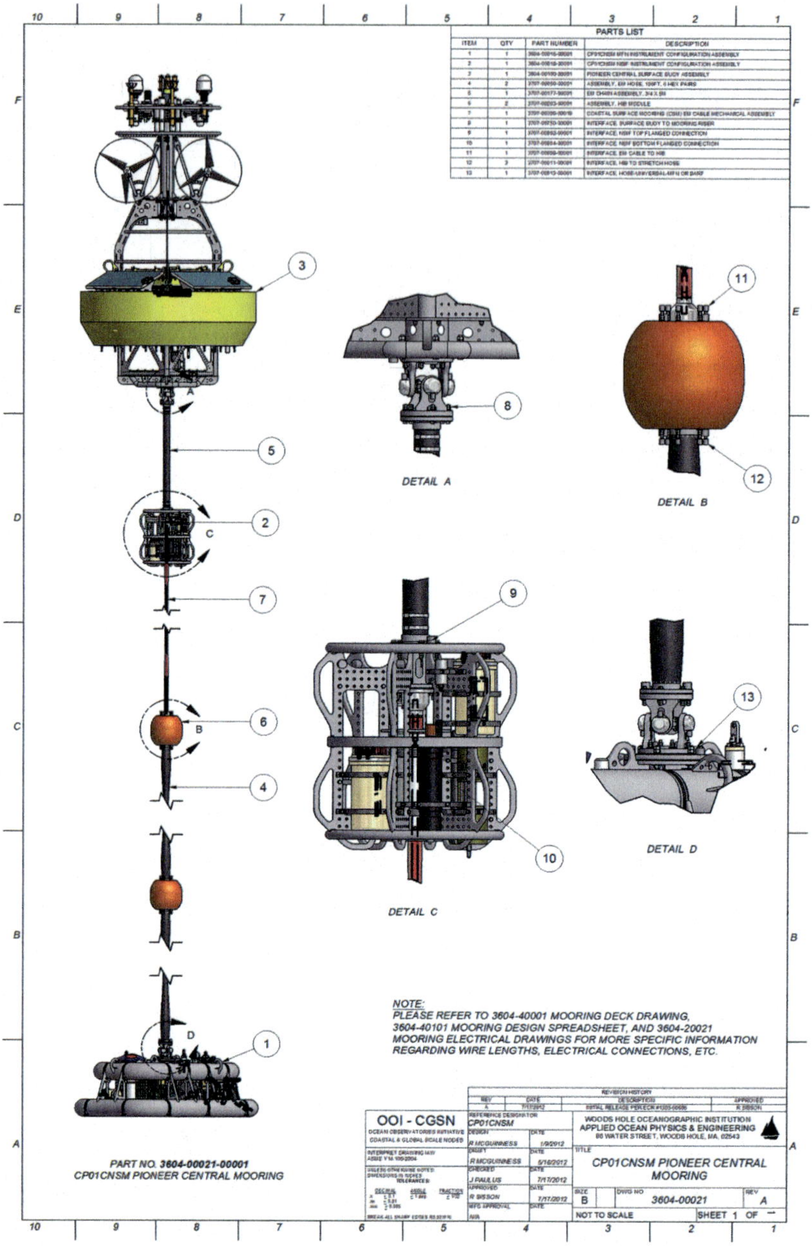

Fig. 7 Mooring schematic of an OOI Coastal Mooring with a Multifunction Node (MFN) Assembly on the seafloor. The MFN collects measurements of velocity, seafloor pressure, pCO_2 water, seawater pH, optical attenuation and absorption, conductivity, temperature, depth, dissolved oxygen, and acoustic backscatter from plankton and zooplankton, and these measurements are transmitted up the mooring to the surface buoy, where they can be transmitted to shore via satellite. The mooring was designed by R. McGuninness

connects the near-surface instrument frame to a surface buoy. The buoy has the capability to transmit data collected on the buoy as well as data from the subsurface instrumentation to shore via satellite. The MFN contains instrumentation to measure velocity, seafloor pressure, pCO_2 water, seawater pH, optical attenuation and absorption, conductivity, temperature, depth, dissolved oxygen, and acoustic backscatter from plankton and zooplankton. The near-surface instrument frame houses sensors to measure velocity, seawater pH, nitrate, spectral irradiance, optical attenuation and absorption, fluorescence, conductivity, temperature, depth, and dissolved oxygen. The buoy has a three-axis motion package, two suites of meteorological instruments, pCO_2 in both air and water, a direct covariance flux sensor, and a surface wave spectra sensor.

Deep ocean surface and subsurface moorings utilize an acoustically triggered release mechanism to completely detach the mooring from its anchor. In these applications, the anchor is left behind. In shallow water, it is possible to recover the mooring and anchor. To recover the anchor, a secondary line connected to the anchor is brought to the surface and used to pull the anchor off the bottom. To simplify the recovery of the anchor, the mooring may be outfitted with two acoustic releases at different depths along the mooring, as shown in Fig. 6. The shallower of the two would be activated first, so that the upper part of the mooring can be recovered. The anchor, a line pack with the second acoustic release and sufficient buoyancy to bring the line pack and release back to the surface, remains on the bottom. The line pack consists of a spool-like framework around which a sufficient length of high strength line is wound. In the center of the line pack is an acoustic release. When the second acoustic release is activated, the backup recovery buoyancy situated above the release brings the line pack to the surface. The line, one end of which remains connected to the anchor on the bottom, spools off the line pack frame as it ascends to the surface. The line pack is recovered and the line is wound onto a winch, which then hauls the anchor to the surface for recovery.

2 Conclusion

Because water blocks the electromagnetic waves used to transmit information to satellites, real-time ocean data telemetry requires that a transmitter be located at the surface. The general challenge of doing telemetry from moorings is that there must be a mechanically, electrically, and/or acoustically robust connection from the instruments at depth, with the transmitter on the surface, and the action of waves, winds, and currents at the surface makes this very challenging. We discussed two approaches used by the US Ocean Observatories Initiative for getting the data from the subsurface instruments: use of surface moorings and use of gliders to "carry" the data from the depth to the surface.

Surface moorings have by definition the platform from which data can be transferred to shore via satellite. To tolerate the conditions on the surface and still stay moored in one location, surface moorings require compliance. Stretch hoses provide both compliance and a means for transferring data; they are well suited for shallow water applications, but can be prohibitively expensive in the lengths that would be required for deep-water surface moorings.

For deep-water surface moorings, the most common approach currently in use, and the one used in OOI, is to use inductive modems to send data up the mooring wire, but it remains an engineering challenge to reliably get data that were collected below the surface up through the air-sea interface. Due to the dynamic environmental conditions at the surface, the mooring components need to be strong and as a result tend to be large and more expensive. The constant buoy motion presents numerous challenges in terms of electrical and mechanical survivability.

Since subsurface moorings do not have a surface buoy, an additional element such as a glider or periodic release of data capsules is required to transfer the data to the surface. Having that additional component adds complexity and expense. The reliability of the telemetry system depends on the reliability of the independently operating element such as a glider.

Subsurface moorings do not require the compliance that is necessary on a surface mooring. Measurements can therefore be made and inductively transferred throughout the entire water column. Since subsurface moorings are not impacted by the dynamic conditions at the surface, their components are generally less robust and therefore less expensive than those required for a surface mooring. Subsurface moorings need to be well below the surface, and therefore cannot make measurements very close to the surface.

While the oceanographic community has met some of the specific engineering challenges for getting data from the subsurface instruments to the surface where it can be transmitted to satellites, these solutions come with appreciable additional cost. One of the most important engineering challenges now is in balancing cost and robustness, and in looking for innovative ways to transfer data from the surface and subsurface oceans, while keeping design complexity and cost to a minimum.

Acknowledgments The authors thank Kathy Ponti, Ruthanne Molyneaux, Sarah Xander, and Jean Taft for assistance with the preparation of the manuscript. The designs and techniques described in this chapter have been greatly influenced by decades of mooring work undertaken by scores of scientists and technicians. Collectively, the authors thank them all for their contributions which have brought them to where they are today. JTF was supported by grant number N00014-13-10453 of the Office of Naval Research and by grant number NNX15AG20G of the National Aeronautics and Space Administration.

References

1. Bingham FM, Li P, Li Z, Vu Q, Chao Y (2015) Data management support for the SPURS Atlantic field campaign. Oceanography 28(1):46–55. http://dx.doi.org/10.5670/oceanog.2015.13
2. Cowles T, Delaney J, Orcutt J, Weller R (2010) The ocean observatories initiative: sustained ocean observations across a range of spatial scales. Mar Technol Soc J 44:54–64
3. Farrar JT, Rainville L, Plueddemann AJ, Kessler WS, Lee C, Hodges BA, Schmitt RW, Edson JB, Riser SC, Eriksen CC, Fratantoni DM (2015) Salinity and temperature balances

at the SPURS central mooring during fall and winter. Oceanography 28(1):56–65. https://doi.org/10.5670/oceanog.2015.06

4. McPhaden MJ et al (1998) The Tropical Ocean-Global Atmosphere observing system: a decade of progress. J Geophys Res 103(C7):14169–14240. https://doi.org/10.1029/97JC02906
5. Wentz FJ (1997) A well-calibrated ocean algorithm for special sensor microwave / imager. J Geophys Res 102(C4):8703–8718. https://doi.org/10.1029/96JC01751

Managing Meteorological and Oceanographic In Situ Data in the WMO Framework

Etienne Charpentier

Abstract The chapter begins with the requirements for marine meteorological and oceanographic (meteo-ocean) data for WMO applications. The meteo-ocean observations are not only used for scientific research purposes, but also allow realizing socioeconomic benefits and addressing the needs of many activities such as the safety of life and property at sea, operations in the open and coastal ocean areas, the protection and sustainable development of the ocean and marine environment, numerical weather prediction and operational meteorology, the monitoring and prediction of seasonal-to-interannual climate variability and climate change, and the efficient management of marine resources. The chapter provides information on the processes used to assess gaps and to provide guidance and World Meteorological Organization (WMO) Member Countries and Territories to address the gaps and make the observing system evolve. The role of the WMO, collaborating with the Intergovernmental Oceanographic Commission of UNESCO (IOC), in making and collecting observations from and over the oceans is explained. In particular, implementation targets for various types of meteo-ocean observing platforms are detailed. The chapter then provides information on meteoceano data management in the WMO framework, including for the collection of data in real-time and delayed mode, quality control, collection of metadata, and feedback of quality information to the observing platform operators. The existing sources of meteo-ocean data are listed, and relevant data policies explained. Finally, the chapter gives information on how to access data and provides an incentive for sharing the data.

E. Charpentier (✉)
WMO Observing Systems Division, WMO Secretariat, July 2016, Geneva, Switzerland
e-mail: echarpentier@wmo.int

R. Venkatesan et al. (eds.), *Observing the Oceans in Real Time*, Springer Oceanography, DOI 10.1007/978-3-319-66493-4_14

1 Requirements for Marine Meteorological and Oceanographic (Meteo-ocean) Data for *WMO Applications*

1.1 The Role of the WMO

The World Meteorological Organization (WMO) is the United Nations (UN) system's authoritative voice on the state and behaviour of the Earth's atmosphere, its interaction with the oceans, the climate it produces and the resulting distribution of water resources. Under WMO leadership and within the framework of WMO programmes, national meteorological and hydrological services contribute substantially to the protection of life and property against natural disasters, to safeguarding the environment and to enhancing the economic and social well-being of all sectors of society in areas such as food security, water resources and transport.

1.2 The WMO Application Areas

To realize these objectives, the WMO is providing a framework for international cooperation in a large scope of activities involving WMO application areas,[1] which are rely heavily on the availability of global environmental observations from observing platforms operating in space (e.g. geostationary and polar orbiting satellites), the atmosphere (e.g. aircraft and weather balloons), on land (e.g. manned and automatic weather stations, weather radars, wind profilers) and in the oceans (e.g. data buoys, ships, profiling floats, tide gauges, offshore rigs and platforms, coastal stations). The making, collection and quality-monitoring of observations are realized and regulated in the framework of the WMO Integrated Global Observing System (WIGOS) and in cooperation with WMO cosponsors. The data are then internationally exchanged in the WMO framework through the WMO Information System (WIS).

[1]There are currently 14 application areas of interest to the WMO, including (1) Global Numerical Weather Prediction (NWP); (2) High-Resolution NWP; (3) Nowcasting and Very Short Range Forecasting; (4) Sub seasonal to longer predictions; (5) Aeronautical Meteorology; (6) Forecasting Atmospheric Composition; (7) Monitoring Atmospheric Composition; (8) Providing Atmospheric Composition information to support services in urban and populated areas; (9) Ocean Applications; (10) Agricultural Meteorology; (11) Hydrology; (12) Climate Monitoring; (13) Climate applications and services; (14) Space Weather. In addition, the Global Cryosphere Watch (GCW) is regarded as an activity cross-cutting between the above application areas.

1.3 The Use of Meteo-ocean Data

Besides the use of meteo-ocean observations for scientific research purposes allowing to better understanding of atmospheric, ocean and cryospheric geophysical processes, such data also allow realizing socioeconomical benefits and addressing the needs of many activities. These include the safety of life and property at sea, operations in the open and coastal ocean areas, the protection and sustainable development of the ocean and marine environment, numerical weather prediction and operational meteorology, the monitoring and prediction of seasonal-to-interannual climate variability and climate change, and the efficient management of marine resources. A number of meteo-ocean products and services provided by national meteorological and hydrological services are therefore delivered to end users such as policy-makers, local authorities, port authorities, public health officials, disaster risk reduction managers, the maritime and oil and gas industries, which are relying on infrastructures deployed at sea and in coastal areas, the fisheries, the tourism industry, insurance companies and the mass media. Meteo-ocean products are essentially based on forecasting products (e.g. wave models, ocean models, Tsunami monitoring). Such products and services rely directly on observations, and in some cases on the analysis and forecast fields from numerical weather and/or ocean prediction models. Services provided to end users include, for example, the provision of maritime safety information (typically marine forecasts) to the shipping industry, assistance to ship routing and assistance to search and rescue operations at sea, all requiring a good knowledge or estimate of current and future weather and marine environment conditions (e.g. currents, wave conditions, wind, precipitation).

Meteo-ocean data are also crucial for climate services, where the WMO is playing a key role through the Global Framework for Climate Services (GFCS). Current and historical marine meteorological and oceanographic climate data allow understanding the global climate, including its variability on timescales ranging from subdaily to weekly to centuries, and are used for the validation of ocean and climate models at various temporal scales. Marine climate information provides the means to deliver the statistical climatological products needed by the end users, helping them to minimize losses due to climate variability and change and to manage natural and human systems effectively. For example, the design and construction of a port or an offshore platform require marine climatological information to estimate the operating conditions and expected frequency of extreme events, both having an impact on the sizing and robustness of the infrastructure.

Besides research activities using meteo-ocean data and aimed at improving the understanding of geophysical processes for their better simulation in forecasting models, a basic and important use of the observational data is their routine assimilation in real time or semi real time in operational numerical prediction models. The modern data assimilation schemes make use of advanced four-dimensional variational (4DVar) techniques, allowing the assimilation in real time of high temporal (hourly and subhourly) resolution observations. One advantage of assimilation in numerical weather and ocean prediction models for the observing platform operators is the availability of quality-monitoring information routinely produced by comparing the observations with the model's first-guess field (typically a 6-h forecast).

1.4 Documenting the User Requirements

Technology-free[2] observational user requirements for all WMO application areas are collected and recorded for more than 200 observed variables in a database called OSCAR.[3] The requirements are expressed quantitatively in terms of space and time resolution, timeliness, uncertainty and stability criteria. For each criterion, three values are provided: (1) the threshold, that is, the value below which observations are worthless, (2) the breakthrough, that is, the proposed practical target for making significant progress and optimal cost/benefit), (3) the goal, that is, the value beyond which improvement gives no additional benefit.

1.5 Gap Analysis

WMO experts, using the results of impact studies (when applicable) and comparing the observing systems' capabilities (also recorded in OSCAR) with the observational user requirements, are regularly performing gap analysis in the view to make recommendations on the most critical variables to observe the gaps to address in each of the WMO application areas.

1.6 Guidance to WMO Member Countries and Territories on the Evolution of Global Observing Systems

In turn, the WMO is reviewing all the gap analyses, taking into account the observing systems' cost-effectiveness, available resources and the priority of the organization to develop and agree on an implementation plan[4] for the evolution of global observing systems. The plan currently includes 115 key actions that the WMO Members and identified actors are requested to undertake to better address the identified observational gaps and better meet the requirements of the WMO application areas. Twelve of those actions directly relate to improving space (e.g. sea surface

[2] That is, not precluding about what observing technology is used.

[3] http://oscar.wmo.int

[4] The WMO Implementation Plan for the Evolution of Global Observing Systems (EGOS-IP) is available from the WMO website at http://www.wmo.int/pages/prog/www/OSY/gos-vision.html#egos-ip

temperature and sea level) and in situ (e.g. data buoys, ship-based observations, profiling floats, ice buoys, tide gauges) meteo-ocean observations.

2 The Role of the WMO in the Making and Collection of Meteo-ocean Data

2.1 International Cooperation Between Meteorologists and Oceanographers for the Making of Meteo-ocean Observations

In the WMO framework, marine meteorological observations (ocean surface variables, and atmospheric variables at the ocean surface) are collected and regulated in the WMO framework as part of the WMO Integrated Global Observing System (WIGOS) and its Global Observing System (GOS). The oceanographic observations (ocean surface and below variables) are essentially collected and coordinated in the framework of the Intergovernmental Oceanographic Commission (IOC) of UNESCO and the IOC-lead IOC-WMO-UNEP-ICSU Global Ocean Observing System (GOOS[5]). Strong cooperation between the WMO and the IOC was put in place through the Joint WMO-IOC Technical Commission for Oceanography and Marine Meteorology (JCOMM) for the coordination of the implementation at the international level of both marine meteorological and oceanographic observations (among other activities).

The WMO and the IOC established JCOMM in 1999. JCOMM's mission is building on the capacities of WMO Member countries and territories, and IOC Member States. This cooperation between the two organizations has particularly allowed the development of a strong partnership between operational meteorological agencies and oceanographic research institutions. The former typically bring operational infrastructure for the making of marine meteorological observations from data buoys and ships, for the exchange in real time of the collected observations, for the routine elaboration of weather forecasting products and the delivery of related marine services. The latter typically bring a strong contribution to the global ocean observing system through sustained research programmes, which deploy hundreds of ocean observing platforms such as drifting buoys, tropical moored buoys, and Argo profiling floats in the world oceans. These research programmes also bring improved scientific knowledge on geophysical processes in the world oceans, allowing a continuous improvement of the numerical ocean prediction models.

[5] http://www.ioc-goos.org/

2.2 *Observing Platforms*

Both WMO and IOC are contributing to the GOOS and the WMO-IOC-UNEP-ICSU Global Climate Observing System (GCOS[6]). Through JCOMM, they cooperate in the design, implementation and evolution of the required meteo-ocean observing systems in the view to respond in a cost-effective way to the common GOOS and GCOS user requirements, and take advantage of the strengths of each type of observing platform, to make their observations complementary and not duplicate. By addressing the observational user requirements for climate monitoring (GCOS), which are more stringent than those of other application areas, JCOMM is also de facto addressing most of the requirements of other WMO application areas (e.g. numerical weather prediction, marine services), hence the focus on GCOS requirements.

JCOMM has set implementation targets for each of the observing networks contributing to the global ocean observing system. Table 1 provides for the implementation targets and other information concerning today's typical meteo-ocean observing platform types. Figure 1 provides an example of deployed instruments (here data buoys and tsunameters). However, at the time of writing, JCOMM is working at proposing new implementation targets and performance metrics, and the ones given in Table 1 may evolve soon.

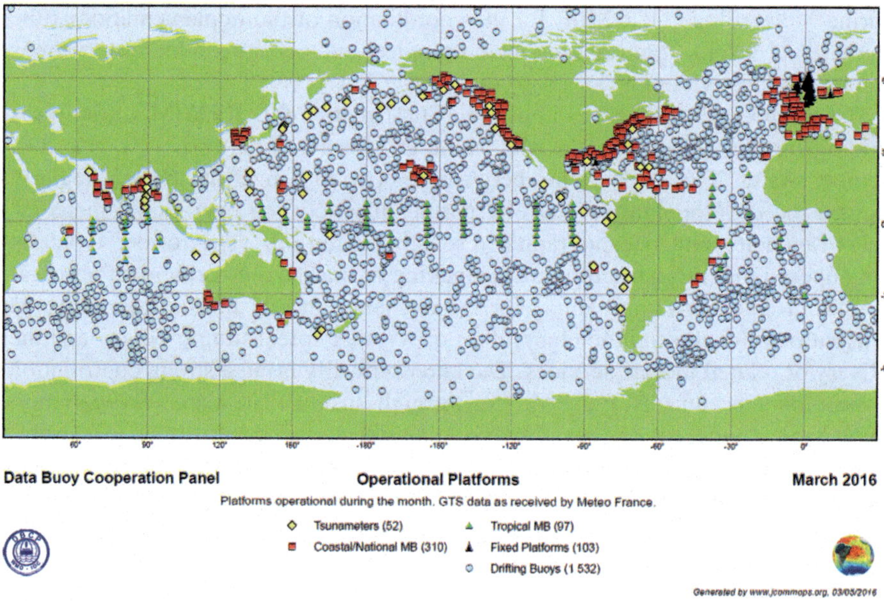

Fig. 1 The status of the data buoy and tsunameter networks in March 2016. The moored buoys shown on the map include both the meteorological and the tropical moored buoys

[6] http://gcos.wmo.int

Table 1 Typical global ocean observing systems monitored through JCOMM

Observing platform type	Key features	Main variables measured	Implementation targets	Current status (2016)
Lagrangian drifters	Small spherical hull with subsurface drogue 15 m below surface	SST, SLP, surface velocity	1250 units	1532 units in March 2016
Meteorological moored buoys	Small to large-scale buoy making surface marine meteorological measurements	SST, SLP, AT, RH, MSWH, wind	No specific target globally	310 units in March 2016
Wave buoy	Small buoy with 3D accelerometers	3D wave spectra	No specific target globally	>270 units
Tropical moored buoys	Relatively small deep ocean buoy, making surface marine meteorological and subsurface oceanographic profile measurements. Network comprised of PIRATA (Atlantic Ocean), RAMA (Indian Ocean) and TAO-TRITON (Pacific Ocean). TAO / TRITON is a central component of the ENSO observing system, deployed specifically for research and forecasting of El Niño and La Niña	SST, SLP, AT, RH, SSS Water temperature and salinity profiles, currents (ADCP). Precipitation, short-wave radiation and long-wave radiation on some surface moorings	PIRATA: 19 units RAMA: 46 units TAO-TRITON: 67 units	PIRATA: 18 units RAMA: 27 units TAO-TRITON: 55 units
OceanSITEs	Deep ocean multidisciplinary time series measurements	Wide range, surface, subsurface, bottom	OceanSITEs is currently defining targets	276 active units

(continued)

Table 1 (continued)

Observing platform type	Key features	Main variables measured	Implementation targets	Current status (2016)
Voluntary Observing Ships (VOS)	Surface marine meteorological observations from recruited maritime industry ships	SST, SLP, AT, RH, waves (visual), wind (visual or measured)	25% of the global active fleet to meet the VOS climate standard	1168 manned VOS, including 338 VOSClim 138 automated VOS, including 100 automated VOSClim
Ships of Opportunity (SOO)	Subsurface oceanographic measurements from recruited maritime industry ships. Typically making temperature profiles on high-density and frequently repeated lines.	Temperature profiles, SST, SSS	Operate 25 frequently repeated lines, and 24 high density-lines for a total of 33,000 XBTs deployed	29 active transects making about 18,000 profiles per year
Research vessels	Wide range of high-quality measurements made available in delayed mode	Wide-range surface and subsurface measurements (e.g. CTD)	No specific target globally	
Profiling floats	Typically making profiles every 10 days down to 2000 m	Temperature and salinity profiles	3000 units	3829 units in April 2016
Gliders	Can be controlled remotely and make measurements in targeted regions	Wide range of subsurface measurements (incl. Profiles)	No specific target globally; JCOMM considering gliders as part of its implementation goals	

(continued)

Table 1 (continued)

Observing platform type	Key features	Main variables measured	Implementation targets	Current status (2016)
Tide gauges	Coastal sea-level observing stations. The Global Sea Level Observing System (GLOSS) GLOSS includes the GLOSS Core Network (GCN), the Permanent Service for Mean Sea Level (PSMSL) and a fast delivery component	Sea level	290 GSN stations	183 stations with updated data in past 5 years
Rigs and platforms	Offshore platforms providing third party data	SST, SLP, AT, RH, MSWH, wind	No specific target globally	>100 units
Coastal stations	Marine meteorological measurements on coast	SST, SLP, AT, RH, MSWH, wind	No specific target globally	
Tsunameters	Tsunami monitoring: ocean bottom pressure measurements transmitted via acoustic communication to surface moored buoy, then by satellite to shore	Bottom pressure	Region-dependant	52 units

2.3 *Satellite Data Telecommunication*

Satellite data telecommunication (Satcom) is key to collect in real-time raw observations from the observing platforms operating in the high seas. Various Satcom systems are being used for such purposes, for example, Inmarsat, Iridium, Argos, Data Collection System – DCS – on meteorological geostationary satellites, etc. Each of these Satcom systems has advantages and drawbacks in terms of cost, geographical coverage, data timeliness, data throughput, electric power consumption and availability of location service. The WMO and IOC have established an International Forum of Users of Satellite Data Telecommunication Systems with the goal of compiling information on user requirements in terms of satellite data collection on one hand, and on the various satellite data telecommunication systems capabilities on the other hand. The information will allow the end users to make informed decision on the best satellite system fitting their user needs.

3 Meteo-ocean Data Management in the WMO Framework

Sharing and exchanging meteo-ocean data on a free and unrestricted basis in both real time and delayed mode are critical for such data to be used by all WMO and IOC applications mentioned above, whether for operational (essentially real time) or research (delayed mode) purposes.

3.1 Real-Time Data Exchange

With the WMO Information System (WIS[7]), the WMO has put in place the necessary infrastructure allowing for the real-time exchange of WMO information, including meteo-ocean data. Observations, data and products can be discovered through the WIS, thanks to availability of data discovery metadata exchanged in ISO-19115 compliant format.

The Global Telecommunication System (GTS[8]) of the WIS provides the protocols and formats used for the real-time exchange of time-critical observations. While alphanumeric codes have been used in the past for such exchange (e.g. FM-13 SHIP, FM-18 BUOY, FM-63 BATHY, FM-64 TESAC formats), more flexible table-driven codes such as FM-94 BUFR[9] are now being used instead.

Ocean observing platforms, the data of which are intended to be distributed on the GTS are identified with WMO numbers (data buoys, profiling floats, gliders, OceanSITEs, ocean rigs and platforms), which can be obtained from nominated national focal points for buoy programmes.[10] Rules for allocating WMO numbers are provided on the WMO website.[11] Ships are identified with ITU call signs or unique identification numbers allocated nationally or regionally by national meteorological services. While WMO numbers can often be reused, the WMO has now agreed to use unique WIGOS identification (ID) numbers to eventually replace the WMO numbers. However, WIGOS ID numbers are not currently being used operationally at the time of writing this chapter.

Observations to be distributed on GTS are encoded in BUFR using predefined sequences or templates specific for each type of observation. There is, for example, a pre-defined BUFR sequence for the distribution of Lagrangian drifter data; another one for Argo profiling floats, etc. The BUFR observations are grouped by type, and then distributed on the GTS into GTS bulletins for their real-time distribution between WMO Member countries and territories. GTS data insertion points are with national meteorological and hydrological services. Details on the GTS, BUFR

---

[7] http://www.wmo.int/wis

[8] http://www.wmo.int/pages/prog/www/TEM/GTS/index_en.html

[9] Binary Universal Form for the Representation of Meteorological Data

[10] http://www.jcomm.info/dbcp-nfp

[11] http://www.wmo.int/pages/prog/amp/mmop/wmo-number-rules.html

and BUFR tables can be found in the Manual on the GTS (WMO No. 386), and the Manual on Codes (WMO No. 306). Further information on WMO codes can be found at the WMO website.[12]

While it is relatively straightforward for observing platform operators working at national and hydrological meteorological services to take the necessary steps for encoding and inserting their data on the GTS in real time, it can actually be challenging for other parties in the oceanographic community. JCOMM Technical Report No. 72,[13] *An Oceanographers' and Marine Meteorologists' Cookbook for Submitting Data and Metadata in Real-time and Delayed Mode* includes practical information in this regard. Practical technical assistance can also be provided in this regard by the JCOMM in situ Observations Programme Support Centre (JCOMMOPS – www.jcommops.org), which can be contacted at support@jcommops.org. The GTS Regional Telecommunication Hubs' focal points can also help. Their contact details can be found on the WMO website.[14]

3.2 *Delayed Mode Data Exchange*

In addition to real-time data exchange, JCOMM is coordinating and evolving marine climatological data management practices. Building on and modernizing the Marine Climatological Summaries Scheme (MCSS), JCOMM is developing the Marine Climate Data System (MCDS) for assuring the flow of in situ delayed mode meteo-ocean observations from the relevant data sources to the long-term archives maintained by Centres for Marine Meteorological and Oceanographic Climate Data (CMOCs). The CMOCs are meant to provide integrated (by observed variables) meteo-ocean data-sets and products based on the availability of delayed mode and long-term records of meteo-ocean data from various types of observing platforms. Key activities of the CMOCs include data rescue, adding value thanks to higher level quality control and bias correction, the compilation of observed variable-based products and the integration of in situ and satellite data.

Today, one CMOC has been established at the National Marine Data and Information Service (NMDIS) of the State Oceanic Administration (SOA) of China. The International Comprehensive Ocean Atmosphere Data-Set (ICOADS[15]) also includes a record of integrated marine meteorological delayed mode data contributed by the MCDS. JCOMM is investigating feasibility of formalizing the ICOADS as a CMOC. In addition, the World Ocean Database (WOD[16]) also provides integrated oceanographic data-sets.

---

[12] http://www.wmo.int/pages/prog/www/WMOCodes.html#Codes

[13] http://www.jcomm.info/index.php?option=com_oe&task=viewDocumentRecord&docID=16012

[14] http://www.wmo.int/pages/prog/www/ois/RTHFocalPts/RTHs_en.html

[15] http://icoads.noaa.gov/

[16] https://www.nodc.noaa.gov/OC5/WOD/pr_wod.html

In the MCDS, the delayed mode data flow to the CMOCs through networks of specialized Data Acquisition Centres (DACs) and Global Data Assembly Centres (GDACs). GDACs are normally specialized by observing platform type (e.g. ship-based observations, drifting buoys, moored buoys, etc.), and they undertake minimum quality control. Observations from these observing networks can also be obtained directly from the GDACs. JCOMM is making efforts to integrate the existing sources (Table 2) of marine meteorological and oceanographic data into the MCDS network of centres.

In the wider meteo-ocean community, JCOMM has published a report on data systems relevant to its activities. The document includes comprehensive information on all such systems, and can be downloaded from the JCOMM website.[17] Practical information on how to submit data in delayed mode can also be found in JCOMM Technical Report No. 72.[11]

3.3 Recent Approaches Regarding Ocean Data Integration

Recently, JCOMM has initiated activities for achieving better interoperability, data integration and accessibility to meteo-ocean data. Integration is being investigated at different levels: within a particular observing network, across networks and by the observed variables. The approach will be to focus on creating interoperable data frameworks rather than data portals. Tools such as the Environmental Research Division's Data Access Program (ERDDAP[18]) can be used for that purpose. ERDDAP can connect to and serve data in many different formats, and users can then use specific Application Programming Interfaces (API) to access and discover data through various interoperable services. Data requests through ERDDAP allow to produce custom data-sets, formatted according to the requirement of the targeted end user application; it can serve metadata, images and analyse results. The goal is to simplify the end user efforts for accessing data from various sources formatted differently, while making them compatible to analysis tools. It is expected to focus on specific variables to begin with (e.g. SST, salinity).

JCOMM has also drafted a WMO and IOC Strategy for Marine Meteorological and Oceanographic Data Management (2018–2021). The draft Strategy, which has yet to be approved by both Organizations, is structured along six key outcomes related to (1) promoting data sharing, (2) improving data collection, (3) achieving better data integration, access, rescue and preservation, (4) improving data dissemination, (5) Making oceanographic and marine meteorological data sets discoverable using WMO and IOC information systems, and (6) capacity development. The strategy was drafted taking into account the over strategic framework of both WMO and IOC Organizations, and particularly the further development of the WMO Information System (WIS) and the WIS Strategy 2.0, as well as the IOC Strategic Plan for Oceanographic Data and Information Management (2017–2021).

[17] http://www.jcomm.info/index.php?option=com_content&view=article&id=331

[18] http://coastwatch.pfel.noaa.gov/erddap/index.html

Table 2 Existing sources of marine meteorological and oceanographic data

Data type	Real-time sources	Delayed mode sources
VOS	GTS/FM-13 SHIP GTS/FM-94 BUFR	Marine Climatological Summaries Scheme (MCSS) Global Collecting Centres (UK: http://www.metoffice.gov.uk/public/weather/marine-observations/#?tab=marineObs Map, and Germany: http://www.dwd.de/DE/leistungen/gcc/gcc.html) (IMMT format) International Comprehensive Ocean-Atmosphere Data Set (ICOADS) – http://icoads.noaa.gov/
SAMOS/underway marine MET and surface ocean data[a]	SAMOS 1.0 (custom ASCII format)	Shipboard Automated Meteorological and Oceanographic System (SAMOS) http://samos.coaps.fsu.edu/html/
Surface underway T&S	GTS/FM-62 TRACKOB GTS/FM-94 BUFR	Global Ocean Surface Underway Data (GOSUD) http://www.gosud.org/
Upper ocean T&S	GTS/FM-63 BATHY GTS/FM-64 TESAC	Global Temperature and Salinity Profile Programme (GTSPP) http://www.nodc.noaa.gov/GTSPP/
Argo profiling float data	GTS/FM-64 TESAC GTS/FM-94 BUFR Argo PIs	Argo Data System – http://www.argo.ucsd.edu/Argo_data_and.html The Argo Data System is comprised of: Argo Global Data Assembly Centres (GDACs) Argo national Data Assembly Centres (DACs)) Argo Regional Centres (ARCs)
Drifters	GTS/FM-18 BUOY GTS/FM-94 BUFR	MCDS trail GDAC for drifting buoys – http://www.meds-sdmm.dfo-mpo.gc.ca/isdm-gdsi/drib-bder/index-eng.htm Global Drifter Programme (GDP) Drifter Data Assembly Centre (DAC) – http://www.aoml.noaa.gov/phod/dac/dacdata.php
Meteorological moored buoys	GTS/FM-13 SHIP GTS/FM-94 BUFR	National Meteorological and Hydrological Services (NMHSs) operating the buoys, e.g. USA: http://www.ndbc.noaa.gov/

(continued)

Table 2 (continued)

Data type	Real-time sources	Delayed mode sources
Tropical moored buoys	Surface data & T profiles GTS/FM-18 BUOY GTS/FM-94 BUFR	Global Tropical Moored Buoy Array (GTMBA) TAO (Tropical Pacific Ocean): http://www.pmel.noaa.gov/tao/ TRITON (Western Tropical Pacific Ocean): http://www.jamstec.go.jp/jamstec/TRITON/real_time/ PIRATA (Tropical Atlantic Ocean): http://www.pmel.noaa.gov/pirata/ RAMA (Tropical Indian Ocean): http://www.pmel.noaa.gov/tao/rama/
Ocean reference sites	GTS/FM-18 BUOY GTS/FM-94 BUFR	OceanSITEs: Deep Ocean Time-Series Multidisciplinary data http://www.oceansites.org/ OceanSItes GDACs at IFREMER (France) and NOAA/NDBC (USA): ftp://ftp.ifremer.fr/ifremer/oceansites/ ftp://data.ndbc.noaa.gov/data/oceansites
Tide gauges		Global Sea Level Observing System (GLOSS) http://www.gloss-sealevel.org/data/
Tsunameters	GTS/FM-94 BUFR FTP/FM-94 BUFR	International Tsunami Information Centre (ITIC): http://itic.ioc-unesco.org/index.php
Gliders	Miscellaneous Distribution via GTS under consideration	Miscellaneous http://www.ego-network.org/dokuwiki/doku.php
High-resolution SST data from satellites	Space agencies GTS/multiple sources for in situ data	Group for High-Resolution Sea Surface Temperature (GHRSST) Regional Data Assembly Centres https://www.ghrsst.org/

[a]Beyond standard VOS, flow water system can be more than T & S

3.4 *Quality Control and Feedback to the Observing Platform Operators*

Controlling the quality of the meteo-ocean observations at various steps in the data processing chain is important for assuring that the quality of the data reaching the end users is best and known. For example, knowing the uncertainty of the data allows giving appropriate weight to the data from different sources in data assimilation schemes of numerical weather prediction, or for simply making decision on what data to use and what data to reject. Knowing the quality of the data requires not only undertaking some quality control checks but also having access to information about the observing platforms and their instruments (see Sect. 3.5). These allow obtaining estimates of the expected uncertainty of the measurements and knowing the conditions under which the measurements were taken (e.g. siting of the instruments and representativeness of the observations). Quality control is made at various levels:

- Simple automatic QC checks at the data acquisition platform (e.g. removal of spikes in the sampled raw data). One may refer, for example, to WMO No. 8,[19] Guide to Meteorological Instruments and Methods of Observation.
- Simple automatic QC checks at the data processing centre before actually inserting the data in real time on the GTS (e.g. comparison of the data with gross error or climatological limits, data consistency check, etc.). See, for example, the DBCP Technical Document No. 37,[20] Guide to Buoy Data Quality Control Tests to Perform in Real Time by a GTS Data Processing Centre.
- Higher level delayed mode QC checks, based on human scientific expertise. Such tests can involve comparing the data with nearby or colocated observations from different sources.
- Automatic delayed mode monitoring statistics performed by the operational centres using the data. For example, data can be compared with the Numerical Weather Prediction (NWP)'s first-guess field (i.e. a 6-h forecast) and statistics of differences produced in terms of standard deviation and bias.

3.5 *Instrument and Platform Metadata*

Instrument and platform metadata, that is, information about the observing platforms and their instruments, are required for a number of activities, including climate applications and research, and operational applications. They allow, for example:

[19] http://www.wmo.int/pages/prog/www/IMOP/CIMO-Guide.html

[20] http://www.jcomm.info/index.php?option=com_oe&task=viewDocumentRecord&docID=5657

1. To interpret the data correctly by knowing the measurement uncertainty and the conditions under which the measurements are made (e.g. siting of the instruments, representativeness). Specific weighting can, for example, be applied to different types of observing platforms (e.g. drifting buoys are regarded as providing higher quality sea surface temperature data than Voluntary Observing Ships).
2. To ensure traceability to standards by knowing calibration information, including what standards have been used for that purpose in the traceability chain.
3. To enhance coherence of data records, for example, through filtering and bias correction.
4. To facilitate quality-monitoring activities by knowing, for example, who operates the platforms, for what programme and by what country. Feedback on the quality of their observing platform data can then be provided to the platform operators.

In the WMO framework, the WIGOS metadata standard described in the WIGOS Manual must be followed by WMO Members when recording and sharing their observing platforms' metadata. WMO is collecting such metadata from its Members, recording them and making them publicly available through the OSCAR[21] database. For meteo-ocean observing platforms, JCOMMOPS is playing a key role in collecting the corresponding metadata from the operators of these platforms, and then for making them available through OSCAR.

3.6 Data Discovery Metadata

JCOMM is encouraging meteo-ocean data centres to make the historical and recent data-sets held by them discoverable by the wider WMO and IOC user communities through the WMO Information System (WIS) or the IODE Ocean Data Portal (ODP[22]). ISO-19115[23] standard for describing geographical information is used for that purpose. Both WIS and ODP are interoperable; so, when a data-set is discoverable through one, it should also in principle be discoverable through the other. However, this may not be the case in all instances as WMO and IOC use slightly different profiles of ISO-19115.

In case of the WIS, the metadata and data can be made available through either of the following WIS infrastructure elements: a 'National Centre' (NC) or 'Data Collection or Production Centres' (DCPC). Those centres providing data through WIS have to comply with the WIS compliance specifications of Global Information System Centre (GISC), DCPC and NC.[24]

[21] http://oscar.wmo.int

[22] http://www.oceandataportal.org/

[23] http://www.iso.org/iso/catalogue_detail.htm?csnumber=53798

[24] http://www.wmo.int/pages/prog/www/WIS/ref_docs_en.html

3.7 Data Policies

For meteo-ocean data, WMO and IOC data policies are applicable in the JCOMM framework. These include:

- WMO Resolution 40 (Cg-12),[25] WMO policy and practice for the exchange of meteorological and related data and products including guidelines on relationships in commercial meteorological activities. The Resolution particularly includes the following elements: (i) *as a fundamental principle of the World Meteorological Organization (WMO), and in consonance with the expanding requirements for its scientific and technical expertise, WMO commits itself to broadening and enhancing the free and unrestricted1*[26] *international exchange of meteorological and related data and products*;(ii) *WMO Members shall provide on a free and unrestricted basis essential data and products which are necessary for the provision of services in support of the protection of life and property and the well-being of all nations, particularly those basic data and products, as, at a minimum, described in Annex 1 to this resolution, required to describe and forecast accurately weather and climate, and support WMO Programmes.* Marine meteorological data are regarded as essential data in the framework of the Resolution.
- WMO Resolution 60 (Cg-17), WMO Policy for the International Exchange of Climate Data and Products to Support the Implementation of the Global Framework for Climate Services. The Resolution particularly urges WMO Members *to provide the additional GFCS relevant data and products that are required to support and sustain the operational climate services as the core element of the Framework and WMO initiatives at the global, regional and national levels and, further, as mutually agreed, to assist other Members to enhance access to GFCS relevant data and products and in the provision of climate services in their countries; such additional GFCS relevant data and products are listed in the annex to the present resolution and could have conditions associated with their use, such as restrictions on their use for commercial purposes, attribution of their source or licensing.*
- Resolution IOC-XXII-6,[27] IOC Oceanographic Data Exchange Policy. The preamble of the Resolution says: *The timely, free and unrestricted international exchange of oceanographic data is essential for the efficient acquisition, integration and use of ocean observations gathered by the countries of the world for a wide variety of purposes including the prediction of weather and climate, the operational forecasting of the marine environment, the preservation of life, the*

---

[25] https://www.wmo.int/pages/about/Resolution40_en.html

[26] 'Free and unrestricted' means non-discriminatory and without charge [Resolution 23 (EC-XLII) – Guidelines on international aspects of provision of basic and special meteorological services]. 'Without charge', in the context of this resolution means at no more than the cost of reproduction and delivery, without charge for the data and products themselves.

[27] http://www.iode.org/index.php?option=com_content&view=article&id=51&Itemid=95

mitigation of human-induced changes in the marine and coastal environment, as well as for the advancement of scientific understanding that makes this possible. Clause 1 of this Resolution states that *Member States shall provide timely, free and unrestricted access to all data, associated metadata and products generated under the auspices of IOC programmes.*

3.8 How to Access Data

Historical meteo-ocean climate data can be obtained from the CMOCs, ICOADS and WOD. In addition, more recent data-sets can be obtained from the sources listed in Table 2.

Other sources of marine meteorological and oceanographic data include:

- The World Data System (WDS[28]) and the former World Data Centres for Oceanography (WDCs);
- The World Ocean Atlas (WOA[29]) operated by the US National Oceanographic Data Centre
- The World Ocean Database (WOD[30])
- The Ocean Data Portal (ODP[31])
- The National Oceanographic Data Centres (NODCs[32]) of the International Oceanographic Data and Information Exchange (IODE)

3.9 Incentive for Sharing the Data

Principal investigators and those in charge of meteo-ocean observing programmes are encouraged to share their data and to make them available in real time and delayed mode through the WMO and IOC data systems. Sharing and exchanging meteo-ocean data with WMO and IOC users in real time and delayed mode offer the following advantages:

- Our data contribute to global programmes and allow improving the products and services provided to the public, for example, operational weather and ocean forecasting, climate monitoring and marine services.
- The quality of our data is to be automatically monitored, thanks to modern data assimilation techniques and their comparison with the numerical model's first-guess field (typically comparing colocated data with the model's 6-h forecast).

[28] https://www.icsu-wds.org/

[29] https://www.nodc.noaa.gov/OC5/indprod.html

[30] https://www.nodc.noaa.gov/OC5/SELECT/dbsearch/dbsearch.html

[31] http://www.oceandataportal.net/portal/

[32] http://www.iode.org/datacentres

Monitoring statistics and feedback on the quality of our observing platforms are then provided to us by JCOMMOPS;

In addition, JCOMMOPS can assist us to realize the sharing and the exchange of our data through the WMO and IOC data systems. This will include assistance to obtain WMO identification numbers for our observing platforms and practical technical assistance on available formats, protocols, available software and contact information in data centres which could further assist.

4 Conclusion

Meteo-ocean data are essential to estimate the state of the atmosphere and the ocean in support of operational and research applications undertaken in the WMO and IOC frameworks. In this way, they contribute to weather and ocean forecasting, marine services, climate monitoring and climate services. Such data are available to end users through WMO and IOC data systems. Scientists and operators of meteo-ocean observing platforms are encouraged to share and have their data exchanged in real time and delayed mode with the WMO and the IOC. If needed, JCOMMOPS (support@jcommops.org) can provide them with practical technical assistance in this regard.

Acronyms

ADCP	Acoustic Doppler Current Profiler
AT	Air Temperature
CTD	Conductivity Temperature and Depth
ERDDAP	Environmental Research Division's Data Access Program
ENSO	El Niño Southern Oscillation
GCOS	WMO-IOC-UNEP-ICSU Global Climate Observing System
GFCS	Global Framework for Climate Services
GOOS	IOC-WMO-UNEP-ICSU Global Ocean Observing System
ICSU	International Council for Science
IOC	Intergovernmental Oceanographic Commission of UNESCO
JCOMM	Joint WMO-IOC Technical Commission for Oceanography and Marine Meteorology
Meteo-ocean	Marine Meteorological and Oceanographic
MSWH	Mean Significant Wave Height
NWP	Numerical Weather Prediction
OSCAR	Observing System Capability Analysis and Review platform – oscar.wmo.int
PIRATA	Prediction and Research Moored Array in the Tropical Atlantic

RAMA	Research Moored Array for African-Asian-Australian Monsoon Analysis and Prediction
RH	Air Relative Humidity
SLP	Sea Level Pressure
SOO	Ship of Opportunity
SSS	Sea Surface Salinity
SST	Sea Surface Temperature
TAO	Tropical Atmosphere Ocean
TRITON	Triangle Trans-Ocean Buoy Network
UN	United Nations
UNEP	United Nations Environment Programme
UNESCO	United National Educational, Scientific and Cultural Organization
VOS	Voluntary Observing Ship
WIGOS	WMO Integrated Global Observing System
WIS	WMO Information System
WMO	World Meteorological Organization
XBT	Expendable Bathythermograph

Part VI
Societal Applications

Applications of Ocean In-situ Observations and Its Societal Relevance

M. Ravichandran and M.S. Girishkumar

Abstract The present status of ocean observation networks, especially in-situ, and their potential applications and societal relevance are summarized here. In-situ ocean observations are imperative to understand dynamics and thermodynamics of the ocean and its near-surface atmosphere, and they enhance our knowledge about weather and climate. Moreover, in-situ observations are directly assimilated into ocean and atmosphere models to support operational forecasts of ocean and atmospheric conditions. They complement the extensive data sets gathered by satellites, and they augment and validate the parameter estimates provided by satellites and other remote sensors through precise, direct measurements of ocean and atmospheric conditions. Global, national, and local ocean observational networks are a key foundation of operational oceanography. They underpin services of broad societal importance and economic value. These include the forecast of weather conditions, including seasonal and subseasonal monsoon forecasts; the provision of warnings of extreme weather and ocean events, such as tropical cyclones, storm surges, high waves and tsunamis; and information services in support of other ocean or coastal activities such as ocean transport and search and rescue operations. These services deliver direct and indirect benefits to a wide spectrum of society.

1 Introduction

Apart from a central role in global transport, the ocean is a source of food, medicines, minerals, and energy resources, and it significantly affects every human life through modulation of weather and climate. Hence, information on the past, present, and the likely future state of the ocean is important for many stakeholders including climate science and weather forecasters; the maritime industry and fisheries; agencies concerned with national security, public health, and the environment; the tourism sector; and the offshore mining and oil industry. The physical, chemical,

---

M. Ravichandran (✉)
National Centre for Antarctic and Ocean Research (NCAOR), Goa 403 804, India
e-mail: mravi@ncaor.gov.in

M.S. Girishkumar
Indian National Centre for Ocean Information Services (INCOIS), Hyderabad 500090, India

R. Venkatesan et al. (eds.), *Observing the Oceans in Real Time*, Springer Oceanography, DOI 10.1007/978-3-319-66493-4_15

and biological processes within the oceanic system operate over various spatiotemporal scales and are in constant dynamic interaction with the overlying atmosphere, posing a demand for continuous observation of ocean.

Ocean observation systems consist of (a) in-situ measurements, using sensors mounted on ships, buoys, and moorings, or the ocean floor and (b) remote sensing systems such as satellites, aircraft, radar, etc. Remote sensors in general and satellite measurements in particular reveal the spatial distribution of ocean surface variables, such as temperature, sea surface height, and ocean color, as well as several meteorological parameters for the calculation of air-sea momentum, heat, and freshwater fluxes. The temporal and spatial coverage of satellite measurements is extraordinarily valuable. However, satellites cannot see below the sea surface, and their reliance on parameter estimates made through indirect sensing place limits on the absolute measurement accuracy for some ocean characteristics, compared with in-situ observations. Hence, an integrated in-situ ocean observing system is required in addition to remote sensing to monitor the range of ocean conditions needed to understand ocean and atmospheric dynamics, and to apply that understanding to a range of societal services.

At present, various national and international oceanographic agencies maintain an in-situ ocean observation network consisting of different observation platforms. A sustained observation program to detect, track, and predict changes in physical, chemical, geological, and biological systems and their effects is needed to measure the impacts of humans on the ocean as well as the impact of the ocean on human endeavors. Just as continuous measurements of weather and climatic conditions are maintained on land, similar sustained measurements of the ocean are required to monitor change and to assist in understanding and predicting its impacts.

2 The Current Status of the Ocean Observations Network in the Indian Ocean

The Indian Ocean was undersampled and poorly understood compared to other tropical oceans before the twenty-first century. However, during the past one-and-a-half decades, there have been sustained and dedicated efforts to monitor conditions in the Indian Ocean region. An extensive in-situ observation (Fig. 1) network has been developed through the coordinated efforts of various oceanographic institutions in India, complemented by the contributions of companion networks from neighbor countries and international agencies and programs. A large number of observation platforms were deployed and maintained in the coastal and interior ocean areas of the Indian Ocean by the Indian National Centre for Ocean Information Services (INCOIS), the National Institute of Ocean Technology (NIOT), and the

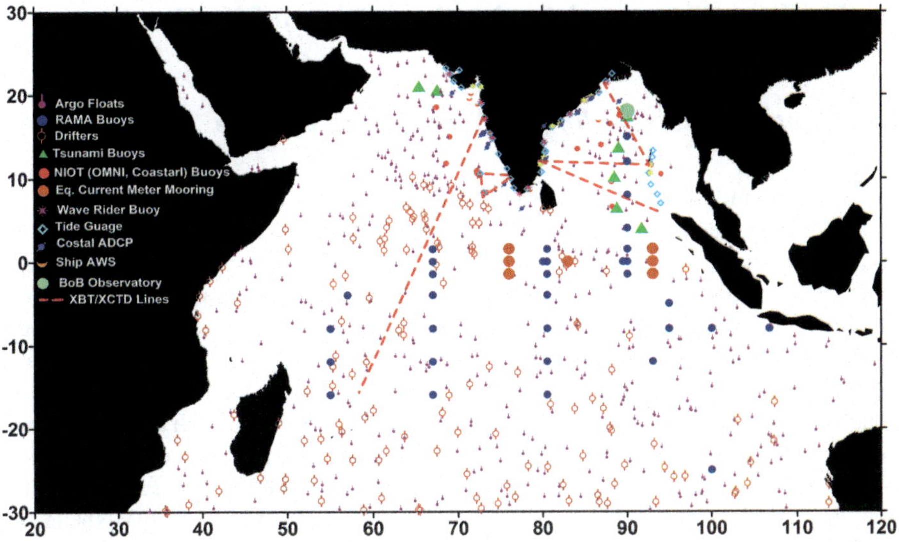

Fig. 1 The status of in-situ observation network in the Tropical Indian Ocean during 2017

National Institute of Oceanography (NIO) through Ministry of Earth Sciences (MoES) funding under the project Ocean Observation Network (OON).

Some platforms, such as automated weather stations onboard ships, the coastal acoustic doppler current profiler (ADCP) network, tsunami buoys (or bottom pressure recorders), tide gauges, the Ocean Moored buoy Network for the Northern Indian Ocean (OMNI) [35], the High frequency (HF) radar network, the Bay of Bengal Observatory, and the coastal Waverider buoy network were indigenously proposed, designed, and implemented by India to support its own ocean information services. The observation platforms such as Argo floats, eXpendable Bathy thermograph/eXpendable Conductivity Temperature Depth probes (XBTs/XCTDs), satellite-tracked surface drifting buoys, and the Research Moored Array for African-Asian-Australian Monsoon Analysis and Prediction (RAMA) mooring networks are parts of the Indian Ocean Observing System (IndOOS), which is a regional contribution to the Global Ocean Observing System (GOOS). GOOS is an international collaborative effort led by the Intergovernmental Oceanographic Commission (UNESCO-IOC) to establish ocean observations and collect real-time oceanographic data across the world's oceans. As a member country of IndOOS, India plays a lead role in maintaining the above-mentioned network in the Indian Ocean region in collaboration with various international agencies. These in-situ observations have succeeded in providing near real-time oceanic and meteorological data in different spatiotemporal scales, and they support ocean research and development of ocean state forecast services in the Indian Ocean.

3 The Importance and Application of the In-Situ Ocean Observation Network

3.1 Better Understanding of Weather and Climate

Ocean observations help to answer fundamental research questions, such as the role of the ocean in weather and climate, quantifying the air-sea exchange of heat, water, momentum and gases, the ocean's role in global warming, understanding the patterns and controls on biological diversity in the oceans, etc. Moreover, to forecast oceanic and atmospheric conditions through ocean, atmosphere, and coupled models, it is imperative to understand these processes and to model them. To characterize and understand the underlying dynamics and physics at various spatiotemporal scales, it is important to monitor the ocean state in real time, for both the surface and the interior ocean, through a continuous and long-term system of ocean observations. The enhancement of in-situ observations in the Indian Ocean region, particularly through the Argo and moored buoy networks (e.g., RAMA and OMNI), leads to a better understanding of the dominant interannual and intraseasonal modes in the tropical climate such as the Indian Ocean Dipole (IOD), the Madden-Julian Oscillation (MJO), and the northward propagating monsoon intraseasonal oscillation (MISO) [19, 28, 33, 38]. These all have the potential to modulate the weather and climate of India and other parts of the globe [2, 7, 9, 14, 20, 26]. In countries like India, food security and a major part of the national employment base rely on rain-fed agriculture, which primarily depends on rainfall in June–September, during the Indian summer monsoon. This has significant interannual variability in both intensity and spatial distribution, leading to extreme events like floods and droughts. It has been shown that different interannual and intraseasonal modes, such as the El Niño-Southern Oscillation (ENSO), the IOD, and the MISO influence the Indian summer monsoon rainfall [1, 13, 38]. Hence, better understanding of the factors that modulate these modes is scientifically and socioeconomically important. Subsurface oceanic and near-surface meteorological observations from the in-situ ocean observation networks provide an excellent opportunity to examine air-sea interaction process that modulate these signals and in turn to a better understanding of the monsoon system.

The ocean observation networks also enable the collection of high-quality in-situ measurements of physical and biogeochemical conditions during extreme weather events such as tropical cyclones, which is not feasible from other means. For example, using moored buoy observation, the air-sea interaction processes and near-surface temperature and salinity structure in the Bay of Bengal was examined during Tropical Cyclone *Nargis* [25], *Jal* [10], and *Phalin* [36]. Further, the study by D'Asaro et al. [6] examined the three-dimensional, time-dependent response of the ocean using profiling floats and surface-drifting buoys during tropical cyclone *Frances*. The inferences from these studies have enhanced our knowledge on air-sea interaction processes during the tropical cyclone period. Even in-situ observation of wind and air pressure from a platform during the passage of a cyclone can provide vital information for an operational forecaster to confirm the direction and intensity of the tropical cyclone in the data-void region [15].

Further, recent studies have shown that sudden tropical cyclone intensification has been linked with high values of upper ocean heat content contained in mesoscale features, particularly warm ocean eddies [17]. Moreover, a recent study by Sandery et al. [31] has shown that reanalyzed sea surface temperatures significantly improve the prediction of tropical cyclone intensity change in the intensification phase. This emphasizes such real-time information from in-situ observation network to be available and readily assimilated into a forecast model during a future cyclone event; it could deliver an improved tropical cyclone forecast, particularly in the intensification phase. Moreover, different operational agencies have used the atmospheric output, especially wind and pressure field, to force their wave and storm surge forecast models. The availability of better atmospheric data during the tropical cyclone period can lead to better atmospheric forecast during tropical cyclone period, and it can facilitate better advisory services of wave and storm surge to the coastal and maritime community.

Recently, there have been efforts to extract new details about ocean turbulence at the submesoscale (100 m–10 km) process from these observation networks. Such processes are extremely important for tracer transport and dispersion, such as freshwater from the Bengal delta, and energy transfer from the mesoscale (~10–100 km) to microscale (1–100 m). Satellites and mooring arrays do not have the required resolution to observe and characterize the behavior of the turbulent flows at these scales. Data collected from drifting buoys and Argo floats with appropriate deployment strategies, such as releasing in pairs and triplets, will enable us to measure relative spreading rates which are governed by the turbulence. Such studies will help us to characterize the scale-dependent kinetic energy distribution and relative dispersion/diffusivity [5, 18].

Coastal current observation from HF radar network and moorings can provide better opportunity to enhance our knowledge on coastal current systems and can have significant operational value. The HF radar network can provide a detailed view to mesoscale coastal ocean surface currents with high spatiotemporal resolutions. The in-situ current observation along with modeling tools provide excellent support for operational agencies for search and rescue operation and monitoring hazardous material spills response. For instance, the real-time current and wave data from the coastal and island-based network of HF radar systems are extensively used for oil and hazardous material spills mitigation and search and rescue operations particularly in the coastal region by operational agencies. The real-time current information from this system will help to determine the oil and hazardous material spreading rate, and this allows spill and search and rescue team to focus their activity on more threatened areas.

3.2 Better Understanding of the Ocean Biosphere

In the last couple of years, enhanced biogeochemical observation, particularly from profiling floats equipped with biogeochemical sensors (e.g., biogeochemical-Argo), has provided unprecedented data sets to help understand the response of biogeochemical parameters to the ocean's physical and dynamic processes. It has also extended the knowledge that was achieved through satellite ocean color measurements to subsurface levels. Moreover, the biogeochemical data from this source provide an opportunity to examine the spatiotemporal evolution of the marine habitat and its behavior under a global warming scenario. For example, information on the depth of the oxycline is critical for understanding the magnitude and extent of the hypoxic zone, and it may help to designate potential fishing zones on an operational basis, as suggested by recent studies [22, 30, 37]. The study by Piontkovski and Al-Oufi [22] in the northwestern Arabian Sea showed that seasonal shoaling of the oxycline leads to compressing of the habitat of marine species into the near-surface layer, which in turn favors the accumulation of pelagic predators and increases marine fish landings. This indicates that enhanced biogeochemical observation may provide an opportunity to better understand seasonal and interannual evolution of subsurface conditions and may help to provide better potential fishing zone advisories/forecasts.

Recent studies have used biogeochemical measurements from biogeochemical-Argo floats to study the complex biophysical interactions during a strong tropical cyclone forcing event, which is not feasible from ship-based observation due to the violent wind and wave conditions associated with a tropical cyclone [34]. Argo floats equipped with biogeochemical sensors provide unmatched and distinctive advantage to enhance our understanding of the impact of tropical cyclone on biogeochemical variability. Better understanding of the upper ocean biogeochemical response to tropical cyclones is worth investigating and crucial to the well-being of the economy and fisheries.

3.3 Validation of Ocean Model Output and Satellite Data to Support the Development of Operational Ocean Systems

Hazardous weather-related events associated with the ocean cause thousands of deaths and extensive property damage each year. These events also impact fisheries, coastal biodiversity, marine transportation, water resource management, etc. Since the ocean state significantly affects society and the economy, decision-makers have come to rely on weather forecasts provided by the operational ocean community. The development of better operational ocean forecast systems and the use of in-situ observation systems to develop and verify those systems are summarized in this section.

In the development of an operational ocean state forecasting model, it is important to evaluate the performance of the model simulation against independent in-situ observations to examine the ability of the model to resolve the variability of interest. The in-situ observations generated by different ocean observation platforms provide a rich data source to qualify the parameters which are simulated by ocean models. They help to understand the sources of error and to rectify operational models. The in-situ data from Argo, moored buoy platforms, and the Acoustic Doppler Current Profiler (ADCP) network provide excellent data for the validation of physical parameters simulated by the ocean model, and data from the Waverider buoy network assist the validation of wave models. Further, the near real-time data from these networks enable operational agencies to refine forecasts and to build user credibility in their systems.

Sea-level monitoring networks of tide gauges and tsunami buoys are crucial components of a tsunami warning system. They measure the changes in water level in real time with high accuracy at coastal sites and in the open ocean to confirm the generation of tsunami waves and to monitor the progress of a tsunami. They also facilitate the near real-time update of tsunami advisories. The real-time sea-level data from these networks help to avoid false alarms and unnecessary public evacuations and to cancel alerts when a threat has passed. Moreover, the data from real-time tide gauges and tsunami buoys can be useful to estimate a tsunami source through sea-level data inversion techniques, which can be useful information for tsunami initiation models.

Ocean models require accurate forcing fields of heat, momentum and water vapor to estimate ocean state. The meteorological data from moored buoy networks and ship-based automated weather stations facilitate the generation of bias-corrected ocean model forcing fields (TropFLUX, OAFLUX etc.) [23, 24, 39]. Moreover, in-situ observations help validate the satellite remote sensing of meteorological and oceanographic parameters [8, 11, 21, 29] and help to improve existing retrieval algorithms and to develop new algorithms to derive geophysical parameters from satellite observations. These data also help to identify and examine deficiencies in the ocean model forcing field and they lead to the generation of a better model forcing field [12].

The development of a marine ecosystem model is a key goal for many organizations and agencies across the world [3]. The validation of these models primarily depends upon the satellite-based chlorophyll measurements and ship-based observations. However, ocean color satellite measurements (primarily visible and infrared band) are incapable of measuring through clouds and below the ocean surface. One of the key advantages of biogeochemical-Argo floats measurements is their ability to provide subsurface information on key parameters, which was previously only possible through ships. However, ship-borne observations have significant bias toward areas that are more easily sampled. Argo floats and moored buoys equipped with biogeochemical parameters (dissolved oxygen, chlorophyll fluorescence, nitrate, and pH) provide an excellent source of data to validate the marine ecosystem models. Ultimately, this will lead to improvement in ecosystem model simulation and to better marine fisheries forecast services, particularly species-specific forecast services (e.g., tuna). Studies have reported that potential fishing zone information can help the fishermen to minimize search time by up to 70% and significantly improve the catch-per-unit-effort economics by saving fuel [4, 32].

3.4 Global Ocean and Atmospheric Analysis/Reanalysis

At present, ocean reanalysis/analysis is a recognized activity in several research and operational centers. The basic idea behind the data assimilation technique is to produce a regular, physically consistent, four-dimensional representation of ocean and atmospheric state from in-situ and remote sensing data sources that sample on different spatiotemporal scales. In operational services, real-time assimilation of observational data is used to improve the estimation of ocean and atmospheric state simulated by numerical models. The main purpose of the ocean/atmosphere analysis product is to provide initial conditions for the seasonal and subseasonal forecasts using coupled models. Data from remote sensing satellites provide extensive spatiotemporal coverage of surface data, but they do not have the ability to provide subsurface information. However, the data collected from the platforms such as Argo, XBT, and moored buoy networks can provide surface and subsurface information, which are useful for assimilation in ocean general circulation models [27]. Further, it was demonstrated that current measurements from HF radar data can be used to improve forecasts of currents and other oceanographic variables in coastal areas [16]. Moreover, near-surface meteorological data collected from moored buoy networks, ship-based automatic weather stations, and surface drifters are assimilated into atmospheric models. Presently, both in-situ and satellite data are being assimilated into the ocean models to provide ocean initial conditions for seasonal forecasts for monsoon prediction. Ultimately, all these efforts lead to better seasonal and subseasonal forecasts. So, it is of immense socioeconomic importance to forecast these variations in advance to devise better policies to mitigate possible disasters and plan for suitable crops. Ocean observations are critical for providing such ocean analysis, and the quality of ocean and atmospheric state forecasts through ocean data assimilation is heavily reliant on the maintenance of ocean observation networks.

In reanalysis, historical observations are assimilated into ocean and ocean/atmosphere models to capture a full representation of past ocean and climate conditions including both measured and model-generated estimates of parameters across the full spatial range of the model, and over timescales that may extend back decades or more. This can provide a high-quality historical record of the ocean variability. Such reanalysis creates a reference data set for the current climate state, and helps in the study of past climate conditions. It is also used to validate seasonal and decadal forecasts. Ocean model simulations can provide some insight into ocean variability, but they are subject to errors associated with initial conditions, forcing fields and model numeric. The combination of ocean models with data assimilated from ocean observations can provide more accurate information than model-only estimations. One of the important future modeling goals of various research and operational centers is the assimilation of biogeochemical parameters in the marine ecosystem models.

4 Summary

In this chapter, the current status of the in-situ ocean observation network in the Indian Ocean region and potential applications of an in-situ ocean observation network and its societal relevance are summarized. To characterize and understand the ocean dynamics and physics at various spatiotemporal scales, it is important to monitor the ocean state in real time, for both the surface and the interior ocean. The data collected by these observation networks also support a variety of operational services and system developments that deliver great benefit to various stakeholders in their day-to-day activities affected by the ocean. The in-situ data collected from ocean observation networks are extensively used to validate simulated ocean parameters from model hindcasts/forecasts and from satellite-derived parameters. Access to observation data also builds our understanding of ocean models' abilities and limitations and assists the fine-tuning of models for optimal performance, which in turn leads to improved forecasting systems. An ocean observation network fuels the growth of scientific knowledge of the ocean environment and its interactions with weather and climate, while at the same time underpinning forecasting systems that inform human decision-making across a number of sectors of society, and provide improved early warning for ocean-related hazards such as waves and storm surges. Today, the in-situ ocean observation networks has emerged as one of the strongest pillars of operational oceanography that delivers direct and indirect benefits to a wide spectrum of society. The evolution of ocean observing systems to encompass biogeochemical monitoring, backed by models of the marine habitat, will deliver even more scientific understanding and societal benefits into the future.

References

1. Ashok K, Guan Z, Saji NH, Yamagata T (2004) Individual and combined influences of ENSO and the Indian Ocean dipole on the Indian summer monsoon. J Clim 17(16):3141–3155. https://doi.org/10.1175/1520-0442(2004)017<3141:IACIOE>2.0.CO;2
2. Cai W, Pan A, Roemmich D, Cowan T, Guo X (2009) Argo profiles a rare occurrence of three consecutive positive Indian Ocean dipole events, 2006–2008. Geophys Res Lett 36:L08701. https://doi.org/10.1029/2008GL037038
3. Chakraborty K, Gupta A, Lotliker A, Gavin T (2016) Evaluation of model simulated and MODIS-aqua retrieved sea surface chlorophyll in the eastern Arabian Sea. Estuar Coast Shelf Sci 181(2016):61–69. https://doi.org/10.1016/j.ecss.2016.08.002
4. Choudhury SB, Jena B, Rao MV, Rao KH, Somvanshi VS, Gulati DK, Sahu SK (2007) Validation of integrated potential fishing zone (IPFZ) forecast using satellite based chlorophyll and sea surface temperature along the east coast of India. Int J Remote Sens 28(12):2683–2693
5. Cole ST, Wortham C, Kunze E, Owens WB (2015) Eddy stirring and horizontal diffusivity from Argo float observations: geographic and depth variability. Geophys Res Lett 42:3989–3997. https://doi.org/10.1002/2015GL063827
6. D'Asaro EA, Sanford TB, Niiler PP, Terrill EJ (2007) Cold wake of Hurricane Frances. Geophys Res Lett 34:L15609. https://doi.org/10.1029/2007GL030160

7. Drushka K, Sprintall J, Gille ST, Wijffels S (2012) In situ observations of Madden–Julian oscillation mixed layer dynamics in the Indian and Western Pacific Oceans. J Clim 25:2306–2328. https://doi.org/10.1175/JCLI-D-11-00203.1
8. Ebuchi N, Graber HC, Caruso MJ (2002) Evaluation of wind vectors observed by QuikSCAT/SeaWinds using ocean buoy data. J Atmos Ocean Technol 19:2049–2062
9. Girishkumar MS, Ravichandran M (2012) The influences of ENSO on tropical cyclone activity in the Bay of Bengal during October–December. J Geophys Res 117:C02033. https://doi.org/10.1029/2011JC007417
10. Girishkumar MS, Suprit K, Chiranjivi J, Udaya Bhaskar TVS, Ravichandran M, Venkat SR, Pattabhi Rama Rao E (2015) Observed oceanic response to tropical cyclone Jal from a moored buoy in the south-western Bay of Bengal. Ocean Dyn 64(3):325–335
11. Goswami BN, Rajagopal EN (2003) Indian Ocean surface winds from NCMRWF analysis as compared to QuikSCAT and moored buoy winds. Proc Indian Acad Sci Earth Planet Sci 112:61–77
12. Goswami BN, Sengupta D (2003) A note on the deficiency of NCEP/NCAR reanalysis surface winds over the equatorial Indian Ocean. J Geophys Res 108:3124. https://doi.org/10.1029/2002JC001497
13. Goswami BN, Xavier PK (2005) ENSO control on the south Asian monsoon through the length of the rainy season. Geophys Res Lett 32:L18717. https://doi.org/10.1029/2005GL023216
14. Goswami BN (2005) South Asian monsoon. In: Lau WKM, Waliser DE (eds) Intraseasonal variability in the atmosphere–ocean climate system. Springer Praxis, Berlin, Heidelberg, pp 19–61
15. Harikumar R, Balakrishnan Nair TM, Rao BM, Prasad R, Phani R, Nagaraju C, Ramesh Kumar M, Jeyakumar C, Shenoi SSC, Nayak S (2016) Ground-zero met–ocean observations and attenuation of wind energy during cyclonic storm Hudhud. Curr Sci 110(12):2245–2252
16. Iermano I, Moore AM, Zambianchi E (2016) Impacts of a 4-dimensional variational data assimilation in a coastal ocean model of southern Tyrrhenian Sea. J Mar Syst 154:157–171. https://doi.org/10.1016/j.jmarsys.2015.09.006
17. Lin I-I, Goni GJ, Knaff JA, Cristina F, Ali MM (2012) Ocean heat content for tropical cyclone intensityforecasting and its impact on storm surge. Nat Hazards. https://doi.org/10.1007/s11069-012-0214
18. Lumpkin R, Elipot S (2010) Surface drifter pair spreading in the North Atlantic. J Geophys Res 115:C12017. https://doi.org/10.1029/2010JC006338
19. Madden RA, Julian PR (1994) Observations of the 40–50-day tropical oscillation—a review. Mon Weather Rev 122:814–837
20. Parampil SR, Gera A, Ravichandran M, Sengupta D (2010) Intra-seasonal response of mixed layer temperature and salinity in the Bay of Bengal to heat and freshwater flux. J Geophys Res 115:C05002. https://doi.org/10.1029/2009JC005790
21. Pickett MH, Tang WQ, Rosenfeld LK, Wash CH (2003) QuikSCAT satellite comparisons with nearshore buoy wind data off the U.S. West Coast. J Atmos Oceanic Technol 20:1869–1879
22. Piontkovski SA, Al-Oufi HS (2014) Oxygen minimum zone and fish landings along the omani shelf. J Fish Aquat Sci 9:294–310. https://doi.org/10.3923/jfas.2014.294.310
23. Praveen Kumar B, Vialard J, Lengaigne M, Murty VSN, McPhaden MJ (2012) TropFlux: air-sea fluxes for the global tropical oceans - description and evaluation. Clim Dyn 38:1521–1543. https://doi.org/10.1007/s00382-011-1115-0
24. Praveen Kumar B, Vialard J, Lengaigne M, Murty VSN, McPhaden MJ, Cronin MF, Pinsard F, Gopala Reddy K (2013) TropFlux wind stresses over the tropical oceans: evaluation and comparison with other products. Clim Dyn 40(7–8):2049–2071. https://doi.org/10.1007/s00382-012-1455-4
25. McPhaden MJ, Foltz GR, Lee T, Murty VSN, Ravichandran M, Vecchi GA, Vialard J, Wiggert JD, Yu L (2009) Ocean-atmosphere interactions during cyclone nargis. EOS Trans Am Geophys Union 90(7):53–54. https://doi.org/10.1029/2009EO070001

26. Nyadjro ES, McPhaden MJ (2014) Variability of zonal currents in the eastern equatorial Indian Ocean on seasonal to interannual time scales. J Geophys Res Oceans 119:7969–7986. https://doi.org/10.1002/2014JC010380
27. Ravichandran M, Behringer D, Sivareddy S, Girishkumar MS, Chacko N, Harikumar R (2013) Evaluation of the global ocean data assimilation system at INCOIS: the tropical Indian Ocean. Ocean Model 69:123–135
28. Saji NH, Goswami BN, Vinayachandran PN, Yamagata T (1999) A dipole mode in the tropical Indian Ocean. Nature 401(6751):360–363. https://doi.org/10.1038/43854
29. Satheesan K, Sarkar A, Parekh A, Ramesh Kumar MR, Kuroda Y (2007) Comparison of wind data from QuikSCAT and buoys in the Indian Ocean. Int J Remote Sens 10:2375–2382
30. Stramma L, Johnson GC, Sprintall J, Mohrholz V (2008) Expanding oxygen-minimum zones in the tropical oceans. Science 320:655–658
31. Sandery PA, Brassington GB, Craig A, Pugh T (2010) Impacts of ocean–atmosphere coupling on tropical cyclone intensity change and ocean prediction in the Australian region. Mon Wea Rev 138:2074–2091. https://doi.org/10.1175/2010MWR3101.1
32. Solanki HU, Prakash P, Dwivedi RM, Nayak S, Kulkarni A, Somvamshi VS (2010) Synergistic application of oceanographic variables from multi-satellite sensors for forecasting potential fishing zones: methodology and validation results. Int J Remote Sens 31(3):775–789
33. Tziperman E, Cane MA, Zebiak SE, Xue Y, Blumenthal B (1998) Locking of El Niño's peak time to the end of the calendar year in the delayed oscillator picture of ENSO. J Clim 11(9):2191–2199. https://doi.org/10.1175/1520-0442(1998)011<2191:LOENOS>2.0.CO;2
34. Udaya Bhaskar TVS, Chiranjivi J, Rama Rao EP, Rao KH (2016) Spatio-temporal evolution of chlorophyll-a in the Bay of Bengal: a remote sensing and bio-argo perspective. Proceedings SPIE 9878, remote sensing of the oceans and inland waters: techniques, applications, and challenges, 98780Z 7 May 2016. doi:https://doi.org/10.1117/12.2223880
35. Venkatesan R, Shamji VR, Latha G, Simi M, Rao RR, Muthiah A, Atmanand MA (2013) In situ ocean subsurface time-series measurements from OMNI buoy network in the Bay of Bengal. Curr Sci 104(9):1166–1177
36. Venkatesan R, Simi M, Vimala J, Latha G, Arul Muthiah M, Ramasundaram S, Sundar R, Lavanya R, Atmanand MA (2014) Signatures of very severe cyclonic storm Phailin in met–ocean parameters observed by moored buoy network in the Bay of Bengal. Curr Sci 107(4):589–595
37. Vaquer-Sunyer R, Duarte CM (2008) Thresholds of hypoxia for marine biodiversity. Proc Natl Acad Sci U S A 105:15452–15457
38. Webster PJ, Moore AM, Loschnigg JP, Leben RR (1999) Coupled ocean-atmosphere dynamics in the Indian Ocean during 1997–98. Nature 401(6751):356–360. https://doi.org/10.1038/43848
39. Yu L, Weller RA (2007) Objectively analyzed air-sea heat fluxes for the global oce-free oceans (1981–2005). Bull Ameri Meteor Soc 88:527–539

Index

R. Venkatesan et al. (eds.), *Observing the Oceans in Real Time*, Springer Oceanography, DOI 10.1007/978-3-319-66493-4

P

Printed by Printforce, the Netherlands